Benchmarking – Theory and Practice

Benchmarking – Theory and Practice

Edited by

Asbjørn Rolstadås

Faculty of Mechanical Engineering
University of Trondheim
Norwegian Institute of Technology
Norway

Published by Chapman & Hall on behalf of the
International Federation for Information Processing (IFIP)

CHAPMAN & HALL

London · Weinheim · New York · Tokyo · Melbourne · Madras

Published by Chapman & Hall, 2–6 Boundary Row, London SE1 8HN, UK

Chapman & Hall, 2–6 Boundary Row, London SE1 8HN, UK

Chapman & Hall GmbH, Pappelallee 3, 69469 Weinheim, Germany

Chapman & Hall USA, One Penn Plaza, 41st Floor, New York NY 10119, USA

Chapman & Hall Japan, ITP-Japan, Kyowa Building, 3F, 2-2-1 Hirakawacho, Chiyoda-ku, Tokyo 102, Japan

Chapman & Hall Australia, Thomas Nelson Australia, 102 Dodds Street, South Melbourne, Victoria 3205, Australia

Chapman & Hall India, R. Seshadri, 32 Second Main Road, CIT East, Madras 600 035, India

First edition 1995
Reprinted 1996

© 1995 Chapman & Hall

Printed in Great Britain by Antony Rowe Ltd, Chippenham, Wiltshire

ISBN 0 412 62680 2

A catalogue record for this book is available from the British Library

∞ Printed on permanent acid-free text paper, manufactured in accordance with ANSI/NISO Z39.48-1992 and ANSI/NISO Z39.48-1984 (Permanence of Paper)

CONTENTS

PREFACE

IFIP working group for computer -aided production management (WG5.7) organized a workshop on "Benchmarking - Theory and Practice" in Trondheim, Norway, June 16 - 18, 1994. This book contains revised and enhanced papers from that workshop.

Benchmarking is a modern buzz-word. Many enterprises and consultants are interested in this topic, and there are a number of opinions on what benchmarking is, why it is necessary, how it should be done, and what results to obtain. The purpose of the workshop was to explore this emerging topic by looking both at theoretical fundamentals of benchmarking and at the same time learn from some of the recent applications in industry world wide.

The workshop was organized by inviting researchers and industrialists to present position papers. These papers were the basis of extensive discussions during the workshop. After the workshop, the authors have been invited to submit full papers based on their position papers, the discussion during the workshop and on later findings.

During the workshop group work on three topics was carried out. The topics were:

- Implementing benchmarking
- Modeling for benchmarking
- Performance measurement

The group work was split into sessions with a plenary presentation and discussion after each session. Reports from the group work are included at the end of the book.

The book has been divided into parts as follows:

One   Management Issues
Two   Applications
Three   Modeling
Four   Tools and Techniques
Five   Performance Indicators and Measurement
Six   Group Work

The management issues focus on the extended enterprise and benchmarking as a tool for reengineering processes. Applications include a number of interesting industrial cases. Modeling, tools and techniques as well as performance indicators and measurement discusses important aspects of performance management - both from a practical and theoretical viewpoint.

The workshop, being held in Trondheim, has been sponsored by the research program TOPP, which aims at improving productivity in the Norwegian industry. Without the support of TOPP, this workshop may not have been organized.

I will extend warm thanks to all the authors that spent much time and effort to produce the papers of this book. I will also thank MSc. Bjørn Andersen and Mrs. Solfrid Sørensen who has helped organizing the workshop and developing this book.

I hope the book will stimulate productivity-related work in enterprises all over the world.

Trondheim, October 15, 1994

Professor Asbjørn Rolstadås, Editor

PART ONE

Management Issues

The Rank Xerox Experience: Benchmarking Ten Years On

R. Cross[a] and A. Iqbal [b]

[a] Rank Xerox Limited, Parkway, Marlow, Buckinghamshire SL7 1YL, United Kingdom

[b] Rank Xerox A/S, PO Box 905, N-1301 Sandvika, Norway

1. INTRODUCTION

1994 represents the tenth anniversary of the introduction of Leadership Through Quality to Rank Xerox. Leadership Through Quality led to a fundamental change in the modus operandi of Rank Xerox. The Rank Xerox Quality journey, documented in the award winning submission to the European Foundation for Quality Management in 1992 is a compelling case study of how to achieve business excellence. As an international company, a joint venture between the Rank Organisation and Xerox Corporation, the company was originally a monopoly. There was a rags to riches rise to fame. Protected by patents, Rank Xerox was one of the fastest growing post war businesses in the sixties and seventies.

When the patents expired in the mid-seventies the Japanese challenged the supremacy of Rank Xerox. The company was introspective and like many monopolies bureaucratic. By the early 1980s as market share slumped survival was at stake. Our benchmarking initiative, pioneered in 1979 and further developed as part of the search for a recovery strategy, has been credited as a major contributor to the turnaround of Rank Xerox. However, success in one arena is not a recipe for success elsewhere. Benchmarking does not carry a total satisfaction guarantee. The Quality journey is paved with good intentions as well as fashionable concepts. This year's flavour in Europe appears to be benchmarking.

1.1 Aims

The aims of this paper are to provide a background context on benchmarking and to address the following two issues. Firstly, what are the core components of a very successful benchmarking exercise. Secondly, how can benchmarking efforts be best prioritised by an organisation. In this respect the paper will cite major lessons learned by Rank Xerox. Additionally there is a summary of some practical benefits gained by Rank Xerox, the company many observers cite as the founders of modern benchmarking.

The paper starts with a brief history of benchmarking in Rank Xerox, then there is a definition of terms, including a typology of benchmarking exercises and outline of the research undertaken. Following this is a review of the characteristics of very successful benchmark studies. Finally, there is a review of lessons learned and an overview of how Rank Xerox prioritises benchmarking activities along with a summary of some key benefits.

1.2 History

Benchmarking was a catalyst for change in Rank Xerox. Originally termed competitive benchmarking, the first efforts were in our Manufacturing Operations. The results from our early benchmarking exercises were like shock therapy for senior management. We discovered our unit manufacturing costs were the same as the competitors' selling price. The widespread belief prior to benchmarking was that the competitors' machines were poor quality. This was proved by benchmarking to be wrong. And to drive the point home they were making profit!

Following further study of the successes of our joint-venture Fuji Xerox in collaboration with Xerox, we used the information gained to develop our Leadership Through Quality strategy. Increasingly benchmarking became seen as a tool to assist the total company. Originally there had been a quantitative emphasis: a focus on targets or standards to achieve. As described in the seminal work on the subject by Bob Camp we realised that benchmarking could be applied to all aspects of our business, and comparing ourselves against companies outside our immediate competition or in some cases, industry, was also valid. Gradually the focus has changed from cost comparison with competitors to include best in class and the *processes* that lead to superior results. Now benchmarking has become a lifestyle and is applied across all aspects of our business - from environment, purchasing, human resources through to software design, the audit function and so on.

1.3 Definitions

The formal definition of benchmarking used by Rank Xerox is: "a continuous systematic process of evaluating companies recognised as industry leaders, to determine business and work processes that represent best practices and establish rational performance goals". In operational terms this is frequently condensed to "the search for industry best practices that lead to superior performance." By "best practices" is meant the methods used in work processes that best meet customer requirements. Benchmarking is then not simply about what we want to achieve; the benchmarks or the measurements of best performance, but also how they are achieved, the processes that are used.

2 RESEARCH APPROACH AND FINDINGS

It is important to stress that an underlying tenet of benchmarking is that it is a learning experience, an experience that has as its end product positive change. Given then that benchmarking is an organisational learning experience, what are the approaches used and what makes for success. These are two of the questions posed when on behalf of Xerox Quality Solutions, the Total Quality consultancy arm of Rank Xerox, a survey based on the Xerox Ten-step approach (shown in table 1 on the next page) was conducted. This survey was sent to one hundred and twenty of the top 1000 companies in the UK and the response rate was eighty-four per cent. Seventy-four per cent of these companies described themselves as using benchmarking. Twenty per cent stated their exercise was very successful, forty-two per cent stated the exercise was successful and thirty-eight per cent concluded their exercise was moderately successful. Incidentally none described their benchmarking as unsuccessful.

It goes without saying that it is good news that the benefits of benchmarking are such that there were no unsuccessful exercises. That being so, moderate success is insufficient in the current competitive environment. Benchmarking can be resource intensive and takes time. It is critical to understand how to make benchmarking exercises very successful. As this was a pilot research project we are presented with a number of hypotheses for further exploration. That said, the findings confirm our experience and present a number of areas which in our view merit further exploration. At a more pragmatic level they present a checklist for success.

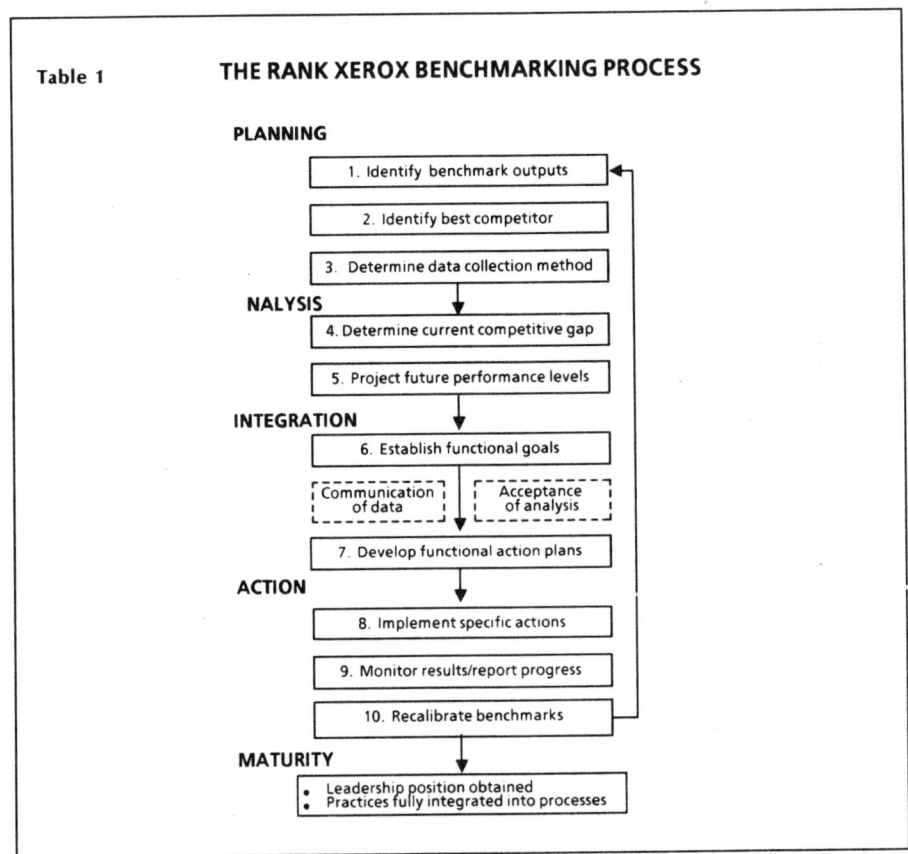

Table 1   **THE RANK XEROX BENCHMARKING PROCESS**

PLANNING

1. Identify benchmark outputs

2. Identify best competitor

3. Determine data collection method

NALYSIS

4. Determine current competitive gap

5. Project future performance levels

INTEGRATION

6. Establish functional goals

Communication of data   Acceptance of analysis

7. Develop functional action plans

ACTION

8. Implement specific actions

9. Monitor results/report progress

10. Recalibrate benchmarks

MATURITY

• Leadership position obtained
• Practices fully integrated into processes

2.1 Benchmarking types

Three main types of benchmarking were identified by the survey: competitive internal and functional. Within these types there tended to be a focus on metrics and targets (tending to be more strategic benchmarking) or processes. The approaches taken were identified as ranging from informal to formal. A formal appproach is defined as use of a disciplined and structured methodology. Informal is defined as "common sense" or in some cases muddling through. Based on the research findings and experience at Rank Xerox as hosts of and participants in numerous studies, a classification emerges. On the one hand there is the change-agent tour which uses a structured approach. These are strategic studies where top management of a company reviews issues of strategic significance. There are project plans developed as a consequence, and actions take place. Such tours generate teamwork and vision, a sense of what is possible and commitment to change. Then there are the process professionals, these are generally process owners. They are experts in their field, have an agreed approach, understand their subject and there is two-way sharing. These first two studies can be very successful. However not all formal studies are successful, there can be the unfinished doctorate. This team loses their way, perhaps obsessed in detail the project takes too long to complete. En route the reason for the benchmark study or sponsor disappears.

On the less structured side, with the more informal approach there are the Business tourists. This is a group that arrives, asks general questions and has not done their homework. At one extreme such tours can be characterised as shopping expeditions, at their worst the host organisation may perceive them to be shoplifting. Given that the above classification of benchmarking exists, the list by the way is not exhaustive. What makes for a *very* successful benchmarking exercise? Ten years on the challenge was to establish whether application of the ten-step process devised by Xerox was valid. Were there, for example, any steps in the process that were more significant for success. The next section reviews the key findings.

2.2 Phase 1 : Planning

This phase is frequently identified as the most important, as well as time-consuming phase in the process. The objective of this phase is to prepare the plan for benchmarking. And the first three steps are deceptively simple. The questions addressed are what is the subject to be benchmarked, who are the best "competitors", and what is the best data collection method?

The survey identified that from the outset very successful studies had a stronger focus to exceed best practices than others. But it wasn't just the initial focus of the study that appeared to be important. The guidelines to the first step refer to the need to review the subject of the study with the internal customer(s) and users of the study. The strength of this advice correlates with the finding that *one hundred per cent* of the very successful studies analysed to a large degree who the main users would be, and *the effects* it would have on a specific function within their organisation. Conversely fifteen per cent of the moderately successful studies had not done this at all and only thirty-five per cent identified to any great degree who the users would be. Related to this, on what is called the buy-in or commitment factor, *fifty-seven per cent* of the very successful studies had extensively consulted management and customers of the study - only fourteen per cent of the moderately successful studies had done so.

A further question was posed, relating to whether the study had a sponsor, a senior manager committed to the exercise. It is striking to note that *all* the very successful studies had a sponsor. By contrast, rather like characters in search of a play, forty-six per cent of the moderately successful studies had no sponsor. A *strong process orientation* was also evident and an attribute of the very successful studies. *Fifty-seven per cent* of the very successful studies documented their internal processes *to a very great degree* compared to only seventeen per cent of the moderately successful studies. It was also revealed that in the data preparation phase the importance of reviewing questions beforehand was crucial. Sixty-two per cent of the very successful studies pretested questions to a very great extent, whereas only twenty six per cent of the moderately successful did so to the same degree. It should be axiomatic that documentation of processes and pretesting of questions are prerequisites for effective benchmarking. This step allows for a focus on what is to be benchmarked and subsequently facilitates comparison between organisations. Comparison is after all the essence of benchmarking. Again this reinforces the guidelines associated with the first step of the process. The guidelines state that when you consider what is to be benchmarked, before you look externally you need to *understand your internal process*. As the research suggests this advice is ignored at the peril of the benchmarker.

2.3 Phase 2: Analysis
This phase is concerned with understanding the competitors' strengths. When assessing a company against the strengths it is important to address two questions. Firstly, is the competition better and secondly, why? In this phase the process orientation of the very successful studies is once more underlined. Seventy per cent of the very successful studies ensured that processes were fully understood before concentrating on metrics. This was not the case for the moderately successful studies where instead thirty per cent concentrated on metrics.

2.4 Phase 3: Integration
During this phase functional goals and action plans are developed. The requirement is to communicate the data and gain acceptance to the analysis. The survey results indicate that the commitment to communicate the findings and win support was highest for the very successful exercises. One might hazard a guess that they had more to communicate! Major gaps will require significant efforts. Two enemies of change are encountered at this stage. There is the Not Invented Here syndrome and associated with this the *shoot the messenger* tactic. Bringers of bad news are not always welcome, even though they may have the facts and data. This reinforces the need to have sponsors involved at the planning stage of any study. Furthermore it suggests that emphasis on the planning step is in itself part of the unfreezing process which creates widespread acceptance and commitment to change. The resistance to change challenge has to be tackled at the outset of a study.

2.5 Phase 4: Plannning
The extent to which change is achieved may be a key yardstick of a successful study. Twenty-one per cent of the very successful studies initiated change to a very great extent, sixty-four per cent to a great extent. In comparison all the moderately successful studies implemented change to some extent. More significantly *ninety two per cent* of the very successful studies incorporated benchmarks into the management process and found the study helpful in setting targets. This was not consistently the case for the moderately successful studies where *only* twenty-five per cent found the exercise helpful in setting targets. One additional finding was that the commitment to do something with the results was higher for the very successful studies: they *all* developed action plans.

2.6 Conclusions and the Rank Xerox experience

The picture which emerges from this external research tends to support our own experience from 1990. Then a multi-national team reviewed the effectiveness of benchmarking across the corporation. Combined with the findings from the above

---

Table 2   **LESSONS LEARNED (1990 QUALITY ASSESSMENT)**

- Inadequate understanding of benchmarking and the resource required
- Leadership an issue, variable knowledge and few champions
- Linkages to other Quality initiatives not understood
- Little evidence of strategic benchmarking
- Incomplete preparation for benchmarking studies
- Benchmarking conducted primarily for problem solving
- Results of benchmarking studies from the 'not invented here' syndrome
- Not integrated into the planning process

---

survey the lessons learned in table two demonstrate some of the key challenges for any organisation keen to maximise the utilisation of benchmarking as a management tool. Our experience is that these are some typical issues faced by many organisations in the early stages of benchmarking. The survey findings do not mean that an informal approach to benchmarking cannot work, rather the implication is that for optimum success a high degree of discipline and adherence to elements of the process is important. There is an educational challenge pertaining to the nature of benchmarking, the extent of preparation required and the skills to document processes. Allied to this the commitment of senior management is paramount. There must be champions or *sponsors* of studies. Benchmarking is not a quick fix and there need to be realistic expectations before embarking on any exercise. Finally there has to be a link between benchmarking and a company's planning process. To maximise effectiveness benchmarking should not be planned in isolation of corporate priorities.

3 IMPROVEMENT ACTIONS AND IMPLEMENTATION OF BENCHMARKING

As a consequence of the 1990 internal assessment two decisions were made at top level in Rank Xerox. Firstly there was an integration of policy deployment with benchmarking, secondly intensive training workshops were developed and targeted to groups who were implementing benchmark studies. Subsequently in 1992 as part of our own Business Excellence Certification process (where each unit assesses itself against the attributes of a world class company) benchmarking was highlighted as an item for special attention. The actions taken may not be appropriate to all organisations but in our case we have evidence to prove that they have worked.

3.1 Policy Deployment

Policy Deployment is designed to link the four corporate priorities of Customer Satisfaction, Employee Motivation and Satisfaction, Market Share, and Return on Assets with those of all employees. Launched in 1989 Policy Deployment is used as the annual process for turning strategic direction into operational business plans. Implementation of Policy Deployment provides a framework for identifying and implementing the key improvement actions (*called the vital few*) necessary to achieve the strategic intent relative to the four corporate priorities. Starting from the top of the organisation a hierarchy of priorities, objectives and vital few actions is cascaded through each Operating unit, where it becomes the framework for development of their objectives and activities. In each unit, functional, departmental, and individual objectives are negotiated and agreed. A diagnostic tool as well as a process for setting objectives, Policy Deployment is bottom-up as well as top-down. It has helped focus on the vital few actions.

As an outcome of the internal assessment the decision was made to formalise the link between benchmarking and Policy Deployment. Now benchmarking supports Policy Deployment at a strategic level, where benchmarks are established for the four corporate priorities. At a tactical level once vital few actions have been agreed there is examination of the benefits available from benchmarking in the context of the business priorities and the allocation of resources against these priorities. Through the Policy Deployment process benchmarking, where deemed appropriate, forms an integral part of the improvement plans to deliver breakthroughs.

In this way a unit or function can apply knowledge of their own operations and market conditions in order to focus on resourcing the benchmarking exercises that will *yield the most significant business impact*. This linking of benchmarking to policy deployment has refocussed the nature of benchmark studies and ensures that there is congruence with business priorities before a benchmark study is implemented. The organisation is now more selective in the choice of benchmarking studies. We are also able to agree pan-European studies on areas of common need.

3.2 Training Intensification

An application driven training programme for sponsors, project leaders and benchmarking study team members was also introduced. Sponsors of studies are taken through an exercise which highlights the key elements of successful benchmarking, the resources required and how to inspect the progress of benchmarking studies. Benchmarking project leaders and team members attend a one day workshop which trains them in detail on all steps of the benchmarking process. After the first workshop, the teams attend a half day start-up workshop where the the study starts in real time with emphasis on the first step in the process. These training courses and workshops are designed and conducted by internal specialists from our own quality network or consultants from Xerox Quality Solutions (XQS).

3.3 Benchmarking and Business Excellence

The progress of benchmarking and its application throughout Rank Xerox is monitored and reviewed through the internal self-assessment process called Business Excellence Certification. The concept underpinning Business Excellence is that each unit self-assesses its progress against the attributes of a world class company. Each unit rates itself along a seven point scale from nil to world class on all thirty seven attributes. One of these attributes is benchmarking and it has been classified as an area where we want to ensure we are at world class standards. This means that benchmarking receives senior management scrutiny on an ongoing basis. Each unit is aware of where benchmarking is utilised, the strengths and the areas for improvement. Accordingly we have evidence of where benchmarking has been applied effectively, and improvements made in use over the last three years.

4 CONCLUSION: BENCHMARKING APPLICATION AND BENEFITS

There are numerous examples of the application of benchmarking in all areas of the Rank Xerox business. A widely publicised success has been in the evolution of Just in Time. Over a period of ten years the Manufacturing and Logistics people in Rank Xerox have achieved major process changes which have raised customer satisfaction, reduced assets and costs. Benchmarked against the best in the field - IBM, DEC, Apple and others - the activity was run on Quality principles with cross functional project teams. This work started in 1983 and comprised of the introduction of JIT principles, reduction in the supplier base, quality certified suppliers, the elimination of waste creating activities and so on. Most importantly the following results were achieved: raw materials stocks went down eighty per cent, production lead times down forty-six per cent, and component defects down ninety-nine per cent.

As mentioned in the introduction benchmarking is part of our lifestyle rather than an esoteric tool. Most recently there has been a series of internal benchmarking programmes between the twenty different operating companies across Europe. In this work the process established to *transfer* the best practices is crucial to success and documentation of best practices mandatory. As an example, the Austrian operating unit currently has the highest customer retention rates in Rank Xerox. This is the result of a concerted programme on the part of senior managers and the sales force. The initiatives they use are now being adopted across all countries. The company also examined the role of information technology and a team of managers - cross-functional and cross-site - benchmarked different approaches internally. As a result, the cost of running the international data centre has been significantly reduced with associated improvements in process performance.

In conclusion there is a placebo-like quality to benchmarking. Rather like a therapy, the effectiveness and success of benchmarking has a lot to do with how ready and receptive a company is to change. Rank Xerox started benchmarking when the company was in a crisis. The success of our initial benchmarking has led us to believe in its curative power. However it must be stressed that it is not a panacea. To be very successful in benchmarking a combination of *commitment, critical self-analysis* and *discipline* at *key steps* in the process, are essential.

REFERENCES:

1. Camp R C, 1983, 'Benchmarking: the Search for Industry Best Practices that lead to Superior Performance"
    ASQC Quality Press, Milwaukee.
2. The Rank Xerox European Quality Award Submission Document 1992
    XBS, Uxbridge.
3. Cross R and Leonard P, 1994, 'Benchmarking: a Strategic and Tactical Perspective' in B Dale (ed.), 'Managing Quality' 2nd edn:
    Prentice Hall, Hemel Hempstead.
4. Haworth R, 1994, 'Benchmarking in British Industry'.
    Unpublished dissertation, De Montfort University, Leicester.

The Extended Enterprise - A Context for Benchmarking

J. Browne[a], P.J. Sackett[b] and J.C. Wortmann[c]

[a]CIMRU, University College, Galway.
[b]CIMI, Cranfield University
[c]Eindhoven University

1. INTRODUCTION

In his introduction to the 1992 European Manufacturing Futures Survey, De. Meyer suggests that manufacturing must "see itself as a link in an *integrated value added chain*, whose goal is to serve the customer" (my emphasis). Coming from a very different perspective, namely that of environmentally benign production, Tipnis (1993) also suggests a similar view when he subtitles his paper "How to design products that are environmentally safe to manufacture/assemble, distribute, use, service/repair, discard/collect, disassemble, recover/recycle and dispose?".

It is clear that the manufacturing function must look beyond "the four walls of the manufacturing plant". Global competition and the emerging pressures to develop environmentally benign products and processes, force manufacturing professionals to take a broader view. We term this broader view the *"Extended Enterprise" and we believe that it provides the context within which Benchmarking takes place*. In this paper we explore the concept of the Extended Enterprise and show how benchmarking can be an important tool to drive performance in the Extended Enterprise.

2. WHAT IS BENCHMARKING?

According to the PA Consulting Group, quoted in Kleinhans, Merle and Doumeingts (1994) "Benchmarking is the ongoing task, at all levels of our business, of finding and implementing world best practice in the key things we do that deliver customer satisfaction". The question arises as to what constitutes customer satisfaction. In the past, customer satisfaction was seen in terms of a quality product, delivered on time and at the right price. Today the situation is somewhat more complex. Customers, influenced by access to global information and increasingly environmentally aware, are taking a more sophisticated view. Environmental regulations and consumer preference for "Green" products are forcing corporations to design products that are environmentally safe to manufacture, use and dispose of. (See Tipnis 1993). Companies now gain competitive advantage because customers prefer their "Green" products and processes. Thus when we talk about benchmarking, defining best or "world class" practice in manufacturing, we are not thinking in terms of manufacturing performance as measured by cost, due date or quality metrics, rather we must consider the total product life cycle, from concept through design and manufacture and on to distribution, end of life disposal or refurbishment and reuse.

An important issue in benchmarking is the definition of appropriate measures of performance. Measures of performance are important. Clearly they provide milestones against which performance can be evaluated. However, measures of performance also serve to influence, often times to direct behaviour. Therefore the measures of performance must reflect very well the objectives of the system they purport to measure. In the 1960s and 1970s, the focus within manufacturing was on the cost of production. Labour productivity became an important measures of performance. Industrial companies, and indeed industrial economies benchmarked their performance in terms of labour productivity. In the 1980s, the emphasis shifted on to quality. Zero defects as suggested by Japanese customers, became the benchmark. Today the appropriate measures are less clear. Certainly cost, quality and timely delivery are important performance measures against which industrial companies can benchmark. However, other issues, thrown up by global competition and the increased emphasis on environmental concerns must also be considered.

3. THE EMERGENCE OF THE GLOBAL MARKET

Changes in the international trading relationships, manufacturing and information technology and infrastructure developments are leading to the realisation of the global market. In the past, companies operated in local markets, were subject to local cost and pricing structures and competed with local competitors. In many cases companies were effectively restricted to local suppliers, since information on remote suppliers was unavailable. The communications and transport costs associated with doing business with distant suppliers and customers were prohibitive. Trading conditions, customs controls and tariffs provided further barriers to global trade.

In recent years, the transportation problem has been radically reduced through vastly superior infrastructure and more efficient forms of transport such as economical air freight for perishable goods. This has greatly reduced the importance of manufacturing products close to their point of consumption. Miniaturisation of many product has also contributed to this trend. Communications technology has provided the customer with the ability to choose among a far greater range of vendors, operating both locally and in distant locations. In response to customer demands, trade restrictions have been greatly reduced, evidenced most recently by the signing of the new GATT Agreement by over 100 countries in 1994.

Such changes have certainly led to pressures on industry. Competitors may operate from a radically different cost base existing in a different country or continent, with cheaper labour rates, and can therefore offer more attractive prices to the customer. Other firms may have a technological advantage, or an extensive marketing apparatus, and so on. Industry must therefore strive for both cost competitiveness and product differentiation in order to compete. Conversely, successful firms can be rewarded well with a greatly expanded market in which to sell their products. The increasing sophistication of products, and pressures to reduce costs through efficient manufacture, has also led to increases in the capital outlay required to manufacture them. This can restrict manufacture to those companies with large, often global, markets where sufficient product sales can cover such high initial expenditure. Smaller firms are increasingly outsourcing and sub-contracting to specialist firms who operate in a different part of the world.

3.1 The Concept of Mass Customisation.

Up to the mid-nineteenth century, manufacturing in so far as it existed at all was confined to the production of low value customised products – in effect craft production. With the development of hard automation (the transfer line etc.) the age of mass produced standard products began. The relatively low cost of such products allowed significant proportions of the population of the western world to acquire consumer goods. As the technology of automation and mass production developed further, costs were further reduced. The development of information technology, and it's application to manufacturing technology in the 1960s and 1970s changed everything. Automation based on information technology is inherently different to the older "hard" automation in that it is flexible. Computer based automation facilitates the production of a wider range of products in lower volumes economically. Today we are approaching the stage where "mass customisation" is becoming a reality.

Mass customisation requires significant alterations to the traditional organisational, human and technological support systems of an enterprise. Existing companies have for many years carried out "continuous improvement" programmes which aim at continuously improving the efficiency and quality of the manufacturing process, and gradually optimising that process by increasing flexibility, employee skills, eliminating waste, and so on. Those employees who were responsible for operating the process were encouraged to find methods of improving it, continuously, in small increments. It has been assumed by some companies that a natural progression from continuous improvement would lead to the process and human flexibility required to support mass customisation. However, a critical difference between these two approaches is that with continuous improvement the product is stable, employees can assume this is what the customer requires. With mass customisation, the product specification cannot be taken for granted.

Mass customisation requires a dynamic network of relatively autonomous units. Each unit handles a specific process or task, such as manufacturing a particular component, or performing a particular type of assembly. These units are not linked together in a serial manner. They are organised to be able to link together in any sequence which may be required to satisfy a customers requirements.

This approach implies a fundamental change in the way an enterprise plans for the future. Rather than ensuring it has the capability to produce a limited range of products quickly, efficiently and with zero defects, it must plan to produce any product variation the customer requires. Therefore, product and process design must concentrate on not only a limited number of known products, but on the capabilities required to design and manufacture customised products as required.

3.2 The Environmental Imperative.

Society is putting pressure on manufacturers, in order to create production systems which are neutral with respect to the environment. This pressure acts through government regulations and customer requests. It may take the form of legal regulations, economic and marketing requirements. These pressures constitute a challenge for companies, to develop new technology and materials. In the long run, environmentally benign production may become a competitive edge. Environmentally benign production requires a shift of paradigm for engineers, accountants, government agencies, and many other parties. There is one line of development which is clearly emerging from the environmental requirements for manufacturing as a system: *the object of study should not be restricted to a single plant or production facility, but should include chains or networks of production and physical distribution.* Issues such as design for recycling, refurbishment, environmental costing, and many legal issues can only be studied if the scope of study is enlarged to the chain of value-adding activities, including ultimately, end of life disassembly and refurbishment.

Today progressive manufacturing companies are developing a total life cycle approach to their products. One multinational supplier of telecommunications equipment and services to the European market has developed a "Product Life Cycle Management" program which currently includes six major activities, namely: design and technology, purchasing of supplies and materials, manufacturing processes, waste reduction and energy management, packaging and post consumer materials management. This company now includes environmental considerations as part of it's supplier qualification process. The "post consumer materials management" activity represents a long term challenge but is based on the following ideas : in the near future, manufacturing will refurbish and reuse products and recycle as much of their contents as possible; finally companies will source new markets for the recycled material and safely dispose of residual materials when necessary.

For some manufacturing companies the "recycle, refurbish and resale" process has already begun. A recent article in the Economist (October <~> November 5th, 1993, Page 99) reported that a UK company is producing and selling recycled and reconditioned motorcycles.

A major European based computer manufacturer has developed a "Design for Sustainable Development" programme. This programme was developed in the early 1990s in response to three external influencing factors viz, emerging international legislation, the need to maintain and increase customer satisfaction and finally in an attempt to secure competitive advantage in the market. This company found that certain key customers were refusing to purchase consumables from it until it had implemented a return process. (Examples include laser printer toner kits). Furthermore, other customers were requiring the reuse and return of product accessories in order for the company to qualify as a key supplier. The company is convinced that it will develop a competitive advantage over it's rivals through it's "Design for Sustainable Development" programme. This programme focuses on "waste management", through waste reduction at source, design for disassembly and maximising the opportunity to recover materials through reuse, recycling, reclamation, resale, reconditioning and re manufacturing. This company has already developed a "Waste Management" infrastructure, which includes an operating special purpose facility in Europe to take back computers from key customers and recycle, reclaim, refurbish and resell them.

Alting (1993) suggests that the research and developments community will have to come up with a life-cycle concept, which individual companies can use to tailor their own specific concepts. Based on this life-cycle concept guidelines must be developed for each of the phases, that is design for environmentally and occupational health friendly production, distribution, usage and disposal/recycling. The important point that we can no longer see manufacturing as an end in itself; rather as one in a long chain of value adding activities. Figure 1 outlines this perspective.

Tipnis makes reference to the "expanded responsibility" of the manufacturer over the entire life cycle and talks about the design of products for sustainability. In his view design for sustainability provides specific targets for design for manufacture, assembly, service, disassembly, and recycling. Again we have evidence of the expanding role of the manufacturing <~> beyond the four walls of the manufacturing plant.

We must also recognise the pressure of legislation forcing industrial companies to take responsibility for the environmental impact of the products and processes.

In Europe, the Council of the European Union has issued several directives regarding the environmentally sound production, distribution, use and disposal of products. These directives include the principles of the civil liability of the manufacturer for

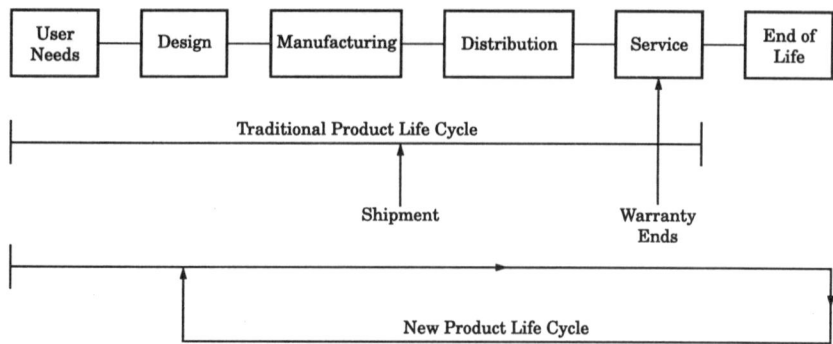

Fig. 1   Product Life Cycle as Seen by Producer

environmental damage caused by their products, financial incentives to achieve effective protection of the environment such as taxes on the use of damaging materials (for example Cholor-Floro-Carbons <~> CFCs), material disposal charges, independent environmental audits and charges on non-biodegradable materials such as certain plastics. It is likely that in the near future manufacturers will be required to recycle obsolete products themselves or face high public disposal charges.

In Germany, government ordinances have been produced covering electronic waste regulation. These require that the manufacturer takes back all electronic equipment produced. If they have no recycling programme in place, then they are forced to bear the cost of "stockpiling" themselves, or pay large landfill charges. There is also legislation in place which requires that all plastic parts are marked to International Standards Organisation (ISO) 1042 standard which will facilitate recycling in the future.

In the United States, government bodies have defined recycling goals, source material and energy reduction goals, incentives for take back programmes, and credit for existing programmes. Certain taxes already exist to discourage the emission of toxic waste. The Environmental Protection Agency (EPA) provides a product life-cycle assessment service, which can be used by manufacturers to pinpoint the occurrences and causes of environmental pollution in the life of their products. Individual states have also determined requirements for recycling and usage bans on specific materials and components.

4. THE EXTENDED ENTERPRISE

How can a manufacturing company respond to the pressures placed on it by global competition? In the past the response was to make the factory more efficient and responsive by developing CIM (Computer Integrated Manufacturing) solutions. Now, however, the challenge is greater and requires that *a degree of integration takes place across the whole value chain*. The United States Department of Defence initiative, CALS <~> Computer-aided Acquisition & Logistics Support, has recognised this.

It is likely that, together with the European initiatives, CALS will influence the standards and the pace for inter organisation technical communication. The market place to which manufacturing businesses, business integration researchers, their teaching and their facilities must respond includes:-

- Business processes which cross enterprise boundaries to interface with functional areas in *other* companies, for example, product design or manufacturing process definition.

- Supplier/customer integration (people and processes) through interchange of commercial/technical data.

- Ability to function effectively as links for information and product in unbuffered supply/distribution chains.

The ability to network the activities of a number of entities to produce and sell manufactured products profitably depends on the relationship of these entities and the communication that passes between them (figure 2).

We are accustomed to thinking about this in the context of a single enterprise with different departments, Sales, Design, Engineering, Manufacturing, Distribution etc. However, within a global market-place, entities from many different enterprises, or entities which in themselves are nominally independent enterprises, relate via a single product to produce a designed result. An example might be a merchandising entity recognising a business opportunity and requesting :-

- a design entity to design it;

- a manufacturing entity to build it;

- a distribution entity to distribute it and;

- a marketing entity to sell it.

The implication of such an example is that all the entities can be considered as "flexible" or "programmable" within their expertise envelope.

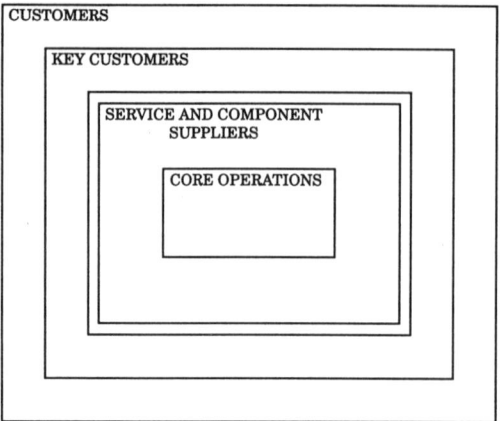

Fig. 2   The Extended Enterprise

We must also recognise that developments in computing and telecommunications facilitate the realisation of the networked Extended Enterprise. Referring to electronic data interchange and to the emerging integration of computing and telecommunications technologies, Keen uses the terms "Reach" and "Range". Reach is the extent to which one can interact with other communication nodes <~> in the limit it becomes anyone, anywhere. Range defines the information types that can be supported from simple messaging between identical platforms to any computer generated data between any operating platforms (figure 3). Electronic mail available to all members of a single department offers the lowest level of technology integration. Until recently, extending reach across other parts of the company and particularly beyond the company has been both technologically difficult and expensive.

Extending range requires a good definition of business processes. Tools and organisational attitudes are now becoming available that make range extension viable. The driving force is the enhancement of business degrees of freedom to respond to the volatile market place. The level of change in the market place requires that we have flexibility not only in product but in business structure and business processes. A high level of reach and range provides the business freedom to operate in the Extended Enterprise. The mapping of a company onto the range/reach chart gives a good indication of its scope for innovative business improvement through the use of integration technologies (figure 4). In the Extended Enterprise the bringing together of core competencies from many different organisations to provide a short manufactured life product means regular enterprise business process restructuring. The pressure on time scales means interactive decision support is required. Companies without appropriate reach and range will not be able to participate in this Extended Enterprise

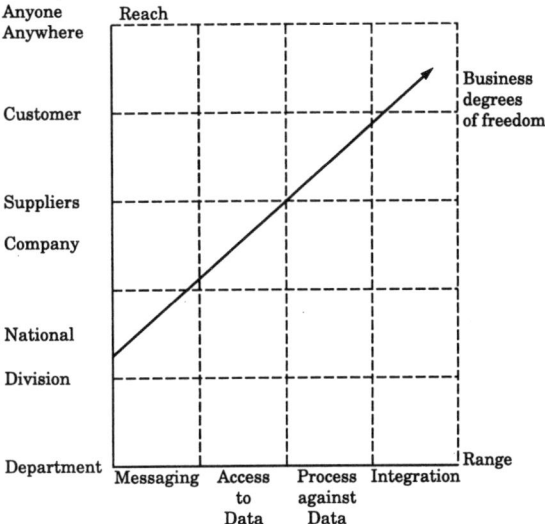

Fig. 3 Business and Technology Integration

business system. Technology integration price/performance improvements will greatly reduce price entry barriers to all companies world wide. In the past high reach/range has been recognised as valuable but has been only available in specialist applications.

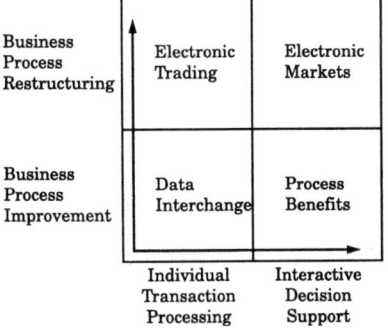

Fig. 4 Scope for Business Improvement through Integration

5. BENCHMARKING AND THE EXTENDED ENTERPRISE

The institutionalisation of the Extended Enterprise will have far reaching outcomes. This will involve major structural change in business organisation. The basis of partnership within the Extended Enterprise is not yet well understood but alternative operations models are likely to be industry and market sector specific. Concurrent Engineering is becoming accepted but understanding of best practice on how, when and in what order to implement it is needed. The extension of tools to embrace environmental issues could offer significant benefit to small and medium sized enterprises. Organisational learning realisation offers the utilisation of industy's most significant business resource, its highly educated labour force. This will involve a re-assessment of the value of knowledge and the development of organisation based tools to use it. The appraisal of manufacturing business options must be developed to match the changes in the business operations environment described previously. Manufacturing information systems need to provide task defined specific solutions embracing emerging multimedia advances.

In Figure 4 we suggested that the development of electronic trading and indeed electronic markets will require business process restructuring. Already we see examples of this happening. The advanced use of EDE (Electronic Data Exchange) and EDI (Electronic Data Interchange) systems has fundamentally changed the nature of the interaction between many large customers (say final assemblers) and their component suppliers. Completely new ways of doing business have been developed. The authors are aware of one example where a component supplier sends his dispatcher to the site of his customer (a major telecommunications systems assembler) to identify the customers needs for components and is paid for these components based on the quantity of finished products shipped by the customer to his customers! Certainly this is an example of the "Extended Enterprise" in practice. It also indicates the need to redesign business processes to cope with the new reality. The redesign of business processes in turn is aided by the use of benchmarking which allows a manager or analyst to understand the potential for improvement in his own process by comparing them to world class and best practice.

6. CONCLUSIONS

In this paper, we have considered the emergence of the Extended Enterprise as a response to the pressures now impinging on manufacturers. We considered these pressures in terms of the development of the global market and the need to develop environmentally benign processes and products. The Extended Enterprise represents a new way of doing business and requires that we redesign and reengineer appropriate business processes. Benchmarking can support this activity by allowing an understanding of best practice. The Extended Enterprise allows us to provide for local customers while meeting global standards. Benchmarking facilitates the identification of global standards.

REFERENCES

1 Browne, J., Sackett, P.J., Wortmann, J.C., "The System of Manufacturing: A perspective study", Study for DGXII of the EU, November, 1993.
2 Alting, L. "Life-Cycle Design of Products: A New Opportunity for Manufacturing Enterprises", in "Concurrent Engineering: Automation, Tools & Techniques", Editor: A. Kusiak, J. Wiley & Sons, 1993.
3 Kleinhans, S., Merle, C., Doumeingts, G., "Determination of What to Benchmark: a customer - oriented methodology", IFIP WG 5.7 Working Conference on Benchmarking, Trondheim, Norway, June 1994.
4 Tipnis, V.J., "Evolving Issues in Product Life Cycle Design", Annals of the CIRP, Volume 42, No.1, 1993.
5 O'hEocha, C., "A Framework for Concurrent Engineering", Master of Engineering Thesis, U.C.G., 1994.
6 De Meyer, A., "Creating the Virtual Factory" - Report of the 1992 European Manufacturing Futures Survey", INSEAD, France, December, 1992.
7 Keen, P.G.W., "Shaping the Future : Business Design through Information Technology", Harvard Business School Press, 1991.

3

The Role of Competency Questions in Enterprise Engineering

Michael Grüninger and Mark S. Fox *

Department of Industrial Engineering, University of Toronto,
4 Taddle Creek Road, Toronto, Ontario M5S 1A4
{gruninger, msf}@ie.utoronto.ca

Abstract

We present a logical framework for representing activities, states, time, and cost in an enterprise integration architecture. We define ontologies for these concepts in first-order logic and consider the problems of temporal projection and reasoning about the occurrence of actions. We characterize the ontology with the use of competency questions. The ontology must contain a necessary and sufficient set of axioms to represent and solve these questions. These questions not only characterize existing ontologies for enterprise engineering, but also drive the development of new ontologies that are required to solve the competency questions.

1.0 Introduction

Market competition is forcing firms to reconsider how they are organized to compete. As a basis for change, they are exploring a variety of concepts, including Benchmarking, Time-based Competition, Quality Function Deployment, Activity-Based Costing, Quality Circles, Continuous Improvement, Process Innovation, and Business Process Re-Engineering. Regrettably, most of the concepts are descriptive, if not ad hoc, and lack a formal model which would enable their consistent application across firms. Consider business process re-engineering [Davenport 93], [Hammer & Champy 93]. It is very much in the "guild" mold of application; management consultants are the "masters" and they impart their knowledge through "apprenticeship" to other consultants. The knowledge of business process re-engineering has yet to be formalized and reduced to engineering practice.

The goal of the Enterprise Engineering Project at the University of Toronto is to:

• Formalize the knowledge found in Enterprise Engineering perspectives such as Benchmarking, Time-based Competition, Quality Function Deployment, Activity-Based Costing, Quality Circles, Continuous Improvement, Process Innovation, and Business Process Re-Engineering. By formalize, we mean the identification, formal representation and computer implementation of the concepts, methods and heuristics which comprise a particular perspec-

*. This research is supported, in part, by the Natural Science and Engineering Research Council, Digital Equipment Corp., Micro Electronics and Computer Research Corp., and Spar Aerospace.

tive. This not only enables a precise formulation of the intuitions implicit in practice, but it is also a step towards automating the execution of certain tasks involved in enterprise engineering.

• Integrate the knowledge into a software tool that will support the enterprise engineering function by exploring alternative organization models spanning organization structure and behaviour. The Enterprise Engineering system allows for the exploration of a variety of enterprise designs. The process of exploration is one of design, analysis and re-design, where the system not only provides a comparative analysis of enterprise design alternatives, but can also provide guidance to the designer.These ideas are formalized in the notion of advisors (cf. [Grüninger & Fox 94]) that are able to analyze, guide, and make decisions about the current enterprise and possible alternatives.

• Provide a means for visualizing the enterprise from many of the perspectives mentioned above. The process of design is performed through the creation, analysis and modification of the enterprise from within each of the perspective visualizations.

Enterprise modelling is an essential step in defining the tasks and functionality of the various components of an enterprise.The goal is to create generic, reusable representations of Enterprise Knowledge that can be applied across a variety of enterprises. Towards this end, the TOVE (Toronto Virtual Enterprise) ontology [Fox et al 93] has been developed and applied to enterprise engineering [Fox et al 94], enterprise integration, and integrated supply chain management. An ontology is a formal description of entities and their properties; it forms a shared terminology for the objects of interest in the domain, along with definitions for the meaning of each of the terms. The TOVE ontology currently spans knowledge of activity, time, and causality, resources, and more enterprise oriented knowledge such as cost, quality and organization structure. The TOVE Testbed provides an environment for analyzing enterprise ontologies; it provides a model of an enterprise and tools for browsing, visualization, simulation, and deductive queries.

In this paper we present a logical framework for the TOVE ontology. We also present a set of tasks that arise in enterprise engineering and the requirements on any ontology that is used to represent the tasks and their solution. These requirements, which we call competency questions, are the basis for a rigorous characterization of the problems that the enterprise model is able to solve. The enterprise model must be able to represent the tasks specified by the competency questions and their solution. The questions are also those tasks for which the enterprise model finds all and only the correct solutions. Tasks such as these can serve to drive the development of new theories and representations and also to justify and characterize the capabilities of existing theories for enterprise modelling.

2.0 Common Sense Enterprise Modelling

The basic entities in the TOVE model are represented as objects with specific properties and relations. Objects are structured into taxonomies and the definitions of objects, attributes and relations are specified in first-order logic. An ontology is defined in the following way. We first identify the objects in our domain of discourse; these will be represented by constants and variables in our language. We then identify the properties of these objects and the relations that exist over these objects; these will be represented by predicates in our language.

We next define a set of axioms in first-order logic to represent the constraints over the objects and predicates in the ontology. This set of axioms constitutes a microtheory ([Lenat & Guha 90]) and provides a declarative specification for the various tasks we wish to model. Further, we need to prove results about the properties of our microtheories in order to provide a characterization and justification for our approach; this enables us to understand the scope and limitations of the approach. We use a set of problems, which we call competency questions, that serve to characterize the various ontologies and microtheories in our enterprise model. The microtheories must contain a necessary and sufficient set of axioms to represent and solve these questions, thus providing a declarative semantics for the system. It is in this sense that we can claim to have an adequate microtheory appropriate for a given task, and it is this rigour that is lacking in previous approaches to enterprise engineering.

The competency questions are generated by requiring that the ontologies and microtheories be necessary and sufficient to represent the tasks and their solutions for the various components of the system. Within enterprise engineering, these include:

• Temporal projection -- Given a set of actions that occur at different points in the future, what are the properties of resources and activities at arbitrary points in time? This includes the management of resources and activity-based costing (where we are assigning costs to resources and activities).To solve this problem, we need to define axioms that express how the truth of a proposition changes over time. In particular, we need to address the frame problem and express the properties and relations that change or do not change as the result of an activity. We will use this task to characterize the ontologies in this paper.

• Planning and scheduling -- what sequence of activities must be completed to achieve some goal? At what times must these activities be initiated and terminated?

• Benchmarking -- Can activities from one enterprise be used in another while still satisfying the constraints that exist within the enterprise's environment and achieving the goals of the enterprise?

• Hypothetical reasoning -- what will happen if we move one task ahead of schedule and another task behind schedule? What are the effects on orders if we buy another machine?

• Execution monitoring and external events -- What are the effects on the enterprise model of the occurrence of external and unexpected events (such as machine breakdown or the unavailability of resources)?

• Time-based competition -- we want to design an enterprise that minimizes the cycle time for a product [Blackburn 91]. This is essentially the task of finding a minimum duration plan that minimizes action occurrence and maximizes concurrency of activities.

Claiming that any ontologies are adequate for enterprise modelling requires proving that the ontologies can represent and solve these competency questions.

3.0 Ontologies and Microtheories

In this section we present the ontologies and microtheories in TOVE for time, activity, and cost. These ontologies will then be used to specify the tasks addressed by the components of the enterprise engineering system; the final section of the paper will present the competency questions that serve to characterize the ontologies and microtheories.

3.1 Time and Action

The problem of benchmarking requires that an enterprise understand its processes and the processes of another enterprise in order to determine whether there are any comparable processes that can be adopted. This requires an adequate representation for processes. In the following sections, we present the TOVE ontology for activity, state, causality, and time, and define the semantics of the constructs in the ontology using the situation calculus.

The intuition behind the situation calculus is that there is an initial situation, and that the world changes from one situation to another when actions are performed; the function $do(a,\sigma)$ is the name of the situation that results from performing action a in situation σ. There is a predicate $Poss(a,\sigma)$ that is true whenever an action a can be performed in situation σ. The structure of situations is that of a tree; two different sequences of actions lead to different situations. The tree structure of the situation calculus shows all possible ways in which the events of the future can unfold. Thus, each branch that starts in the initial situation can be understood as a hypothetical future.The work of [Pinto & Reiter 93] extends the situation calculus by selecting one branch of the situation tree to describe the evolution of the world as it actually unfolds. This is done using the predicate $actual(\sigma)$.

Situations are assigned different durations by defining the predicate $start(s,t)$. Each situation has a unique start time; these times begin at 0 in σ_0 and increase monotonically away from the initial situation.Time is represented as a continuous line on any branch in the tree of situations; on this line we define time points and time periods (intervals) as the domain of discourse. We define a relation < over time points with the intended interpretation that $t < t'$ iff t is earlier than t'. Using this relation, we can define the temporal relations of [Allen 84] over intervals.

To define the evaluation of the truth value of a sentence at some point in time, we will use the predicate $holds(f,\sigma)$ to represent the fact that some ground literal f is true in situation σ. Using the assignment of time to situations, we define the predicate $holds_T(f, t)$ to represent the fact that some ground literal f is true at time t. A fluent is a predicate or function whose value may change with time. Another important notion to represent is the occurrence of actions at points in time. To represent this we introduce two predicates: $occurs(a,\sigma)$ (action a occurs in situation σ), and $occurs_T(a,t)$ (action a occurs at time t) defined as follows:

$$occurs(a,\sigma) = actual(do(a,\sigma)) \tag{EQ 1}$$

$$occurs_T(a,t) = occurs(a,\sigma) \wedge start(do(a, \sigma), t) \tag{EQ 2}$$

3.2 Activities and States

At the heart of the TOVE Enterprise Model lies the representation of an *activity* and its corresponding enabling and caused *states* ([Sathi et al. 85], [Fox et al 93]). In this section we examine the notion of states and define how properties of activities are defined in terms of these states. An activity is the basic transformational action primitive with which processes and operations can be represented; it specifies how the world is changed. An enabling state defines what has to be true of the world in order for the activity to be performed. A caused state defines what is true of the world once the activity has been completed.

An activity, along with its enabling and caused states, is called an *activity cluster*. The state tree linked by an *enables* relation to an activity specifies what has to be true in order for the activity to be performed. The state tree linked to an activity by a *causes* relation defines what is true of the world once the activity has been completed. Intermediate states of an activity can be defined by elaborating the aggregate activity into an activity network (see Figure 1).

There are two types of states: *terminal* and *non-terminal*. In Figure 1, *es_fabricate_plug_on_-wire* is the nonterminal enabling state for the activity *fabricate_plug_on_wire* and *pro_fabricate_plug_on_wire* is the caused state for the activity. The terminal conjunct substates of *es_fabricate_plug_on_wire* are *consume_wire*, *consume_plug*, and *use_inject_mold* since all three resources must be present for the activity to occur; the terminal states of *pro_fabricate_plug_on_wire* are *produce_plug_on_wire* and *release_inject_mold*.

In TOVE there are four terminal states represented by the following predicates:*use(s,a)*, *consume(s,a)*, *release(s,a)*, *produce(s,a)*. These predicates relate the state with the resource required by the activity. Intuitively, a resource is used and released by an activity if none of the properties of a resource are changed when the activity is successfully terminated and the resource is released. A resource is consumed or produced if some property of the resource is changed after termination of the activity; this includes the existence and quantity of the resource, or some arbitrary property such as color. Thus *consume(s,a)* signifies that a resource is to be used up by the activity and will not exist once the activity is completed, and *produce(s,a)* signifies that a resource, that did not exist prior to the performance of the activity, has been created by the activity. We define use and consume states to be enabling states since the preconditions for activities refer to the properties of these states, while we define release and produce states to be caused states, since their properties are the result of the activity.

FIGURE 1   Activity-State Cluster

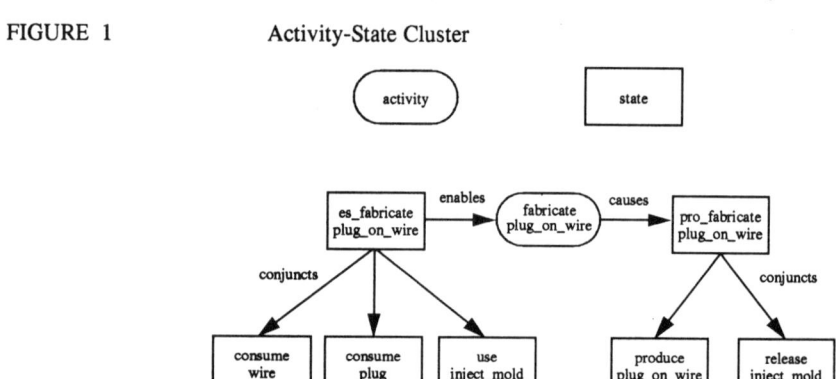

Terminal states are also used to represent the amount of a resource that is required for a state to be enabled. For this purpose, the predicate *quantity(s,r,q)* is introduced, where *s* is a state, *r* is the associated resource, and *q* is the amount of resource r that is required. Thus if *s* is a consume state, then *q* is the amount of resource consumed by the activity, if *s* is a use state, then *q* is the amount of resource used by the activity, and if *s* is a produce state, then *q* is the amount of resource produced.

In this section, we formalize the relationship between states and activities. First we examine the notion that an activity specifies a transformation on the world; this requires that we introduce fluents for states and activities, and the actions that change these fluents. The axioms presented adequate for solving the temporal projection problem for these properties of states and activities.

To formalize the notions of nonterminal states and aggregate activities, we introduce occurrence axioms for a set of actions.

3.3 Successor Axioms for Status of Terminal States

The primary fluents we will consider are the values assigned to states to capture the notion of the status of a state. We define a new sort for the domain of the status with the following set of constants:{ *possible, committed, enabled, completed, disenabled, reenabled*}. The status of a state is changed by one of the following actions:*commit(s,a), enable(s,a), complete(s,a), disenable(s,a), reenable(s,a)*. Note that these actions are parametrized by the state and the associated activity.

The next step is to define the successor axioms that specify how the above actions change the status of a state. These axioms provide a complete characterization of the value of a fluent after performing any action, so that we can use the solution to the frame problem in [Reiter 91]. Thus if we are given a set of action occurrences, we can solve the temporal projection problem (determining the value of a fluent at any point in time) by first finding the situation containing that time point, and then using the successor axioms to evaluate the status of the state in that situation. For example, we present two of the successor axioms in the microtheory:

The status of a state is committed in a situation iff either a commit action occurred in the preceding situation, or the state was already committed and an enable action did not occur:

$$(\forall s,a,e,\sigma)\ holds(status(s,a,committed),\ do(e,\sigma)) \equiv (e = commit(s,a) \wedge holds(status(s,a,possible),\sigma)) \vee \neg(e = enable(s,a)) \wedge holds(status(s,a,committed),\sigma)$$ (EQ 3)

The status of a state is enabled in a situation iff either an enable action occurred in the preceding situation, or the state was already committed and a complete action or disenable action did not occur:

$$(\forall s,a,e,\sigma)\ holds(status(s,a,enabled),\ do(e,\sigma)) \equiv (e = enable(s,a) \wedge holds(status(s,a,committed),\sigma)) \vee \neg[(e = complete(s,a) \vee e = disenable(s,a)) \wedge holds(status(s,a,enabled),\sigma)]$$ (EQ 4)

Using the successor state axioms, we can derive occurrence axioms that make the relationship between the occurrence of the actions that change the status of a state and the preconditions for these actions:

$$(\forall s,a,\sigma)\ occurs(commit(s,a),\sigma) \supset holds(status(s,a,possible),\sigma)$$ (EQ 5)

$$(\forall s,a,\sigma)\ occurs(enable(s,a),\sigma) \supset holds(status(s,a,committed),\sigma)$$ (EQ 6)

How are these incorporated into the activity-state clusters, which only represent the causal relationships among states and activities? The occurrence of a commit action is not explicitly given in the specification of an activity. However, since the status fluents can only be changed by the above set of actions, the following sentence can be derived from the axioms:

$(\forall \, s,a, \, \sigma) \; occurs(enable(s,a), \, \sigma) \supset (\exists \sigma') \; occurs(commit(s,a), \, \sigma')$ (EQ 7)

Similarly, the precondition for the *commit* action is that the state be *possible*. In [Fadel et al. 94] it is shown how the *possible* status is defined in terms of the availability of a resource for the activity. This includes the configuration or setup of a resource as well as capacity constraints for the concurrent execution of activities with a shared resource. Axioms similar to those above would be used to express the occurrence of the appropriate setup activities for some activity. This is necessary for formalizing time-based competition, where the occurrence of setup activities is minimized.

3.4 Status of Non-Terminal States

In TOVE, non-terminal states enable the boolean combination of states. We will consider four non-terminal states:*conjunctive, disjunctive, exclusive, not.* What precisely does it mean for a non-terminal state to be a boolean combination of states? For example, how do we define the status of a non-terminal state given the status of each substate? To define this notion, we must refer to the occurrence of the actions that change the status of the states.In this way we can define arbitrary nonterminal states as occurrence axioms.

Disjunctive states are used to formalize the intuition of a resource pool. We may have a set of resources, such as machines or operators, that can possibly be used by an activity. The activity only requires one of these resources, so the activity only needs to nondeterministically choose one of the alternative resources in the pool. Thus, the status of the disjunctive state changes if one of the resources has been selected and its status has been changed. For example, we have

$(\forall \, s,s_1,...,s_n,a, \, \sigma) \; disjunctive(s,a) \wedge substate(s_1,s) \wedge ... \wedge substate(s_n,s) \supset occurs(enable(s,a), \, \sigma) \equiv occurs(enable(s_1,a), \, \sigma) \vee ... \vee occurs(enable(s_n,a), \, \sigma)$ (EQ 8)

The successor axioms for the other values of status are defined in the same way. In other words, the occurrence of an action for a disjunctive state is equivalent to a disjunctive sentence of occurrence literals for each disjunct substate.

Similarly, we have the following constraints on conjunctive states:

$(\forall \, s,s_1,...,s_n,a, \, \sigma) \; conjunctive(s,a) \wedge substate(s_1,s) \wedge ... \wedge substate(s_n,s) \supset occurs(enable(s,a), \, \sigma) \equiv occurs(enable(s_1,a), \, \sigma) \wedge ... \wedge occurs(enable(s_n,a), \, \sigma)$ (EQ 9)

The occurrence of an action for a conjunctive state is equivalent to a conjunctive sentence of occurrence literals for each conjunct substate. Note that we make the assumption that all conjunct substates change their status at the same time.

3.5 Ontology of Cost

The ontology for activity-based costing is a formal specification of the assignment of costs to activities based on costs for the resources utilized by these activities [Tham et al. 94]. Each resource is assigned a unique cost depending on the status of its terminal state; these are represented by the predicates *committed_res_cost_unit(a,r,q,v)*, *enabled_res_cost_unit(a,r,q,v)*, *disenabled_res_cost_unit(a,r,q,v)*, *reenabled_cost_unit(a,r,q,v)*, for some activity *a* and resource *r*. The parameter *v* represents the cost metric for a unit *q* of the resource. It is assumed that the values for these costs are completely known and that they are unique. Based on the duration of a particular status value, the axioms in the ontology of cost assign a unique cost for the state at a point in time. The cost assigned to an activity at a point in time is the aggregation of the costs for the states of the activity at that point. In this sense, the task addressed by the ontology of activity-based costing is a special case of temporal projection. We thus use successor state axioms similar to those in earlier sections. For example, we have the following successor axiom for computing the cost associated with the enabled status of a terminal state, where t,t' are the endpoints of the interval over which the state is enabled:

$(\forall\ a,r,s,t,t',c,c')\ holds(enabled\_res\_cost(s,r,a,c'),\ do(e,s)) \equiv (e = disenable(s,a) \lor e = complete(s,a)) \land enabled\_res\_cost\_unit(r,a,q,v) \land holds(enabled\_res\_cost(s,r,a,c)\ ,\ \sigma) \land c' = c+vq(t'-t) \lor \neg[(e = complete(s,a) \lor e = disenable(s,a)) \land holds(enabled\_res\_cost(s,r,a,c'),\ \sigma)]$ (EQ 10)

Given the costs computed for each status of a state, the resource cost point (represented by the predicate *cpr*) is computed by summing the costs for each status value of the state:

$(\forall\ a,r,s,t,c,c_1,c_2,c_3,c_4)\ holds_T(cpr(s,a,r,c),\ t) \equiv holds_T(committed\_res\_cost(s,a,r,c1) \land holds_T(enabled\_res\_cost(s,a,r,c1)\ \land\ holds_T(disenabled\_res\_cost(s,a,r,c1)\ \land\ holds_T(reenabled\_res\_cost(s,a,r,c1)\ \land\ c = c1+c2+c3+c4$ (EQ 11)

The cost for an activity at a point in time is the sum of the costs for each of its resources; this is represented by the predicate *cpa(a,c)*.

The ontology for activity-based costing therefore consists of resource cost units, successor state axioms, and axioms defining the aggregation of costs for resources, activities, and orders.

3.6 Competency Questions for Ontologies

In this section we rigorously specify several of the tasks that the various advisors must solve, and claim that the ontologies and microtheories presented earlier in this paper are necessary and sufficient to represent these tasks and their solutions. We can express these as the following theorems; let T_{succ} be the set of successor axioms and let $T_{occurrence}$ be a complete specification of action occurrences and the times at which the actions occurred.

Theorem 1: At any time point t, state s, and activity a there exists a status value X such that

$$T_{succ} \cup T_{occurrence} \models holds_T(status(s,a,X),\ t)$$

In other words, the status of a state is completely determined at any point in time.

Let T_{cost} be the set of successor axioms for cost and the complete set of resource cost units for every resource, activity, and status value.

Theorem 2: At any time point t, state s, resource r and activity a there exists a cost c and a cost c' such that

$$T_{succ} \cup T_{occurrence} \cup T_{cost} \models holds_T(cpr(s,a,r,c)\ ,\ t) \land holds_T(cpa(a,c')\ ,\ t)$$

Thus the costs assigned to a resource and activity are completely determined at any point in time.

We can further show that the axioms are necessary and sufficient to prove these theorems in the sense that if any of the axioms are removed then we can no longer prove the theorem. Thus these temporal projection problems serve as benchmarks for any theories of processes and activity-based costing.

Competency questions can also serve to drive the development of appropriate microtheories. For example, the goal of time-based competition is to find the enterprise model with the minimum cycle time. Within the ontology of activity, this is equivalent to finding the ordering of activities with the minimum duration. The first step in solving this task is to define the conditions under which a set of activities may be completely assigned a unique minimum duration; this competency question serves a characterization for any theory of time-based competition. In order to do this, we must also define the conditions for the existence of bottlenecks and other limitations of concurrency within an enterprise model, such as computing the maximum number of activities that may be supported by a resource. This in turn provides a competency question for the ontology of resources in [Fadel et al. 94].

4.0 Summary

In this paper, we presented a logical formalization of the TOVE ontology of activity and time that has been designed to specify the tasks that arise in enterprise engineering. To this end, we have defined the TOVE ontologies for activities, states, time, and cost within first-order logic. This formalization allows deduction of properties of activities and states at different points in time by formalizing how these properties do or do not change as the result of an activity (temporal projection). The representation of aggregate activities, and the role of temporal structure in this aggregation, is accomplished through axioms that allow us to reason about the occurrence of actions.

Competency questions are used to characterize each of the ontologies and microtheories; these questions present tasks such that the microtheories are a necessary and sufficient set of axioms for representing and solving these tasks. Furthermore, the use of competency questions serves two roles -- they characterize the ontologies and microtheories that have been designed for each task and they also provide direction for the development of new ontologies and microtheories.

The ontologies for activities, states, and time defined in this paper have been implemented on top of C++ using the ROCK knowledge representation tool from Carnegie Group. The successor state axioms and occurrence axioms have been implemented using Quintus Prolog.

5.0 References

[Allen 83] Allen, J.F. Maintaining Knowledge about Temporal Intervals. *Communications of the ACM.* 26:832-843, 1983.

[Blackburn 91] Blackburn J. *Time-based Competition.* Business One Irwin, 1991.

[Davenport 93] Davenport, T.H. *Process Innovation: Reengineering Work through Information Technology.* Harvard Business School Press, 1993.

[Fadel et al. 94] Fadel, F. , Fox, M.S., Grüninger, M. A generic enterprise resource ontology.*Third Workshop on Enabling Technologies: Infrastructure for Collaborative Enterprises*, Morgantown, West Virginia, 1994.

[Fox et al. 93] Fox, M.S., Chionglo, J., Fadel, F. A Common-Sense Model of the Enterprise, *Proceedings of the Industrial Engineering Research Conference 1993.*

[Fox et al 94]Fox, M. S., Grüninger, M., Zhan, Y.. Enterprise engineering: An information systems perspective. *Proceedings of the Industrial Engineering Research Conference 1994).*

[Grüninger & Fox 94] Grüninger, M. and Fox, M.S. An advisor-based architecture for enterprise engineering. *Workshop on Artificial Intelligence in Business Process Reengineering*, AAAI 94, Seattle.

[Hammer & Champy 93] Hammer, M. and Champy J. *Reengineering the Corporation.* Harper Business, 1993.

[Lenat & Guha 90] Lenat, D. and Guha, R.V. *Building Large Knowledge-based Systems: Representation and Inference in the CYC Project.* Addison Wesley, 1990.

[Pinto & Reiter 93] Pinto, J. and Reiter, R. Temporal reasoning in logic programming: A case for the situation calculus. In *Proceedings of the Tenth International Conference on Logic Programming* (Budapest, June 1993).

[Reiter 91] Reiter, R. The frame problem in the situation calculus: A simple solution (sometimes) and a completeness result for goal regression. *Artificial Intelligence and Mathematical Theory of Computation: Papers in Honor of John McCarthy.* Academic Press, San Diego, 1991.

[Sathi et al 85] Sathi, A., Fox, M.S., and Greenberg, M. Representation of activity knowledge for project management. *IEEE Transactions on Pattern Analysis and Machine Intelligence.* PAMI-7:531-552, September, 1985.

[Tham et al. 94] Tham, D., Fox, M.S., Grüninger, M. A cost ontology for enterprise modelling. *Third Workshop on Enabling Technologies: Infrastructure for Collaborative Enterprises*, Morgantown, West Virginia, 1994.

4

Developing performance networks to improve the benchmarking process
- Action Research for Productivity Improvement

Thomas A. Hansen and Jens O. Riis

The Department of Production, University of Aalborg
Fibigerstraede 16, 9220 Aalborg, Denmark

Benchmarking is related to many areas of interest when improving the competitiveness of an industrial enterprise or any other activity center which functions under market economic conditions. An important element in gaining full benefit from a benchmarking process is the ability to implement cross functional activities and manage change. This ability may be improved by the implementation of performance networks which are aligned according to tasks and core competences. This paper discusses a framework for implementing performance networks and assessment of the performance network maturity stages, which is important if feasible improvement actions are to be initiated. The paper is based on results from a Danish research project in an industrial enterprise.

1. IMPLICATIONS OF BENCHMARKING

The benchmarking process may be defined as follows (**Camp, 89**):

"Benchmarking is the continuous process of measuring products, services, and practices against the toughest competitors or those companies recognized as industry leaders".

This definition holds important implications as word 'continuous' indicates that benchmarking should not be thought of as a one time event. Furthermore the word 'measure' indicates that evaluation of performance based on derived performance measures has to be carried out. Benchmarking activities may be divided into two major groups: internal and external benchmarking. Internal benchmarking is application of the benchmarking philosophy (comparing yourself to the best) within your own organization, i.e. comparing business processes and performance of one organizational units to that of others. It is not only establishing benchmarks (performance measurement). As to external benchmarking one may distinguish between competitive benchmarking which is focused on direct competitors within the same industry and non-competitive benchmarking which is focused on organizations which are not direct competitors. The benchmarking process may be divided into the following steps (from Camp, 89):

Planning (Identify what is to be benchmarked, Identify comparative companies or activity centers, Determine data collection method and collect data).

Analysis (Determine current performance gap, Project future performance levels).

Integration (Communicate benchmark findings and gain acceptance, Establish goals).

Action (Develop action plans, Implement specific actions and monitor progress, Recalibrate benchmarks).

As discussed in a later paragraph it is important to recognize that most of the activities listed above involves organizational change and that a benchmarking process benefits from dynamic cross functional organization structures.

1.2. Benchmarking and performance assessment

Benchmarking implies performance assessment in at least two different environments, your own and the reference object. When benchmarking is used as a leverage technique, it is important to find cardinal competences which may be subject to change and improvement. This may be accomplished by using the model illustrated in figure 1, in which performance requirements of the market are compared to the specific company performance.

The model may be applied to different activity levels, e.g. company level, function level or production unit level etc. The model essentially enables an illustration of company performance to the market "average" and how the portfolio of selected competences support a continuous competitive strength. When the model is applied it is necessary to select a competence related to one or more of the generic performance areas i.e. features (product) and quality, logistics or productivity (manufacturing). The next steps are to assess the market performance (competition point of view) and the company performance. The market performance is separated in two categories with different characteristics. The first category implies a scale of competences, skills or properties which may be denoted 'Qualifiers'. Qualifiers are the foundation for a competitive strength and 'a must' if long term survival is to be achieved. In addition to qualifying properties a range of innovative capabilities should be possessed, because these are the edge-keeping properties of the company and make out the future competitive profile.

Two categories have been introduced in order to represent the company performance profile. For a skill or competence within any performance area we have chosen to distinguish between those fully possessed and those not possessed. If a skill or competence is possessed it should be mastered to an extent enabling optimal use in business processes or production. A skill or competence which is not mastered must undergo further research, development or implementation and training before it is part of the self propelled business organization.

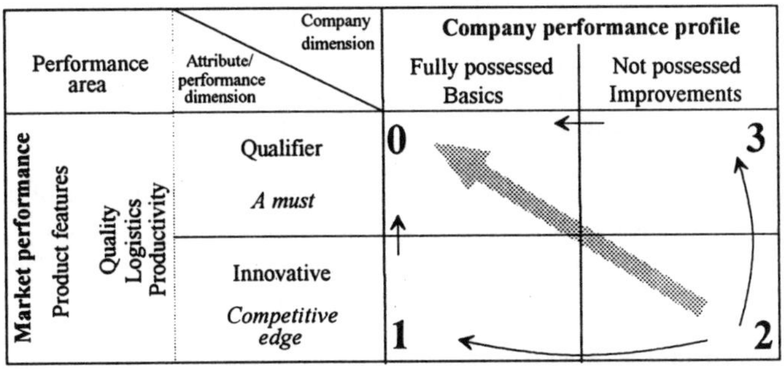

Fig. 1: Capability evaluation model supporting benchmarking efforts and focus.

The different fields of the matrix implies different competitive situations and possibilities. Location in field '0' means that the company masters a competence (e.g. a process, technology etc.) completely which makes it a basic attribute. This attribute does however not provide you with a specific competitive edge as the industry in general offers these attributes at the same performance level. Location in field '1' is only possible if you have either differentiated attributes to offer or if your performance related to these attributes is superior to your competitors because you have managed to make it a basic competence or property of your company. If not, you will be located in field '2', which may be perceived as a future competitive option. Location in field '3' is not desirable, as investments have to be made in order to move from field '3' to '0'.

The path '2' --> '0' is quite neutral, and the one most companies will follow in the long run. The path '2' to '1' is followed by world class performers. Time plays an important role in the model as companies continuously slide towards the '0' field. Thus if the time span going from not possessed to fully possessed is too great it will be very difficult to enter the '1' field. This is caused by the fact that what is new today is not new tomorrow and the customers will either expect the attribute or a competitor will possess the capabilities necessary to offer it.

The capability evaluation model has proven very useful in the high-level business planning process at a Danish industrial enterprise and is believed to be a helpful processing-tool in relation to the planning and analysis phase of benchmark studies. Further more the model helps illustrate the possible effects of the implementation of performance networks. Performance networks play an important role in relation to performance improvement and change and therefore also supports a successful benchmarking study (esp. the integration and action   phases) and also helps companies move directly from field '2' to field '1'.

2. INTRODUCING PERFORMANCE NETWORKS

An organization may be separated into a vertical and a horizontal dimension. The vertical dimension is the most familiar, as investigation of this dimension reveals a typical organization chart with a hierarchical structure limited by functional boundaries. The horizontal dimension appears when a flow-oriented viewpoint is taken. Such a viewpoint reveals the many tasks which cut across functional boundaries such as planning, development etc. This viewpoint enables a new organization chart to be drawn, as the organization members are related to each other by tasks and cross functional objectives.

A good benchmarking study often reveals potential improvement areas and therefore results in change activities. A good mental picture may be a 'change bomb'. It is not likely that the full effect of this change bomb will be obtained unless organizational boundaries are not loosened. A fox-hole organization (cfr. **Savage, 93**) with stiff functional boundaries would minimize the benefits of a benchmarking process as the shock waves are not allowed to spread and give other organizational units an opportunity to initiate change. The horizontal dimension must be perceived and used pro-actively.

The idea of performance networks is based on combining theory and practice of performance measurement and organizational networking (**Hastings, 93**). The results are characteristics of a work-form and some tools centered around performance assessment enhancing implementation of a change oriented organization focused on results and cross functional tasks.

2.1. Defining performance networks

Numerous researchers, consultants and authors interested in organizational development have discussed the perspectives of networks and organizations. It seems however that it is extremely difficult to find a definition of the term including a specification of what the necessary means and conditions for establishment of a networking organization are. Some authors come close to a definition (esp. **Hastings, Savage (1993)**), but never realizes concrete suggestions. In order to carry out a stringent discussion, a work-definition of performance networks is proposed and an investigation of how performance networks should be and are established will be initiated.

A performance network is an organizational cluster established according to recognized cross functional objectives or interdependencies and which aligns improvement efforts to shared and task oriented performance measures.

Performance networks are intended to work cross functionally taking a starting point in interdependencies and interactions (symptoms of cross functional malfunctioning). The basic ideas is that such a starting point is a foundation for analysis and perception of the complex horizontal flow and to identify crucial customer-client relationships. Perception of the horizontal flow and organization enables co-operation and establishment of performance measures focusing on the horizontal organization.

However most organizations are not very good at realizing true horizontal performance networks. Performance measurement has traditionally been very vertically focused. The primary reasons are the complexity of the horizontal system and the traditional functional separation of responsibility. The vertical focus however stimulates sub-optimation and hinders integration. Therefore horizontal performance measurement is a crucial focal point of productivity improvement of the future.

From a theoretical viewpoint, a model implying concurrent development and exploitation would be more suitable, but from an applied viewpoint, taking into account the performance measurement state of the art in most industries, it is suggested that the evolution of performance measurement should progress towards the horizontal dimension by climbing the "Performance Measurement Ladder" shown in figure 2.

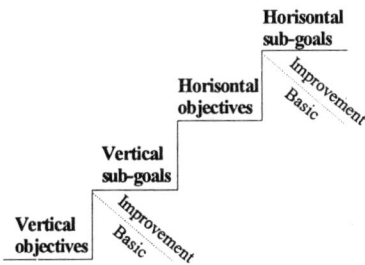

Fig. 2: The performance network ladder

The levels of the performance measurement ladder actually describe the development of a certain work-form along with the establishment of certain structural conditions. The nature of the work form and the structural conditions are described in the following. The implications of the performance network ladder are founded on well known constructs and have been verified during action research at an industrial enterprise.

2.2 Characterizing the levels of the performance ladder

Two factors are important to discuss in relation to performance networks: 1) How tasks and responsibility are defined and handled, 2) How follow-up is established and handled. Each level may be characterized from two viewpoints: a behavioural and a structural. The need for both viewpoints is becoming more and more apparent, as researchers investigate integrated productivity improvement and how implementation of new technology should include parallel work on organizational aspects.

2.3. The level of vertical objectives
Behavioural characteristics
- Functional borders limit task definition and perspective.
- Objectives are received from functional superior but not really decomposed.

Structural conditions
- Performance measures are established within functional borders at responsibility/task level.
- Follow up is initiated by the superior and targets the mass or responsible individuals.

2.4. The level of vertical sub-goals
Behavioural characteristics
- Functional borders limit task definition and perspective
- Objectives received from functional superior are decomposed and operationalized.
- Analysis are to a large extent initiated and carried out by organization members themselves.

Structural conditions
- Performance measures are established within functional borders at sub-task level by organization members themselves.
- Follow up is initiated by organization members and targets sub-task owners.

2.5. The level of horizontal objectives
Behavioural characteristics
- Functional borders are recognized but does not limit cross functional task definition.
- Horizontal objectives are identified and established but not decomposed.

Structural conditions
- Performance measures are established at task level.
- Horizontal follow up is initiated by superior and targets the mass or the responsible individuals.

2.6. The level of horizontal sub-goals
Behavioural characteristics
- Cross functional task definition is widespread, and constantly initiated by organizational members themselves.
- Decomposition of horizontal objectives by organization members themselves
- Cross functional analysis are carried out by task-owners themselves.

Structural conditions
- Performance measures are established at task level but supported by operational sub-task measures established by organizational members themselves.
- Follow up is initiated by organization members and targets sub-task owners.

In todays organizations most units are found on the vertical 'steps'. It seems that two primary reasons influence this: 1) The vertical view of the organization is traditional and easy to cope with. 2) The organizational structures and the management systems (IT) are geared to function in a vertical environment. It is important to seek a two-dimensionally balanced organization where employees are influenced not only to look in a vertical direction but also to work actively in a horizontal dimension which includes participation in decomposition of horizontal objectives and in developing management systems that take these issues into account.

2.7. Differentiating between basics and improvements

In the performance network ladder (fig. 2) it is indicated that there is a need to differentiate between the basic dimension (already possessed tasks or competences) from the improvement dimension. This will allow a continuous improvement process to be initiated and structured. Further discussion of this argument and how it may be combined with the network establishment may be found in a later paragraph.

2.8. Implications of performance networks

Horizontal performance networks implies a measurement system (collection of measures) centered around the horizontal task structure. The purpose of the measurement system is to focus attention to whatever activities or evaluation criteria are important in order to generate improvement of productivity and performance. According to the theory on development of measures and measurement systems, a number of behavioural and structural considerations are to be investigate.

Development of measures should be based on situational analysis and knowledge of the desired relationship between objectives, means (decision variables) and the measurement system itself.

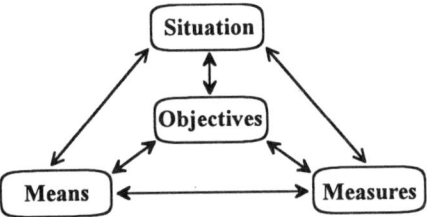

Fig. 3: Central elements of performance measurement and improvement.

Development of a measurement system often will include the following activities: 1. Decomposition of objectives, 2. Analysis of objectives and means, 3. Construction of measures, 4. Implementation of measures and measurement procedures. 5. Development of information and documentation procedures. Experience from applied measurement system elaboration have shown that success is supported heavily by initiating a participary and costumer driven development process (**Bitton, 90 & Brinkerhoff, 90**). Further more the measures developed and established should (**cfr. Hansen, 93 & Christopher, 93**):

1. Be accepted by employees concerned with performance measurement and evaluation (this usually means that they have participated in the development and establishment process).
2. Be established in accordance with the objectives of the system studied.
3. Be easily measured, easily quantified and easily understood.
4. Be situationally adapted in frequency and aggregation.
5. Allow coherence between long term and short term objectives.

If the idea of performance networks and horizontal performance measurement is combined with the knowledge and implications of theory behind performance measurement, a large number of complex questions arise. The questions are easy to identify if the behavioural and structural characteristics of the horizontal sub-goal level mentioned earlier are reviewed. Some of them are:

How may the horizontal dimension be assessed.

- How are task (cross functional) perspectives captured, analyzed and visualized
- How are shared and consistent performance measures established, decomposed an implemented.

How are relevant performance networks identified and established.

- How may information technology be used to support performance networks.

Investigation of these questions is relevant as it often is the basis of performance evaluation in the organization. Thus actions and organizational values are highly influenced by the measurement system. Some tools and considerations concerning establishment of performance networks through performance measurement are presented in the following paragraph.

3. STRUCTURING MEASUREMENT AND IMPROVEMENT EFFORTS

It has been argued that the initial phases of a benchmarking process (planning and analysis) may be supported by applying the model and the underlying mind set of figure 1. This paragraph discusses another model which may support the phases 'Integration' and 'Action' of the benchmarking process and help organizational members establish performance networks by climbing the performance measurement ladder.

3.1. Enabling the organization

The implications of horizontal performance measurement are often very demanding to organizational members, as they are asked to bust boundaries and seek partners outside their own domain. Therefore it is important to provide tools to help them structure the efforts and obtain success. The model in figure 4 is meant to help balance performance improvement in both the vertical and horizontal dimensions and to structure performance measurement. The model has been used in practice with much success in pilot projects at one company, and will be an important instrument in the quest of reaching the upper levels in the performance measurement ladder throughout the organization. The model is extremely simple and may therefore be used by all organizational members.

The model is based on distinguishing between the vertical and horizontal dimension and also between performance areas or measures which focuses on basic activities as opposed to improvement activities. The separation between a vertical and horizontal dimension has already been discussed. The basics are the performance areas or tasks which are mastered well, indicating a good balance between the four elements of figure 3 (situational boundaries, objectives, means and measures). The improvements are performance areas or tasks which are identified but not mastered.

The model enables a continuous and structured increase in performance as organizational members are motivated to force an increasing number of tasks and performance areas into the fields of 'Basics'.

The general patterns of how an organizational unit is placed in the model are somewhat given by the performance measurement ladder. This results in a pattern where organizational units studied often move from working in field '0' or '1' and gradually progress towards field '3'. Operating in field '3' means dealing with and assessing complex problems in a wholistic manner.

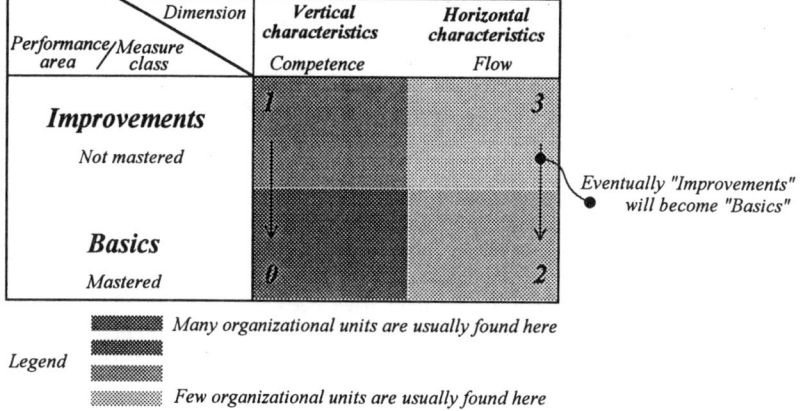

Fig. 4: Generic model for structuring measurements and improvement efforts.

As discussed later, it is possible to evaluate where an organizational unit currently is located. The model has proven itself extremely useful when implementing performance networks in specific production units, but has also been appreciated at higher organizational levels as a cognitive tool.

At higher organizational levels the model was applied in an effort to map the activity span and its characteristics. The objective was to initiate a discussion whether or not the initiatives launched were balanced correctly (as desired) in the different fields of the model. Thus the model was given substance by discussing the characteristics of the results expected given the situation and use of certain means (in terms of basic/improvement, vertical/horizontal).

At lower organizational levels (production units and sections) the model was applied during an effort where the main objective was to structure the forthcoming improvement efforts in a manner that invited to analyze across functions, find partners and establish shared performance measures. Thus a focus was on tasks and appropriate performance measures and their location in the different field of the model. This way a moderate but steady expansion of tasks and performance areas was initiated. At the same time a new way of operation focusing on horizontal dimensions began to appear.

3.2. On the need of horizontally oriented analysis methods

As performance networks are based on horizontal performance measures it is clear that need for cross functional analysis and co-operation appear. However this means that tools supporting these efforts must be developed. When the need of analysis methods is considered, the following points should be kept in mind:

- Employees influenced by the performance measures and the following evaluation should participate in developing and establishing the performance measures needed in order for the measures to be accepted. Final commitment may only be accomplished if employees sense that they have access to some active decision variables which enable them to influence the performance level.

- Organizations are becoming leaner which means that fewer managers and advisers are available in the effort to carry out horizontal analysis. Employees will have to (and should) participate in the analysis effort.

There are major research areas related to these considerations. Important ones are performance measurement systems, consistency analysis, modelling of relationships and interdependencies including the use of information technology (IT) supporting the areas mentioned.

Without going into methodological details, many of the analytical methods and tools known today are most often used consequently by researchers and consultants, and more seldomly managers or operators close to the shop floor. Examples are application of IDEF/SADT diagraming, concept development, activity chain analysis. Even the 'simple' formalisms, tools and techniques usually embraced by the Kaizen-term have been very difficult to implement or effectuate in numerous industrial enterprises in Europe. Thus if a self propelled organization is to be developed there is a need to develop methods and tools useful for horizontal analysis which are simple and yet powerful.

4. THE PERFORMANCE NETWORK MATURITY LEVEL

In several cases during the research project it has been useful to apply evaluation principles given by Kirkpatrick (**Kirkpatrick, 77**) in order to seek an understanding of the current maturity level of the organizational unit studied. The maturity level analysis takes a starting point in the different levels of the performance network ladder and their characteristics. Applying these principles results in an approach illustrated in fig 5.

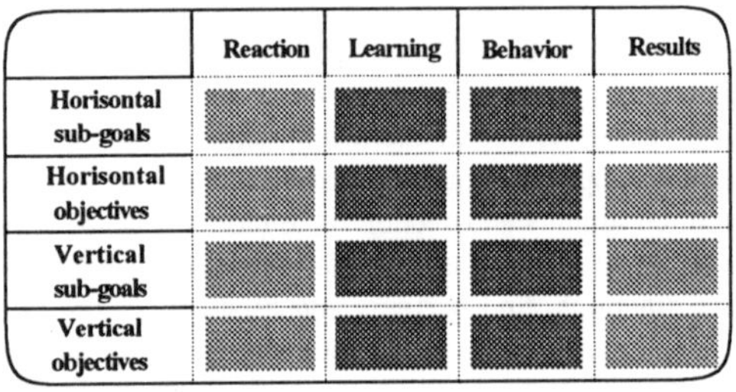

	Reaction	Learning	Behavior	Results
Horisontal sub-goals				
Horisontal objectives				
Vertical sub-goals				
Vertical objectives				

Fig. 5: Organizational performance network maturity assessment.

The basic idea of the evaluation model is to assess the organization by studying selected units at four different evaluation levels: *1)Reaction,* where the objective is to investigate the reactions of the different participants involved when exposed to the ideas and implications of performance networks as presented in this paper. 2)*Learning,* where the objective is to obtain information of the knowledge, skills and attitudes of the participants involved when exposed to the ideas and implications of performance networks. 3)*Behaviour,* where the objective is to obtain information concerning the behaviour of the participants involved. This makes a comparison of desired and actual behaviour possible. 4)*Results,* where the objective is to investigate how results are influenced by existing behavioural characteristics and established structural conditions.

It is important that the investigation targets all evaluation levels of the model shown in figure 5, but experience applying the model has led to the conclusion that the learning and behavioural levels are crucial, as much useful information may be obtained if those levels are assessed correctly.

By evaluating organizational units using the model in figure 5, it is possible to find many signals indicating a feasible path in order to implement performance networks in an industrial enterprise. This has been done in the research project, and both interviews and questionnaires have been applied with much success. The result was development of project ideas and action plans along with initiating a learning process within the units studied.

5. CONCLUSIONS

Benchmarking may support companies in the effort to improve their competitive situation. A benchmarking study is however often centered around a specific organizational unit and a selected business process. The potential benefits of benchmarking increase if the organizational processes in a benchmarking study is not limited by stiff functional boundaries. It is important that the benchmarking process is related to the implementation of horizontal relations as unrealized competitive strength may be found here.

Horizontal relations and interdependencies are central elements in developing performance networks within an organization. This paper discussed why and how performance networks should be established and argued that application of benchmarking along with implementation of performance networks may catalyse realizing the competitive possibilities found in well functioning horizontal relations of any company or corporation. A number of models and tools developed and tested during work with performance networks in an industrial enterprise have been presented.

REFERENCES

Bitton, Marc: ECOGRAI, Méthode de Conception et d'Implementation de Systemes de Mesure de Performance pour Organisations Industrielles. Thèse Docteur, Université de Bordeaux, Septembre 1990.

Brinkerhoff, Robert O. & Dennis E. Dressler: Productivity Measurement - A Guide for Managers and Evaluators. Applied Social Research Methods Series, Vol. 19. Sage Publications, 1990.

Chew, W.Bruce, Timothy F. Bresnahan, Kim B. Clark: Measurement, Coordination, and Learning in a Multiplant Network. Measures for manufacturing Excellence edited by Robert Kaplan, Harvard Business School Series in Accounting and Control, 1990.

Camp, Robert C.: Benchmarking - The search for industry best practices that lead to superior performance. Quality Press, Milwaukee, Wisconsin, 1989.

Christopher, William F. (ed.) & Carl G. Thor, (ed.): Handbook for Productivity Measurement and Improvement. Productivity Press, Massachusetts, 1993.

Frick, Jan & Jens O. Riis: Activity Chains as a Tool for Integrating Industrial Enterprises. Advances in Production Management Systems. Proceedings of the 4th IFIP/TC5/WG5.7 APMS'90 Conference, North-Holland 1991.

Hansen, Thomas A. & Jens O. Riis: Modelling Interactions and Measuring Performance for Productivity Improvement. Proceedings of the 8th IPS Research Seminar, 1993. Report from Dept. of production, University of Aalborg, 1993.

Hastings, Colin: The New Organization - Growing the culture of organizational networking. McGraw-Hill International (UK) Limited, 1993.

Kirkpatrick, Donald L.: Evaluating Training Programs: Evidence vs. Proof. Training and Development Journal, November 1977.

Savage, Charles M.: 5th Generation Management - Integrating Enterprises through Human Networking. Digital Press, 1990.

5

Benchmarking–The neglected element in total quality management

F.W. Swift[a], T. Gallwey[b], and J.A. Swift[c]

[a]Mechanical Engineering, Florida International University
University Park, Miami, Florida, USA

[b]Production and Operations Engineering, University of Limerick, Limerick Ireland

[c]Industrial Engineering, University of Miami, Miami, Fl. USA

1. EVOLUTION OF TOTAL QUALITY MANAGEMENT (TQM)

Long before the twentieth century, merchants and craftsman knew the importance of quality. While the term meant different things to different people, it was valued to be an important, intangible commodity. The concept of specifications, inspection with respect to these specifications and later, the refined aspects of measurements with respect to the field of metrology, has been with us for many years. With the coming of the twentieth century and especially with the advent of the industrial revolution, quality, in many different forms and different names, became the criterion to be "championed". The earliest serious works centered around "quality control" and perhaps quality planning. In the early stages, the emphasis was on inspection, and inferior products (non-conforming products) were reworked or discarded as scrap. The literature is replete with historical and evolutionary quality efforts which were not economically acceptable, and these will not be duplicated here.

What is enlightening, is the fact that the whole field and concept of quality began to change in the late 1940's after World War II. The real explosion of "new" definitions and concepts did not emerge until the 1980's. This revolution is continuing today and will continue to cause change for at least another 10 years.

Of vital importance to this research and this paper, is the changing mosaic which now makes up quality, and TQM's place in society. While some still deny the value and even the existence of its importance, TQM continues to dominate the free market systems in the world today.

Following the lead of the Japanese, as tutored by Dr. Deming, four important elements evolved which revolutionized quality in the market place. First, he tutored upper management to make a commitment to quality and insure that quality was emphasized throughout the organization. Second, all levels and all functions were to receive quality training at some specified level of expertise. Third, quality improvement was to be a continuous process as later defined by the Deming wheel. And finally, the customer was to be the most important concern in the "quality loop". There are many reasons why this has prevailed, and even today the customer is very cognizant of quality and value. Most consumers are more educated in quality, and everyone seems to demand it. But, the battle cry today is "quality at a reasonable cost" instead of quality at any cost. Since many accepted definitions of quality today include a clause or phrase "quality is customer satisfaction", this new concern will receive attention by those who expect to make a reputation by catering to the sophisticated customer.

One difficulty encountered today is the concern related to environmental issues, ergonomics aspects and health and safety concerns. These and other related issues are not normally of a direct concern to the consumer, and therefore not requested by the customer. However, they must be included and they do add incrementally to total cost. Addressing these issues takes a unique manager who can properly balance these concerns while still providing customer wants.

2. HISTORICAL PERSPECTIVE OF BENCHMARKING

Interestingly enough, quality, in its earlier history was at times called benchmarking, or more precisely, benchmarking was called "product quality and feature comparisons". (Camp, p.6) This could account for the similarities and common elements in these two management tools. Even the historical development has some common elements, as seen in this paper.

Since 1979, the concept of benchmarking has taken on radical new meanings. Before that time, benchmarking simply meant comparing various components of the company to the previous year's performance. In the earliest uses of benchmarking, the definition of Webster was sufficient. "...A surveyor's mark...of previously determined position...and used as a reference point." A company would simply decide which measures of performance to follow, and measure themselves from year to year. The most common measures were related to economic parameters. They included profits, sales volume, expenses, some type of performance ratio or some other parameter. Strategic planners would opt for some percentage increase or decrease. If this goal was met or exceeded, then the unit was declared a success for the period in question.

Most researchers and writers suggest that 1979 was the emerging point of benchmarking as a vital management tool. Xerox is given credit for first discovering that it would become vital to use benchmarks to compare itself to other competitors as well as to reference points within their own organization. At that time, Xerox's manufacturing unit decided to compare unit manufacturing costs and features of their copying machines to those of competitor machines. Thus began the concept. While there has been a slow but steady increase in the use of benchmarking, it has been growing rapidly in importance. When one recognizes the value and importance of competitive benchmarking, it is easy to understand the virtual explosion of its use.

Improvement goals based solely on the company's own past performance are no longer sufficient. In fact the definition itself has undergone a revolution. In 1979, Xerox realized that the old Webster definition was no longer sufficient and adopted a new one:

"The continuous process of measuring our products, services, and business practices against the toughest competitors or those companies recognized as industry leaders."

D.T.Kearns, Xerox CEO (Camp, p. 10)

It will be noted by those who work extensively in TQM that this definition has strong similarities with the now currently accepted definitions of quality and quality improvement. There are many other definitions that have evolved, but this paper is not intended as an extensive development of benchmarking. It is intended to show that there is a high correlation between the work being done in TQM and the work being done in benchmarking. There is an inherent synergism which needs to be developed. A systematic development for a common model will advance this management tool to make companies competitive on a world scale.

A Chinese proverb which is attributed to a Chinese general over 2000 years ago states, "If you know your enemy and know yourself, you need not fear the results of a hundred battles". (Tzu, p 84) This proverb (whether true or mythical), can be paraphrased here and used as sound advice for managers. If you know your own product and your own company well enough, you need not fear competition from other companies. It has become a truism that one must study and know more about all aspects of one's company, because in this modern age at the dawn of the 21st century, "knowledge is power".

Benchmarking will become more useful especially in the all important aspect of quality. It will be important to be competitive with the "best of the best" and it needs to be understood that this is a dynamic and continuous process.

3. COMMON ELEMENTS

As noted in the last two sections, there are commonalities in the two management techniques. In all the current definitions for benchmarking and TQM, there are references to: continuous-systematic improvement, meeting customer requirements, performance standards, striving for the best of the best, understanding industries' best practices regardless of the product or service, and many other elements. It is also noted that, the definitions associated with quality and TQM have taken significant, and parallel development with benchmarking concepts since the 1970's.

In the earliest days, conformance to specifications and inspection of final products was considered very important. This began the meaning of quality. Early in the twentieth century, formal definition was given to this concept, and significant importance was given to meeting prescribed specifications as a means of obtaining quality. It was not until the work of Deming, Juran, and Ishikawa in the 1950's and 1960's that the current meaning of quality, and Total Quality Management began to emerge.

With Dr. Deming's concept of continuous improvement using the familiar PDCA continuous improvement wheel, and his success in Japan after World War II; the tremendous focus on customer needs and satisfaction; Dr. Juran's recognition of the Pareto principle and its application to quality problems; we find a great amount of commonality between the ideals of benchmarking and quality improvement. In Quality Planning and Analysis, Drs. Juran and Gryna recognize that competitive analysis is one of the distinguishing features of modern strategic quality management, but they did not emphasize the power of benchmarking functions from dis-similar companies.

A close review of all the writings on benchmarking since 1988 and a review of writing and analysis on total quality management will reveal a closer link than previously realized. Yet, when one studies the criteria for the most prestigious awards, and the criteria for quality certification (ISO 9000), it is curious to note that very little mention is given to benchmarking.

It seems that this omission should be rectified by making an extensive study of the relative merits of benchmarking, and its value to a company's quality efforts, as compared to other topics in the criteria for quality awards and ISO 9000 certification. It will be beneficial to note some specifics especially in the certification criteria.

Of all the certifying criteria, probably the best known and the most widely accepted is the ISO 9000 certifications. A review of the most up-to-date criteria for 9001, 9002 and 9003 would seem to indicate that benchmarking is not essential to receiving certification. While there are sections that relate to the goals of benchmarking, there are no specific criteria that requires that any sort of benchmarking be done.

The three quality awards which have the most prestige today are the Deming Application Prize (DAP), The Malcolm Baldridge National Quality Award (MBNQA) and the Quality Cup (Space limitations do not allow an elaboration of these awards here). Of these three quality awards, the only one which specifically uses benchmarking as a criterion, is the MBNQA award. This award is based on criteria which allocate 1000 points to seven different major categories. Of these categories (which are to receive a total of one thousand points), the only category relating to benchmarking is category two. Category number two is, "Information and Analysis", and sub-category 2.2 is designated "Competitive comparisons and benchmarking". Only 20 points are allocated for this sub-category. Of the sixteen winners for the MBNQA through 1992, only Xerox has been a prominent user of benchmarking. This fact was one of the issues which set it apart from the other winners, and Xerox felt that the extensive use of benchmarking gave it an edge in many of the other categories.

It should be pointed out that winning the award does not guarantee success. Winners do look forward however, to more recognition, and winning this award has certainly opened the door to more advertising potential. As they have all found out, this alone is not sufficient in today's environment. Winners anticipate reduced costs, increased productivity, improved customer satisfaction, and higher profitability. This points out again that knowledge about one's own company, and about the industry in general, can bring rewards far greater than the awards themselves.

If benchmarking could be incorporated to a fuller extent into the awards themselves, it is felt that some essential information could be gained with very little additional expenditure. At the present time, pursuing any of the awards, and/or seeking certification, is costly and time consuming. To independently add the cost of benchmarking onto current management efforts is difficult to justify. Since the aims of both of these management tools are similar, and elements can be found which are common, it would seem prudent to incorporate these studies into one exercise and gain a significant edge in the information and analysis phase of the study.

In very general terms, it could be said that both benchmarking and TQM (both quality awards and certification) are types of performance indicators. As such, they add to the repertoire of tools which can be used to assess company performance. An examination of these indicators needs to be pursued.

4. PERFORMANCE INDICATORS

One of the issues which needs to be studied in detail and perhaps researched with an eye towards completely revamping the benchmarking model, is performance indicators. As pointed out previously, the concept has taken a different focus from old models. Probably the most significant change was the redirection of the reference point from an internal one to an external one. The indicators themselves changed, and the meaning of comparison also underwent drastic revisions.

4.1 Current Practices

In the pre-1970 years, benchmarking was simple and straight forward. Through the 1980's, modern philosophies and techniques have become more comprehensive, complex, and valuable. During these years, the use (or purported use) of benchmarking seemed to have increased dramatically.

On the use of benchmarking, the president of Kaiser Associates reported, "Its earliest citings were in 1979 or 1980, but as recently as 1985 or '86 only half a dozen *Fortune 500* companies were doing it. Now over half of them are" (Industry Week, Nov, 1990). A more recent report claims that a poll carried out by Gallup on behalf of Coopers and Lybrand found that ...two thirds of The New York Times' "Top 1000 companies" claim to use benchmarking. Over 90% of those, claim that benchmarking helps to provide new insight and that it acts as a catalyst for change. Eighty seven percent believe that benchmarking has been successful. A serious study needs to be undertaken to determine the extent of benchmarking and the type that is actually taking place. Competitive benchmarking should be emphasized.

When Xerox originated "competitive benchmarking" in 1979, they made a drastic departure from "internal benchmarking". These early results from competitive benchmarking were so startling, that a corporate vice president announced that competitive benchmarking would become a corporate-wide endeavor. The initial competitive benchmark was conducted in the manufacturing unit, then extended to a few of the operating units. At the 1983 annual shareholders meeting, the CEO announced that his number one priority was to achieve leadership through quality. Benchmarking was one of the three components of that effort. Once again, we learn from history that quality management and benchmarking are tightly interwoven, and should be modeled and researched in the same studies. The benefits have been so great, that Xerox has a program which extends their efforts in benchmarking every year. They are the inventors, the most studied, and unquestionably the experts in this area today. By this initiative, Xerox has demonstrated that an internal benchmark used to measure performance based solely on "last year's" results is of diminished value today.

If a company wishes to begin a "benchmarking" initiative, it should begin modestly using internal markers, with a few important indicators set by management; then move aggressively into competitive benchmarking. To stay compatible with the TQM criteria, it is suggested that the functions of the enterprise should be examined and categorized as inputs to the service or product, the process itself, and all its steps, and finally the outputs. If this can be modeled for a company, or for a class of companies, then the rest of the task becomes straightforward.

An element of vital importance to TQM, is customer satisfaction. This is the element which is of high focus for this paper and which will be universal for every kind of business today and in the future. There are many elements which can be benchmarked in this area and it is receiving a great deal of attention. The customer has a strong desire to obtain "high quality" goods and services, but now the major focus is on "high value". This means that the customer wants what is perceived as high quality, but it must be available for what is perceived as a fair price.

4.2 Integrating Ergonomics, Health and Safety

Of all the benchmarks which receive the most neglect, ergonomics factors have to be at the top of the list. With all the interest in customer satisfaction, all the emphasis on TQM, and the profound discoveries in worker health and morale, this area deserves substantial research.

A search of the literature on benchmarking did not reveal any mention of concern for this vital element. Even in the area of TQM, its depicted role is more limited than one would expect. The role of human factors or ergonomics has primarily been that of inspector for the quality program. Some authors mention human factors for a few products such as automobile design, and furniture design, but none give extensive study to this growing area of human interaction.

If one is to consider ergonomics as a person/product interaction, ergonomics and TQM/benchmarking should play a role in three major areas: worker/product (service) interaction, customer-purchaser/product (service) interaction and customer-user/product (service) interaction. The last two categories must be researched separately, since they appeal to different aspects of a customer. A customer may pay money for something based on its marketing appeal. Emotional interactions, sensory appeal, sex appeal all may enter into a customer making a purchase. The other aspect of a customer then takes over and functional aspects and reliability aspects may become the more prominent concern. In each of these areas, there are opportunities for benchmarking. Improvements in many of these areas could lead to reduced costs, higher quality and greater customer satisfaction.

Looking at the first group, benchmarks could focus on one or more of the following:

> worker and product design interaction
> worker and process design interaction
> worker and machine compatibility
> worker and temperament for the job
> worker and quality inspectors
> > selection/training
> > testing

The next two groups would have some benchmarks in common but each group would also have some unique ones. Where the buyer is concerned, he or she would have to be "excited" about quality features. The marketing specialist would then have to determine factors which would appeal to this "group" of consumers. Quality Function Deployment (QFD) could play a very powerful role in this focus. Most customers would be able to speak in terms of what they want without being encumbered with extensive technical knowledge. Possible benchmarks to be studied would be:

> target group of customers
> emotional interactions
> marketing considerations
> sensory perceptions of customers

For the customer/user, many of the above elements would be important, but functionality may begin to play a more important role. Possible benchmarks to be included would be:

> functional requirements
> reliability
> maintenance considerations
> user friendly aspects

Finally, all groups would have different needs as to the following benchmarks which would relate to human behavior:

> service quality
> product quality
> value perception

If some of these elements could be benchmarked, this would make significant strides in TQM, could contribute to a work force of higher morale, and certainly extend customer satisfaction.

5. SUCCESSES AND FAILURES

There have been attempts to incorporate TQM in many firms, and many companies have made attempts at implementing benchmarking. History is replete with both successes and failures in these endeavors. In so many cases, a corporate executive lays out a strategic plan for TQM which may include an aspect of benchmarking. When the implementation phase comes, benchmarking often takes a very minor role (if any at all), and other aspects of the strategic plan receive more resources and closer attention.

It is felt that this is due, in part, to the criteria laid out for quality certification, or to the criteria laid out for winning one of the awards. Earlier it was pointed out that only one of the major awards directly uses benchmarking as a criterion, and none of the major certification programs include it. Thus, having to develop benchmarking standards, in addition to satisfying criteria for ISO 9000, will be viewed as additional effort.

It would be well to point out that the number of companies seeking awards and/or certification has risen dramatically in the past 6 years. Since this is a fairly time consuming task, complete with serious increases in paper work as well as a significant expense to the company, it cannot be a casual decision. Therefore, unless a way can be found to include the benchmarking model into these undertakings, it will never reach its full potential.

Among the companies who have become very successful in incorporating benchmarking into their corporate philosophy structure are: Federal Express, American Express, L.L. Bean, General Electric, Xerox, IBM, Dow Chemical, Honda Motor, Proctor and Gamble, and Apple Computer. There have likewise been many corporations who have "tried" benchmarking as a component of their standard operating procedures and failed to make it successful. No studies could be found which documented *why* these companies did not find success and this study was beyond the scope of this paper. In the experience of the authors, however, the most significant cause of failure comes from "lack of management commitment" or from the concept of feeling that "if we do it, we have solved the problem and can get on with company business".

Similarities are noted when examining companies which have tried to implement TQM. Both successes and failures are available for study. An article in the October 1993 Quality Newsletter on "The Death of Quality?" described some quality bashers who rejoiced at the demise of companies who were winners of the Malcolm Baldridge award. This only reinforces the point that there is still much work to be done. These management tools are still in their infancy and have not solved all the problems of coping with the new "world environment". They will undergo more growing pains as experience is gained and exogenous factors are better understood. For those fortunate enough to put forth the skill and determination necessary for a new corporate culture, the rewards will be great.

6. ADVANTAGES OF BENCHMARKING AS AN ELEMENT OF TQM

Much of the core of what is to be obtained in benchmarking can be modified to coincide directly with that needed for a TQM project. A review of the current research reveals that in one form or another, the following elements are common to both management tools:

Continuous improvement
Meeting customer requirements
Certain performance standards
Understanding industries' best practices
Concurrent engineering
Measuring of elements (targets)

Using these core elements, benchmarks could be incorporated directly into the TQM model. The real test would come in incorporating these elements into the ISO 9000 criteria. As the authors reviewed these criteria, it seems that some modest research would reveal some common, if not identical, criteria which could easily become a common core.

Currently, the biggest obstacle to overcome would be the idea of getting used to using dissimilar companies as benchmarks. Just as managers had to be educated in achieving customer satisfaction to yield a high quality product, they will need to be educated in the principles of "competitive benchmarking", and will have to be shown its benefits.

7. COMPETITIVE ADVANTAGES

There are many reasons for implementing a benchmarking policy and many of the key reasons are noted in R.C. Camp's book on benchmarking. Only one reason will be noted here. This is "Becoming Competitive". This reason is sufficient to cause companies who compete on the "new world" market to take notice.

7.1 Benchmarking Within The Organization

Mr. Camp states that without benchmarking, a company becomes internally focused, has evolutionary change and low management commitment. With benchmarking, a company develops a concrete understanding of competition, utilizes new ideas of proven practices and technology and has a high level of commitment (Camp, p. 30). While a company will typically begin its external benchmarking by comparing companies in the same industry, much can be learned from companies of a vastly different business.

7.2 Benchmarking Vendors

In recent years, it has become apparent that if a company is to minimize costs, it must be concerned with all aspects of its product or service. In the past, one often neglected element was the relationship with suppliers. It is the responsibility of the vendee initially to select a supplier who meets the requirements and standards of the vendee. The ISO 9000 series has been developed to reassure purchasers that objective quality standards are being met. The ISO Standards Authority certifies that specific suppliers have achieved a minimum standard for a quality process.

REFERENCES

1.   Camp R. C., <u>Benchmarking: The Search for Industry Best   Practices that Lead to Superior Performance</u>, ASQC Quality Press, Milwaukee, Wisconsin, 1989.
2.   Ishikawa, K. <u>Guide to Quality Control</u>, Asian Productivity Organization, Tokyo, 1993.
3.   Juran, J.M. and F. Gryna, Jr.   <u>Quality Planning and Analysis,</u> McGraw-Hill, New York, 1993.
4.   "Malcolm Baldrige 1992 Award Criteria", National Institute of   Standards and Technology, Gaithersburg, MD.
5.   Rothery, Brian, <u>ISO 9000</u>, Gower Press, England, 1993
6.   Tsu, Sun, <u>The Art of War.</u>, Delacorte Press, New York, 1983

6

The role of Benchmarking in the Management of Change process
Some reflections from the TIME GUIDE project

J. COLOM[a], R. SMEDS[b], S. KLEINHANS[c], G. DOUMEINGTS[c], M. BITTON[a]

[a] CLEMESSY SA
18 rue de Thann, BP 2499, 68057 MULHOUSE cedex FRANCE
[b] Faculty of Industrial Management, Helsinki University of Technology
Otakaari 4 A, 02150 ESPOO FINLAND
[c] GRAI/LAP University of Bordeaux
351 cours de la Libération, 33405 TALENCE cedex FRANCE

Abstract:

The use of benchmarking in the management of change is discussed, based on the approach of the EUREKA project TIME GUIDE.
Management of change begins by an essential process: the definition of the target. While it might seem trivial, this process, which needs clear references coming from the outside of the company, is often neglected.
A market study realized during the TIME feasibility phase illustrates this shortcoming in many ways.
In the ongoing project TIME GUIDE, a coherent methodology is developed for the management of enterprise evolution. This methodology contains, as an important part, the benchmarking method and database.

Keywords :

Management of Change, Steering the Evolution, Strategic Planning

1. INTRODUCTION

For almost a decade, modifications in the economic and industrial environment have happened at such an increasing pace that traditional approaches of Manufacturing could not handle them. New models, new ways of thinking are required. In this context, industry as well as academic research pay more and more attention to such concepts as Management of Change including Benchmarking.

Benchmarking is a technique for increasing the knowledge of best practices. It is aimed towards managers wanting to evaluate their own level of performance, according to the achieved best level.

We can ask the question: what are the real objectives of managers using benchmarking?

- Is it to copy what is made by competitors, or even by non competitors in related industries?
- Is it to know what are the weakest points in their own companies, in order to motivate people, working in this area, to enhance their competitiveness?
- Is it on the contrary to know what are the company's strengths, in order to base their competitive strategy on them?

Perhaps all these reasons, and some others, could have a part in the decision of using benchmarking techniques. But the common point within these reasons, is the need to define some direction where the company has to go in its evolution process: *Benchmarking helps to define strategic objectives, in order to steer the evolution and not to suffer it*. It is then an essential step in the more global process of management of change.

Internationally, many projects in this field are emerging: "Agile Enterprise", "Intelligent Manufacturing Systems", etc.

In Europe, such programs as ESPRIT and EUREKA have been created in order to allow a common reflection of both industry and research at a European level, by launching precise projects ending in innovative technologies or management tools.

A group of European industrial and academic partners from Finland, France, Ireland, Italy, Norway, Portugal, Spain and Sweden prepared and proposed an Eureka Project, TIME (Tools and methods for the Integration and Management of Evolution of industrial enterprises). The project aims at developing a methodology and the supporting tools for steering the evolution process of industrial enterprises. Benchmarking is included in this methodology as an important part.

To be able to conduct a thorough market analysis before the specification of the project, it was decided to divide the project in two phases :
- the feasibility study;
- the project itself, called "TIME GUIDE".

This paper gives an overview of the TIME project:

- in the first part, we describe the past feasibility phase of the TIME project (from November 1992 to July 1993) as well as its key results;
- in the second part, we present the TIME GUIDE project stemming from the feasibility phase with special focus on the integration of benchmarking in the overall TIME methodology.

2. RESULTS OF THE TIME FEASIBILITY PHASE: THE MARKET STUDY

2.1. Organisation of the study *[Time 93-2]*

During the TIME Feasibility Study in 1993, a user requirements survey was conducted in the following participating countries: Finland, France, Ireland, Italy, Norway, Portugal, Spain and Sweden. In the survey, industrial enterprises were asked about the evolution needs in their different functions and processes, and about the management of this evolution. Altogether 64 companies answered to the survey.

The targeted enterprises were manufacturing companies in various sectors mainly in automotive industry, machinery and electronics. For each of them, the contact person was a high level manager who had an overall view of the enterprise and could thus fill out most parts of the questionnaire alone.

We developed a "lean" questionnaire which could be completed in no more than two hours. The questionnaire was based on the Norwegian TOPP questionnaire [TOPP 92]. It was divided in three parts:

- The first part is an evaluation of the main functions' performance, compound of approximately 100 questions, each of them split up into 3 elementary sub-questions (performance today, performance in two years, importance for competitiveness). This allows to make projections towards the future, so that trends can be found.

    There are 3 types of questions:
    - overall performance of the function,
    - level of computerization,
    - additional specific questions.
- The second part is related to the practices and interests of the surveyed companies concerning their management of change (strategic issues, use of consulting, etc.).
- Finally, the last part contains background information about the company (name, size, sector, etc.).

The translated lean questionnaires were accompanied by documents explaining the context and objectives of TIME project to motivate the companies to participate to this survey. Each partner distributed this document to companies corresponding to the profile of study. To enhance the return ratio, each partner used its own contacts through the industrial milieu to convince most of them to respond.

A specific software application was developed to support the statistical analysis of the returned questionnaires. To validate and interpret the results further, also a more detailed TOPP analysis at five industrial sites was carried out.

2.2. Main results *[Time 93-2bis]*

The market study showed that most of the surveyed industrial companies have difficulties to define clear objectives in terms of strategic evolution. Indeed, while 77% of the enterprises claimed to have a strategic plan, only 64% had well defined goals and 46% had a clear definition of improvement projects in the strategic plan. This points out important shortcomings in the definition of a global strategy for evolution.

Moreover, the companies in the recorded sample do not have methods to help them to define this strategic plan: only 25% of companies had knowledge about methods and tools for the management of evolution.

However, the consciousness of the importance of a long term vision and of a global reflection is very high. A long term view and management methods are recognised in the companies as key success factors for the evolution process. Nevertheless, they are not considered as sufficient by themselves: without a high motivation of employees, it seems difficult to successfully re-organise a company. Motivation is best achieved through participation. Therefore, it is a major key success factor to achieve a large participation during the change process.

2.3. Functions needing major improvements

According to the results, the greatest need for development in almost all functions was felt in small and medium sized enterprises with less than 500 employees. In general, they lagged behind the big companies in their relative performance. However, the managers of SMEs were aware of their huge need for development.

In those functions that were closely related to the physical system (the manufacturing and assembly functions, key equipment and manufacturing technology), the SMEs felt more competitive than the big companies. In logistics, irrespective of company size, the proximity of the company to the final customer seemed to coincide with higher performance estimates. In contrast to the big companies, the SMEs were however not yet aware of the whole logistic chain and of the importance of suppliers to overall competitiveness.

On the whole, SMEs have been so far focusing on the internal functioning and especially on manufacturing, which they now estimate to be very well equipped and competitive. No big development needs were felt in manufacturing, although quality and lead-time were perceived as important competitive factors. The respondents seemed not to associate development in quality and expedition of work processes with the development of their manufacturing function. The potential for organisational evolution in manufacturing needs still to be uncovered.

The function with the poorest ranking in performance relative to competitors was marketing. However, the perceived development needs in marketing were high. Also the rather low level of computerization in marketing was estimated to rise in the future. The enterprises perceived a need to improve their market knowledge and to react more quickly in order to win the hard competition.

In the researched sample of enterprises, there were problems in the integration of marketing with both production and quality. Also the maintenance function had poor links to quality and to design. Many of the enterprises could not estimate their ability of inter-functional cooperation at all. This is alarming, since the most important areas in evolution often are in the reintegration of differentiated tasks and functions into a more streamlined and efficient process. Concerning the various functions' level of computerisation, attitudes are often ambiguous:

- the overall importance of computerisation for the competitiveness of the enterprise is not recognised,
- for many separate functions however, computerisation is depicted as very important, and moreover, its utilisation will increase in the next years.

Indeed, the strategic potential of IT is still uncovered, and the introduction of IT in industrial management has not yet included the necessary changes in the modes of operation: in work processes and organisation. There is in future a need for training and reorganisation in all business processes to take the full advantage of the already existing technology, and especially to implement most efficiently the huge future IT-investments that are anticipated in the answers.

3. THE TIME GUIDE PROJECT

3.1. Objectives and Domain [Time 93-1]

Accelerating turbulence in the competitive environment forces management to focus on mastering the continuous evolution of the enterprise. However, the methods and tools for this management of change are still insufficient. As a consequence, industrial organisations have evolved slowly and painfully. The evolution know-how has not accumulated in the organisation, and the enterprises' internal innovation potentials for evolution could not be uncovered. The results have been resistance to change, long lead times of transitions, often sub optimal techno-organisational structures and decreasing competitiveness.

TIME GUIDE aims to develop a set of tools for the enterprises' management of change needs. In TIME GUIDE, the evolution path of the enterprise is seen as a succession of intermediate steps - steady states - towards the vision "should be" (c.f. Figure 1). The management of evolution starts from the modelling and analysis of the present situation "as is" and the definition of a future vision "should be", and proceeds then to the modelling of the "next step", which will be the first intermediate steady state towards the vision. These managerial actions are repeated at each "next step", when it becomes the next steady "as is" on the path of evolution. Also the vision itself can get revised at each new "as is".

This continuous evolution through discrete steps has to be monitored and documented effectively, so that the organisation gets feedback from its progress along the path and can learn from its evolutionary actions. The definition of measurable progress factors and the establishment of evolution databases to create an "organisational memory of evolution" are thus important tasks in the management of change.

The self-audit method serves the internal diagnosis of the enterprise in its "as is" state, and contributes to the definition of progress factors. It helps to find the business processes that need to be developed, and to monitor this development. It is closely connected to the benchmarking method, that contributes to the external comparisons between enterprises on process level. The benchmarking results have an impact on the vision "should be": they create the tension for the change and help in the definition of the key performance indicators and the progress factors. An important feature in both the self audit and the benchmarking methods is their participative nature: both analyses are conducted by the personnel of the business processes and create thus commitment to change.

The TIME GUIDE model for the continuous development of enterprises is depicted in Figure 1 below.

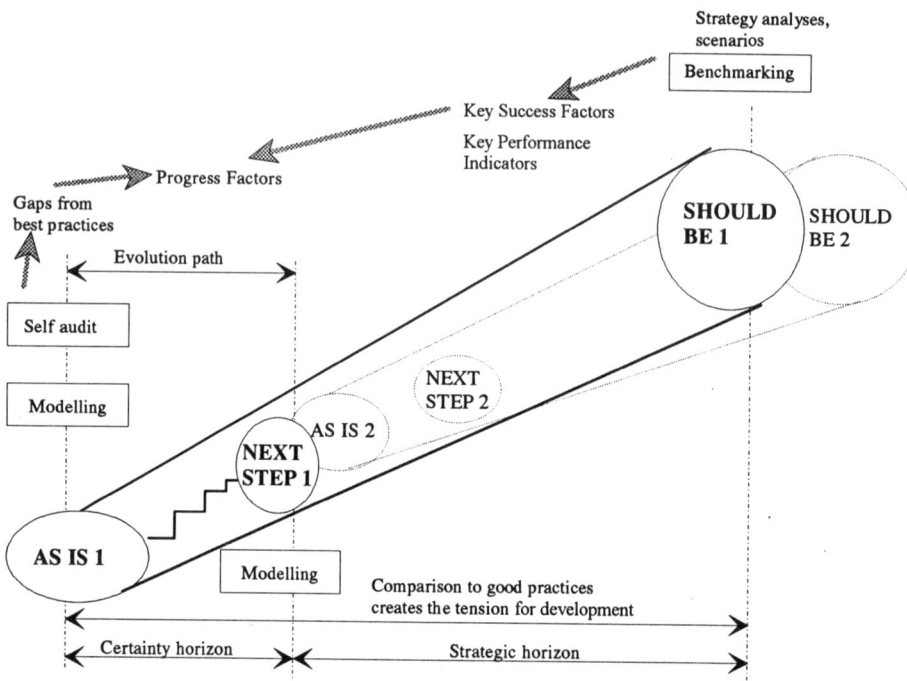

Figure 1: The TIME GUIDE model: Guiding the development of industrial enterprises
[Time 93-3]

3.2. Benchmarking in TIME GUIDE

We would like here to show where and how benchmarking can be used in the overall change process and how the "classical" view of benchmarking can be enlarged and enriched by additional concepts and tools from TIME GUIDE.

3.2.1. Need of tools for benchmarking

Today, benchmarking has reached a stage where everybody agrees more or less on the different steps in the process. Figure 2 hereafter sums up the benchmarking process.

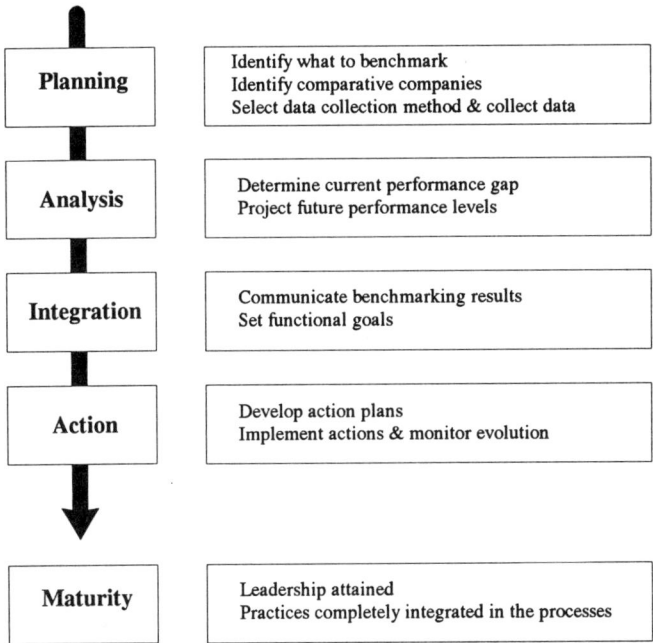

Planning
Identify what to benchmark
Identify comparative companies
Select data collection method & collect data

Analysis
Determine current performance gap
Project future performance levels

Integration
Communicate benchmarking results
Set functional goals

Action
Develop action plans
Implement actions & monitor evolution

Maturity
Leadership attained
Practices completely integrated in the processes

Figure 2: Steps in the benchmarking process [CAMP 89]

Yet, even if the benchmarking tasks are perfectly defined in their functional contents, the tools to support these tasks are missing: *today, benchmarking proposes a process, but it is not enough founded on concrete implementation supports.*

To illustrate this point, a 1992 study (in [Sprow93]) indicates that 95% of American companies do not know how to steer the benchmarking process, even if they are totally convinced of its necessity for the company.

For instance, few tools exist to help "benchmarkers" in the identification of the activity to benchmark. Most authors agree on the fact that this identification should be based on both customer's expectations (the "Order Winning Criteria") and strategic objectives of the company, but none of them proposes a clear methodology to steer this phase.

This is the case for most steps in the overall benchmarking process: how might companies determine the "Best Practices", how can they ensure their personnel's involvement, how can they measure the attainment of the functional goals they have set?

TIME GUIDE intends to provide such tools, and tentative solutions are already proposed :

- by applying *self-audit techniques*, TIME GUIDE grounds the benchmarking process on a strategic diagnosis of the company's strengths and weaknesses allowing to define "those activities that contribute in providing competitive advantage" ([Shetty 93]). Moreover, a self-assessment tool might increase participation within the company, facilitating the final integration of the benchmarking results.

- by using *modelling methods and tools*, TIME GUIDE ensures that all aspects will be taken into account: the system is seen along decisional, informational, physical as well as organisational and human points of view.

The models and data obtained can not only help in identifying the key activities and processes to benchmark, but also:

- in building "generic models" of the benchmarked activity. These models would contain the key elements of a process (objectives, key resources and skills, associated performance indicators, flow of material, information, division of tasks etc.) and could help in finding the right comparative companies.

- in providing visual support (and ensuring a common understanding) to communicate internally the benchmarking results.

- in implementing the change, through the basic principle of participation in modelling and simulating the business processes.

- in documenting the change process. By updating periodically the models, the change actions and achieved improvements can be visualised, and an "organisational memory" of evolution can be created.

3.2.2. *The role of benchmarking in the management of change*

Management of Change is a broader concept than benchmarking: benchmarking is only one of the possible ways to tackle Management of Change. As such, benchmarking is included in the overall TIME GUIDE approach. We shall see hereafter where it takes place in the TIME GUIDE model of evolution: (c.f. Figure 1)

- The main role of benchmarking in our approach is to contribute, in addition to the more general strategic goals of the company, to the *definition of the vision we have of the right direction to follow (the "Should be" system)*. By identifying "Best Practices", we can further determine the "Key Success Factors" leading to excellence in the considered business processes. Obtaining the same level of performance in our company becomes the target to reach, the objective initiating the whole improvement process.

- A second role of benchmarking lies in the determination of the "Next Step" as well as of the "action plans" to get there. By taking into account the situational factors and other "boundary conditions" of the benchmarked activity within our company (for instance, "allocated budget is limited to 1M$", or "Human resources policy forbids any hiring for the next two years"), we can also instantiate generic Key Success Factors and Key Performance Indicators into tangible *progress factors* leading to directly implementable *action plans* and measurable results.

3.2.3. Similarities between the two approaches

The TIME GUIDE project is built in order to develop "Tools and methods for Integration and for Management of Evolution of industrial enterprises" that will take into account the industrial needs collected in the user requirements survey performed during the feasibility study.

The TIME GUIDE key concepts seem to fit well into the benchmarking approach and *vice versa* :

Similarity 1 - **TIME GUIDE as well as Benchmarking aim to steer the evolution process.**
As a matter of fact, the objective of the two approaches is to improve the performance of the firm. The only difference lies in the improvement scope: while TIME GUIDE intends to manage the whole range of changes (structural, organisational, etc.) in an integrated and coherent approach, benchmarking focuses on precise improvement projects.

Similarity 2 - **TIME GUIDE and Benchmarking imply participation and autonomy.**
The two concepts are closely linked : autonomy is given prominence in order to enhance personnel's motivation and to reduce the resistance to change. Participation is therefore a *sine qua non* condition for the change process to be successful. Benchmarking insists on their importance, and TIME GUIDE proposes to embed these principles into all its tools and methods.

Similarity 3 - **TIME GUIDE methodology and Benchmarking have to be applied continuously.**
Most of the time, several improvement actions have to be planned, but also monitored and reviewed simultaneously. Results of these change actions have to be checked permanently, in order to keep coherence in the overall system. Moreover, changes occur permanently in the environment of the system. For adaptation purpose, continuity is required in steering the evolution process.

Similarity 4 - **Both TIME GUIDE methodology and Benchmarking take into account the competitive environment.**
Both approaches end in durable, structural changes in the production system.
That is to say that both approaches deal with strategy, matching internal strengths and weaknesses against the characteristics of the competitive environment.

This is not an exhaustive list of the similarities between TIME GUIDE methodology and Benchmarking, but it shows that both of them are grounded on the same conceptual framework stemming from - and aiming to solve - the difficulties encountered by industry in managing their change.

3.3. The European database for Benchmarking

One of the aims of the TIME GUIDE project is to develop the structure for a European benchmarking database. The database will be built on the results of the TIME GUIDE benchmarking method, and filled with initial data from the TIME GUIDE pilot companies and companies participating in the Norwegian TOPP program. In TIME GUIDE, the feasibility of this database approach is proven, and a structure is settled for its future commercialisation and operational exploitation, which will be tasks of a following project.

3.3.1 *From self audit to benchmarking: experiences from the TIME feasibility study*

In five companies of the participating TIME countries, an in-depth industrial site analysis was carried out during the feasibility study. Researchers in the TIME consortium collected the data from the sites using the full TOPP A questionnaire as an instrument. The aim of these analyses were firstly to test the results of the "lean" market study, and secondly to test the self audit and benchmarking features of the TOPP A approach. The results were analysed at SINTEF using the TOPP method and the already existing TOPP database.

The TOPP A questionnaire is formulated as a self-assessment of the company's own performance compared to the performance of its competitors. Performance estimates are given for each function of the company in its present situation and in the estimated situation in two years. All organisational functions are assessed in principle two times: on a general level by around 20 individual respondents from different units of the organisation, and on a more detailed level by small interdepartmental groups. The assessments are given as relative scores (1...7) of own performance compared to the competitors' performance. Also the importance of each function to the company's competitiveness is evaluated.

The development needs of the five industrial sites could be analysed in a number of ways. The group results highlighted the functions which had biggest gaps between the present performance and the estimated performance in two years. They could also point out the critical development areas, where the importance for competitiveness was assessed to be very high but the present performance was ranked poor. On the other hand, some functions could be detected, where big development needs were expressed, but which were not considered important for competitiveness. The individual responses revealed those functions where the differences of opinion concerning performance were the greatest. This could be interpreted as a sign for development needs in those functions.

3.3.2 *Requirements for the TIME GUIDE benchmarking database*

The TOPP A self audit results can help companies to find, based on internal self-assessment, their most critical functions and processes. When conducted as a continuous self audit activity, the results also show development trends within one company. But the audit scores as such are meaningless for inter company comparison. They are given by the companies themselves as subjective estimates of own performance relative to competitors. For benchmarking, these self images should be complemented by objective measurement data, and a common process modelling structure.

The structure of the self audit hierarchy of measurement criteria will in the TIME GUIDE project be mapped with the structure of the benchmarking database, and links will be established to enable inter company comparison. The objective is to develop a benchmarking procedure and database structure, that allows company comparisons on a process level. A deep benchmarking in the relevant areas of enterprises (in their business processes, in manufacturing, logistics etc.) by the managers and middle managers themselves in a participative manner gives important knowledge about how these enterprises reach their performance levels (not only their performance levels as such).

The TIME GUIDE industrial partners as well as selected TOPP companies will test the benchmarking method and the database prototype in pilot implementations, cooperating with project researchers. Based on this cooperation, pilot cases describing the experiences and results from the benchmarking method will be produced. These cases will later on help further refining the database structure before the final version is developed.

4. CONCLUSION

According to the market survey conducted in the TIME feasibility study, industrial managers feel huge development needs in their companies, and a well managed enterprise evolution is perceived as a competitive advantage. The demand for effective change management tools and methods has awakened and is growing rapidly.

To answer to this need, the TIME GUIDE project will develop a methodology for change management. The TIME GUIDE methodology and benchmarking are closely connected and complementary in at least three dimensions:
- TIME GUIDE fulfils the need for tools that support benchmarking by developing a self audit procedure and modelling techniques
- Benchmarking is perfectly integrated into the overall TIME GUIDE methodology, since it helps in determining the goals to strive for (the "should be" system and the "next steps"); it also helps in the formulation of the progress factors to measure the concrete change actions.
- The objective (improvement of enterprise performance) as well as the underlying concepts (continuity, participation, consideration of the competitive environment) are common to both approaches.

The TIME GUIDE methodology consists of three separate methods: the self audit method, the modelling techniques and benchmarking. All these elements answer to specific requirements in evolution management. Considered separately, each of them has strengths and weaknesses. But built into a coherent TIME GUIDE methodology, the synergy and complementarity of the elements make the whole more than the sum of its parts.

While benchmarking alone is not sufficient to manage enterprise evolution, its integration into the TIME GUIDE methodology leads to a coherent management of change process. The central effort in the TIME GUIDE project will lie in developing this synergy between the three methods.

BIBLIOGRAPHY

[Camp 89]   Robert   C.   Camp,   "Benchmarking   -   the   search   for   Industry
                                 Best   Practices   that   lead   to   superior   performance"   -   ASQC
                                 Quality Press 1989, Milwaukee, Wisconsin.

[Shetty 93]   Y.K.   Shetty,   "Aiming   high: competitive   benchmarking   for   superior
                                 performance" - Long Range Planning, Vol. 26 Nr 1, pp 39-44, Pergamon
                                 Press Ltd, 1993.

[Sprow 93]   E.E.Sprow, "Benchmarking: it's time to stop tinkering with manufacturing
                                 and start clocking yourself against the best" - Manufacturing Engineering,
                                 pp56-69, September 1993.

[TOPP 92]   TOPP "A productivity program for manufacturing industry" - NTNF 1992.

[Time 93-1]   TIME -deliverable D11: "Objectives, concepts and domain of the TIME
                                 project". Last version - 4/10/1993.

[Time 93-2]   TIME -deliverable D21: "Survey methodology". Last version - 6/10/1993.

[Time 93-2bis]   TIME -deliverable D22: "Results of the market survey". Last version -
                                 6/10/1993.

[Time 93-3]   TIME -deliverable D31: "Definition of the deliverables". Last version -
                                 4/10/1993.

7

Benchmarking the Investment Process of New Technology

I. Persson

Dept of Industrial Engineering, Lund University
P.O. Box 118, S-221 00 Lund , Sweden

1. INTRODUCTION

Use of new technology has not always been a success. In fact, rather many heavy capital investments have not provided the expected increase in productivity, and the announced profitability has not always been reached. Osterman (1991) showed that more advanced technology did not mean an increase in productivity, however human resource management was considered to be very important.

The fact that the expected influence from new technology has not been reached has been interpreted as a challenge for improvements in management support when evaluating investments in information technology and advanced manufacturing technology. Many articles have been published. In the streamline of suggested improvements, three different areas have been found:

1. Improvements of techniques used in the evaluation process. There are many articles which discuss improved evaluation techniques. Meredith & Suresh (1986) and Soni et al (1992) have made summaries of the economic and analytical justification methods. Estimated probability (Azzone & Bertele, 1989), or random variables, (Suresh, 1990) are suggested in order to take flexibility into account.

2. Strategic investment appraisal. The link to the strategy is focused (Noble, 1990; Slagmulder & Bruggeman, 1991). The decision to invest is primarily based on an analysis of to what extent the suggested investment can support the actual strategy. This approach relates the investment to the business which will influence the way of working with the investment later on. Slagmulder et al (ibid.) found that the quality of the strategic analysis was the main critical success factor in their study (case studies).

Within this approach, recommendations for which quantitative and qualitative measures are to be used are dependent on the project. Meredith & Suresh (ibid.) and Meredith & Hill (1987) focus on four main approaches at the full integration level: technical importance, business objectives, competitive advantage, and research & development. Though basically in a critical mode concerning misuse of capital budgeting techniques, Clemons (1991) states that "sometimes we can do a net present value analysis... even for strategic investments and capture uncertainty through sensitivity analysis" p33. Sometimes a strategic investment can merely be seen as an option. A strategic approach is suggested for systems that directly contribute to the company's business objectives. (Burstein & Pearson, 1990)

3. Focus on a champion. Ward(1990) emphasises the champion in the implementation process by referring to Schon(1967). Beatty (1992) found that successful implementation requires a skilled champion. Slagmulder et al (1991) found that a champion was not a sufficient condition for success.

2. THE OBJECTIVE OF THE PROJECT

In other studies have been found that new technology is working successfully with the human resource management applied in Japanese companies. (Osterman, 1991). There is, however, no research which penetrates the difference between the investment process of successful projects in new technology and the investment process when the project fails. This study will fill that gap.

The objective is to benchmark successful and less successful investment processes of new technology. Perhaps other countries than Sweden could be involved. Experience has been found to be very important in traditional capital investment processes (Persson, 1990) in Swedish companies as well as manager's perspective. Making successful investments in new technology, however, must not be dependant of experience. Instead, the old process may be an obstacle.

Hypothesis 1. When taking steps in technology or introducing new technology there is a need for new knowledge which is difficult to obtain by traditional educational efforts.

The hypothesis is that when making investments in new technology, experience based knowledge has to be replaced by a process of learning and influence, which creates the corresponding knowledge before the decision to invest is made. If so, this competence would influence, through a deeper analysis to understand what is possible and also make it easier to compare different levels of technology. Instead of making a judgement, the manager has to make the necessary arrangement in order to understand.

Hypothesis 2. The successful capital investments in new technology, have a process which includes learning and influence in the early stages.

The difficulties in the appraisal of new technology are the inabilities to concretise the effects from an investment. Persson, 1994, emphasise the understanding and insight which the managers would like to have when making a decision. Managers need to understand what the technology can offer from a business oriented point of view and what the demands of the investment will be in the different areas, in order for it to be a success. Competence is one example of an important area.

3. METHODOLOGY

To test the hypothesis data on successful and unsuccessful implementation processes of new technology is of interest.

The testing of the hypothesis can be made through surveys.
Cases are the traditional method used for gaining a deeper understanding in a complex environment and will be used in next phase.

3.1. Survey
The hypothesis is that there is a difference between the way people are involved in the investment process, which in itself is a benchmarking process. This hypothesis must first be tested.

In a survey, I want a method that uses the respondents answers. Traditional interviews are one possibility, but involves a lot of interpretative work

The conjoint method is a survey method which can take earlier answers into account by using of computer. There are benefits as well as disadvantages when answers are given through a computer. The benefits, besides the use of earlier data, are the structuring of data and hence less effort is needed in making the analysis. The disadvantages, above all, are that the selection of respondents will be influenced and restricted.

3.2. Cases
Cases is the traditional method used when making in depth studies. Benchmarking of a process also means a study to get understanding. Cases can be carried out as a longitudinal study as well as a reconstructing study. If it is a longitudinal study, the researcher can work passively as well as in an action mode.

Working in an action mode means that the outcome is dependent on the researcher's actions. Based on the hypothesis this could be actual. In order to be meaningful the quality has to reach an acceptable level, which means that the possibility to influence should be important. These actions, however, would be based on important performance indicators before the decision to invest is made. The performance indicators first has to be caught from a survey study.

3.3. Conclusions on methodology
The first goal of research is to test the hypothesis by a survey method. The conjoint method is preferred if possible and the objective is to get which criteria are important.

A second step may be to make deeper studies of the investment process in a company, which has had a successful investment process. The company (companies) will be chosed from the answers in the survey.

REFERENCES

G. Azzone and U. Bertele, Measuring the Economic Effectiveness of Flexible Automation: A New Approach. International Journal of Production Research, Vol.27, No.5, pp735-746, 1989.

C.A. Beatty, Implementing Advanced Manufacturing Technologies: Rules of the Road. Sloan Management Review, Summer, pp49-60, 1992.

M.C. Burstein and G. Pearson, Strategic Justification of Plant-Level Investments. Manufacturing Review, Vol 3, No 3, pp171-177, 1990.

E.K. Clemons, Evaluation of Strategic Investments in Information Technology. Communications of the ACM, January, p25, 1991.

T. Hill, Manfacturing Strategy. MacMillan, 1985.

J.R. Meredith and M.M. Hill, Justifying New Manufacturing Systems: A Managerial Approach. Sloan Management Review, Summer, pp49-61, 1987.

J. Merredith and N.C. Surech, Justification Techniques for Advanced Manufacturing Technologies. International Journal of Production Research, pp1043-1057, 1986.

J.L. Noble, A New Approach for Justifying Computer-Integrated Manufacturing. Journal of Cost Management Winter pp14-19, 1990

P. Osterman, New technology and Work Organization. Paper at the IVA-Conference "Technology and Investment - Critical Issues for the 90´s, 1991.

I. Persson, Use of Information Technology in The Investment Process. WP 1990:20, Institute for Management of Innovation and Technology (within the Information Technology and Management Program), 1990.

I. Persson, The Capital Investment Process and New Technology. WP 1994:60, Institute for Management of Innovation and Technology (within the Information Technology and Management Program), 1994.

D. Samson, Manufacturing and Operations Strategy. Prentice Hall, 1991

D.A. Schon, Technology and Change: The New Heraclitus. Dell, 1967.

R. Slagmulder and W. Bruggeman, The Investment Decision Methodology and its Impact on the Success of Flexible Manufacturing Technologies. In Management and New Production Systems (EIASM) pp637-663, 1991.

R.G. Soni, H.R. Parsaei, and D.-H. Liles, Economic and Financial Justification Methods for Advanced Automated Manufacturing: An Overview. In H R. Parsai and A. Mital (eds.) Economics of Advanced Manufacturing Systems, Chapman & Hall, pp3-19, 1992.

N.C. Suresh, Towards an Integrated Evaluation of Flexible Automation Investments. International Journal of Production Research, Vol 28, No9, pp1657-1672, 1990.

T.L. Ward, Role of Champions in Justification of Computer Integrated Manufacturing Systems. In H.R. Parsai, T.L. Ward and W. Karowski(eds.), Planning, Design, Justification and Costing. Elseiver Science Publishers B.V., pp123-131, 1990.

8

Free Trade and Ecology - Benchmarking as a Method Towards a Sustainable Future

A.-P. Hameri and E. Eloranta

Institute of Industrial Automation, Helsinki University of Technology
Otakaari 1, FIN-02150 Espoo, Finland
Internet: aph@cs.hut.fi

The paper questions whether the prevailing combination of management principles provide ecologically sustainable future. Benchmarking together with the presumptions of free trade and neoclassical economics direct the industrial enterprises to strive towards the best practise and higher profits. To achieve comparative advantage is no longer the main objective of multinational firms, but to obtain supreme advantage through cost efficiency and better exploitation of resources. Free trade with expeditiously moving capital and the increasing use of benchmarking techniques has created a business environment where the legal system is too rigid to provide exogenous and sufficient protection for the nature. We suggest that the external control should itself use benchmarking and shift the obsolete neoclassical world view towards the steady-state economics in order to provide ecologically secured future.

Keywords: benchmarking, free trade, neoclassical and steady-state economics, ecology, globalisation

1 INTRODUCTION - THE GLOBALISATION PROCESS

The Uruguay round of the General Agreement on Tariffs and Trade (GATT) negotiations at the end of 1993 was a partial success. New members were obtained and better prospects for fluent trade were made. This agreement provides the legislative framework for smooth transition from internationalisation to full globalisation of manufacturing and services. GATT is only a part of the global process that reshapes the operating environments of industrial companies by removing trade barriers between nations. International production through world-wide corporate networks with multimodal supply and distribution chains has marked the current era as the period of physically and logically distributed co-operation. The visible proof of this trajectory of development includes the following things (see e.g. OECD, 1992, p. 211):

- the very rapid growth of foreign direct investment;
- the role played by multinational enterprises in world trade and the present volume of intra-firm trade;
- the emergence of highly concentrated international supply structures;
- the increasing number of cross-border mergers and acquisitions;
- the development trends in managerial principles to manage distributed manufacturing and global competition.

The GATT pact between 117 countries aims to secure the economic growth and jobs in export-related industries worldwide over the next decade by lowering the consumer prices and, most important, by preventing protectionism. As the above list of facts indicate that the companies are already exploiting, and have been for some time, the fruits of free trade and comparative advantage through globalisation. The point of this paper is not to argue whether Japan should allow the import of rise, but how the current managerial principles along with benchmarking direct the economy in the future.

One major reason for the emergence of multinational enterprises stems from the obstacles and disparities between countries. Such obstacles and disparities include tariff barriers, wage and interest rate differentials, technological inequalities, and the fact that national economies are at varying stages of development. These differences between nations provide an opportunity for the multinational enterprises to delocate production and other functional parts of the organisation in such manner that overall cost efficiency is achieved. The increasing use of transaction cost theory has also favored the geographical diversification of multinational companies. The phenomenon is of significant scale even if we only consider the flow of intra-triad investment between the three developed market areas in the world (Fig. 1).

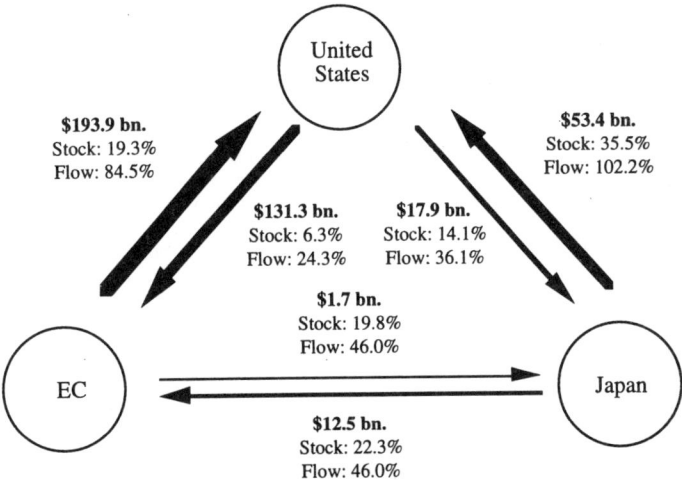

Figure 1. The flow of intra-triad investment, 1985-89 and the 1988 level of stock (source OECD, 1992). Note: Dollar figures show 1988 outward stock; percentages show average annual growth rates, stocks and flows. Stock growth rates are for 1980 to 1988. Flow growth rates are for 1985 to 1989.

The figure shows the evolution inside the developed market areas, which was boosted by the 1980s financial deregulation and globalised monetary markets. Even more rapid has been the development inside the market areas and in the new and virgin areas of trade. The transition process inside the pan-Europe along with the Baltic states, the former Soviet Union and Eastern Central Europe is very vivid because of their superior economical and operational advantages (labour, raw-materials, finance, etc.). Similar process is going on in America where the North American Free Trade Agreement (NAFTA) aims to standardise the basic elements of competition among the northern-American nations. In the far-East the disparities among rivalling nations, e.g. northern-Korea, has been subject even for open trade-wars. It is beyond the scope of this paper to analyze in more detail the globalisation process and, thus, the paper focuses on the implications of the managerial principles pursuing the companies towards ultimate excellence.

2 MANAGERIAL PRINCIPLES - BENCHMARKING

Management of industrial enterprises have faced many themes and methods that provide a quick-fix for competitivity and success. Eccles et al. (1992) describe nicely the continuous change in the prevailing productivity paradigm of consultants, academics and researchers. Eloranta (1993) points out that these hypes (e.g. just-in-time, theory of constraints, time based competition, lean production, activity based management and costing, logistics, benchmarking, kaizen, computer integrated manufacturing, global sourcing, re-engineering, visual management, concurrent engineering, etc.) share the common message to concentrate on the real and value-adding processes of the company. Of all these hypes benchmarking is currently popular among the companies. According to the word definition benchmarking means something that serves as a standard by which others may be measured and judged. On the other hand it may also mean a standardized problem or test that serves as a basis for evaluation or comparison (e.g. tests for computer system performance). In the measurement of industrial performance benchmarking means the continuous process of measuring products, services and practises against the toughest competitors or those companies recognized as industry leaders to achieve superioir performance. As Camp (1989, p. 3) puts it that the essence of benchmarking is to be the best of the best.

There are various levels of benchmarking ranging from strategic and operational benchmarking to the business management benchmarking (Pryor, 1989, p.29). Regardless of the level, the underlying procedure is the same, the emphasis and focus being on different levels or parts of the organisation. By analysing the current performance against the best practise (e.g. market leader) enables to determine the competitive gap. Both negative and positive gaps should be analyzed, yet the negative ones needs to be examined thoroughly to define why differences exist and what factors require change. This analysis phase provides the means to project the future performance levels and the trajectory of the development. The goal is to transform the negative gaps to positive ones and, thus, to gain superior performance over the rivals. Once the goals are set then begins the integration phase, where the benchmarking results are communicated with the organisation. Functional goals and action plans are established for each organisational part that the development affects. Through the understanding of the benchmarking results and the motivation of the personnel the plans are executed. To monitor the development the benchmarks should be recalibrated after certain periods and not only at the end of the project. Therefore benchmarking is a continuous process towards the best practice.

Despite the systemized method of benchmarking (Camp, 1989), it is hard to find good cases of its use in practise. More or less the cases indicate that benchmarking is based on copying methods from other companies, which are doing better. As one consultant declared: "If things are done right, do not hesitate - copy!". Usually the public sources of information provide enough facts on how the rivals are doing certain operations. This is due to the fact that to obtain quantitative and accurate information from competitors is difficult or even impossible. Yet, some world class benchmarking candidates can be easily listed. Pring's (1992) list for the creme de la creme on certain topics is the following:

Purchasing	Honda Motor, Xerox, NCR
Quality process	Westinghouse, Florida Power & Light, Xerox
Manufacturing operation	Hewlett-Packard, Corning Inc, Philip Morris
Sales management	IBM, Procter & Gamble, Merck
Technology transfer	Square D, 3M, Dow Chemicals
Training	Ford, General Electric, Polaroid
Warehousing & Distribution	L L Bean, Hershey Foods, Mary Kay Cosmetics

Table 1. Some world class benchmarking candidates (Pring, 1992).

In one way or an other the many managerial hypes share directly or indirectly the same sincere objective; to provide means for faster throughput times, better quality and improved cost efficiency. Benchmarking enables companies to asses their current performance with respect to other companies. This kind of comparison differs from the traditional customer surveys and other intra-company analyses, which indicate what customers want and how we are doing it. Thus, the learning by doing method is applied. But this kind of analyses may provide one-sided picture of the current situation and the true potential of improvement is not exploited. Benchmarking versatiles the traditional methods, for improvement by enhancing the aspect of learning from others.

3 ECOLOGICAL IMPLICATIONS - THE MASS PRODUCTION PARADIGM

In America Ross Perot has warned the people of the "giant sucking sound" that will take effect along the NAFTA. By this he means the huge transition process during which the American industrial enterprises will move their value-adding facilities on to the Mexican territory. Perhaps there will be a real movement phenomenon, but as we know, the movement has been valid for the past 20 years. Along the increase of labour costs in the US companies have been delocating their activities to the cheaper Mexico. Due to this process the famous Rio Grande has partially turned into sewage. This development is simply the result of the continuous search for better cost efficiency and competivity. Along with these driving measures, the lax environmental standards and poorly enforced regulations usually valid in low cost countries speed up this development.

The mass production paradigm is strongly prevailing the current industrial enterprises where the profit making is still largely based on Fordian wisdom of high volumes and low unit costs. Relying on the huge PIMS data base, Eloranta (1993) argues that this attitude is still prevailing, despite the voluminous debate around the customer oriented production systems. The data base shows that customization is statistically the major determinant of low profitability. Thus, the current managerial principles together with bencmarking largely direct the development towards cost efficiency and high volumes. Taiichi Ohno's (1988) sincere idea of production systems generating productivity of mass production at low volumes has not been achieved yet.

Along with this development the comparative advantage of nations work all the time leading probably towards the absolute advantage. As Daly (1993, p. 51) points out that Racardo's theory of free trade between nations has a critical but often forgotten assumption that factors of production (especially capital) are internationally immobile. In reality this is not true, because capital can be moved from nation to nation almost at the speed of light. The following figure depicts the evolution of free trade towards absolute advantage (ibid., p 52).

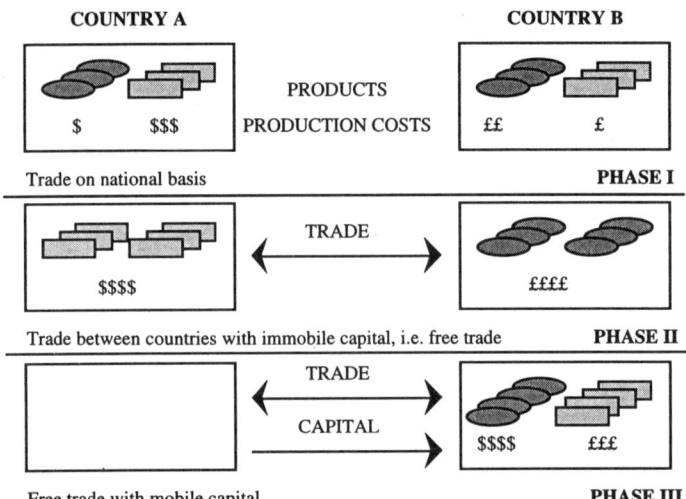

Figure 2. The evolution of trade from national to global through comparative advantage to the absolute advantage (Daly, 1993).

Despite the schematical nature of the picture the rationale is the following:

Phase I   Without international trade each country is limited by its own capital and resources. Some products are comparatively less expensive to produce than others on a per unit basis.

Phase II   Along the free trade countries can specialise on those products with high comparative advantage. All of country's capital can be invested in making certain products. Absolute cost differences between the countries do not matter. The hidden assumption is that capital cannot cross borders.

Phase III   If capital is mobile (as it is), capital may follow absolute advantage rather than comparative advantage. The result may be the one predicted by Ross Perot.

The above dynamics of the development seems to be plausible, yet to verify it is a difficult task. But intuitively without any exhaustive quantitative analysis we can provide evidence of the existence of this development. The following map (Figure 3) depicts the transformation process of Finnish textile companies in the Europe. By viewing the location of factories we may conclude that the driving force has been the cost of labour.

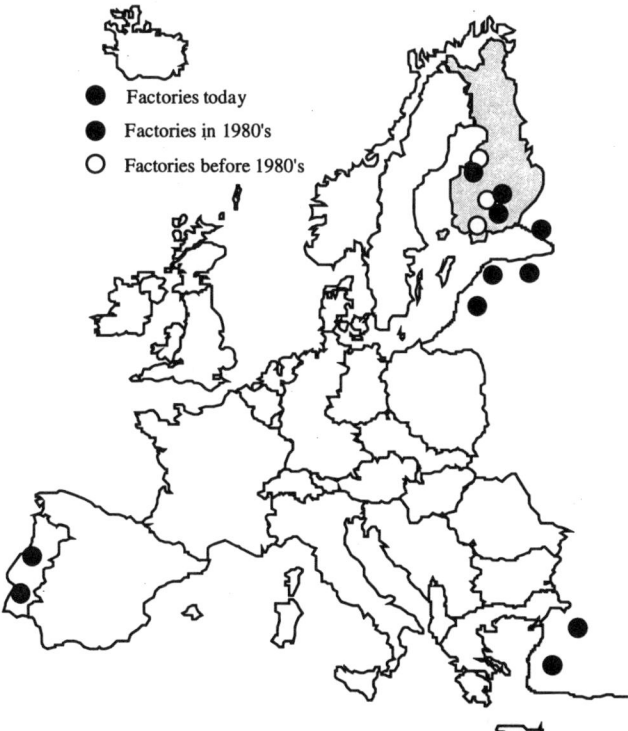

Figure 3. Location of Finnish owned textile factories before, during and after the 1980's.

The map shows nicely the avid pursuit after lower costs. This tendency is emphasised because textile manufacturing is typically very labour intensive industry. Same clustering development could be depicted in other industries and globally the process is very active. Batra (1993) discuses the problems of free trade from the point of view of domestic markets. He proposes that nations should establish strict means to control the trade. As Daly (1993) summarizes it that each country should set the rules of cost internalization in its own market. Whoever sells in a nation's market should play by that nation's rules or pay a tariff sufficient to remove the competitive advantage of lower standards. Thus, tariff policy should consider the environmental preferences and moral judgements between two trading nations.

4 EXOGENOUS CONTROL - STEADY STATE ECONOMICS

So far we have reached the conclusion that free trade and current managerial principles are insufficient to provide ecologically sustainable future. The evolution of industrial enterprises is conducted by the mass production and cost efficiency paradigm. The whole economy is driven by the neoclassical principles, which says that ultimately growth will solve all the problems (immigration, exploitation of raw-materials, ecological problems, etc.). Factors of production are changed into goods and services and this circle of transformation is the whole economy. The system is isolated and thus can be of any size. The unrealistic equilibrium position of an economy with continuous growth expectations is the object of study for neoclassical economics, which has resulted in many "armchair" analyses and mathematical models of this hypothetical situation (for summarizing critique see e.g. Simon, 1983, pp. 13-14; 1986, pp. 22-24). As a former neoclassical economist Boulding (1991, p. 12) hilariously puts it "... we got trapped into what I have called a 'cookbook theory' of production, that we mix land, labour and capital, and out pops potatoes or microwave ovens". To broaden the view of neoclassical economics new approaches have been developed. One mainstrem is the evolutionary economics which have strong emphasis on technological issues as the core of economoc and ecological development.

The neoschumpeterian school (see e.g. Dosi et al., 1988; Freeman & Soete, 1990; Hagedoorn, 1989; Stolper, 1991) have broken the untouched core of the economic system using an empirical approach and delivering real world data of the consequences and nature of technological changes and innovations. The idea that economies should be understood on the basis of the 'evolutionary paradigm', rather than an equilibrium situation or a world of deterministic mechanics has become a prevailing attitude in this research. This work has also created a whole set of new concepts describing technology and technological process. The approach adopted by neoscumpeterian research has also extended the narrow presumptions of the neoclassical models of economic growth. The Nelson-Winter (1982) 'evolutionary' model of technical choice and innovation is one result of the theoretical work of this economic school. In their model, disequilibrium, which is understood as an important characteristic of economic development, plays a central role. Competition is no longer seen as a static phenomenon but a process in which winners and losers are generated. The winning companies have the *carpe diem* characteristics of taking advantage of technological opportunities, while the losers will miss their chances and gradually decline into technological obsolesce. This school has also extended their research on to the ecological issues.

Traditionally, the only methods of appraisal in technology have been based on economic and efficiency criteria, set by the engineers and entrepreneurs. This is due to the obsolete attitude that nature is seen as a finite resource which must be fully utilized. The study of technology assessment is a pursuit to find an appropriate regulatory framework to restrict the risks and costs and to enhance the benefits of technological change. To construct such a framework, all the different and diverse aspects of technologies, must be comprised. As Naschold (1991, p. 73) simply declares, the whole effort of technology assessment is to obtain a better understanding of the complex processes of technological change and, where possible or necessary, to steer or shape these processes. The interest towards technology evaluation has increased, perhaps because of the common awareness of the already realized and the possible negative effects of technological development. Hawthorne (1978, p. 168) specifies more accurately the above general criteria of technology assessment. He classifies technology assessment into categories which include:

- quality of life; impact of technology on individuals,
- social structure; changes in infrastructure and institutional systems,
- external environment; human health and safety together with the natural environment,
- working environment; employment and working conditions,
- alternatives; optional technologies,
- assessment effectiveness; measuring the management performance.

Technology assessment is the tool to guide our economic system towards the steady-state view of economy, according which the economy is only one component of a larger ecosystem in which materials are transformed and energy is converted to heat. As the economy grows larger, its behaviour must conform more closely to that of the total ecosystem. The economy is seen to be one open subsystem in a finite, nongrowing and materially closed system. Being an open subsystem it must be controlled exogenously by providing certain economic incentives, such as pollution taxes or tradeable emission permits. For the policy and regulation makers benchmarking may provide a tool to find the best practises in ecological performance among industrial enterprises. At least the following best practises on each line of business could be tracked by using benchmarking:

- the best recycling performance;
- the best energy efficiency;

- the best production methods and procedures enabling minimum air and water pollution;
- the best organizational modes and wage scales for human friendly work environment.

In addition to these analyses benchmarking could also be used on national basis:

- which country has ecologically the most advanced legislation system;
- which taxation system provides the best incentives for industrial enterprises to develop ecologically sustainable technologies.

As the OECD (1992, p. 208) report concludes that technology policy can become increasingly important instrument in attaining environmental objectives. More cross-sectoral, multi-disciplinary approaches are called for. Thus, the apparent threats of future may be turned to possibilities in a sustainble growth trajectory.

5 CONCLUSIONS

Benchmarking is seen as a method which emphases learning from those who are industry leaders in certain functional performance. This principle along with the traditional and still strongly prevailing mass production paradigm with its managerial principles has initiated the globalisation process. Various free trade agreements will speed up this process, when the search for comparative advantage may change to avid pursuit on absolute advantage. From the ecological point view this development is unfortunate and certainly it is not a sustainable one. It is suggested that the obsolete neoclassical view of economic system is to be changed into steady-state economic approach. To control and direct the economic development towards more sustainable future the policy makers should itself use benchmarking methods to discover environmentally friendly production technologies and to provide legislative means to direct industries to adopt these technologies. From the traditional production management point of view the success factors will remain the same; speed and quality in every operation is the key to prosperity also in the future.

6 REFERENCES

Batra, R., Myths and Misconseptions of Free Trade, Scribner's, 1993.

Boulding, K.E., "What is evolutionary economics?", Journal of Evolutionary Economics, vol. 1, pp. 9-17, 1991.

Camp, R.C., Benchmarking - The Search for Industry Best Practises That Lead to Superior Performance, ASQC Quality Press, Milwaukee, Wisconsin, 1989.

Daly, H.E., "The Perils of Free Trade", Scientific American, pp. 50-57, November, 1993.

Dosi, G., Freeman, C., Nelson, R., Silverberg, G., Soete, L. (eds), Technical Change and Economic Theory, London, Pinter Publishers, 1988.

Eccles, R., Nohira, N., Berkeley, J., Beyond the Hype - Rediscovering the Essence of Management, Harvard Business School Press, 1992.

Eloranta, E., "Value-Adding Factory - A View on the Factory-of-the-Future", paper presented at the Third International IMS Symposium in Vienna, Austria, 30.11.-1.12.1993.

Freeman, C., Soete, L. (eds), New Explorations in the Economics of Technical Change, London, Pinter Publishers, 1990.

Hagedoorn, J., The dynamic Analysis of Innovation and Diffusion, London, Pinter, 1989.

Hawthorne, E.P., The Management of Technology, Maidenhead, McGraw-Hill, 1978.

Naschold, F., "Techno-Industrial Innovation and Technology Assessment - The State's Problems with its New Role", in Hilpert, U. (ed), State Policies and Techno-Industrial Innovation, London, Routledge, pp. 65-82, 1991.

Nelson, R.R., Winter, S.G., An Evolutionary Theory of Economic Change, Cambridge, The Belknap Press of Harvard University Press, 1982.

OECD, Technology and the Economy - The Key Relationships, TEP - The Technology/Economy Programme, Paris, 1992.

Ohno, T., Workplace management, Productivity Press, Cambridge Mass., 1988.

Pring, P., "Benchmarking Strategy for Business Improvement - a 3M View", the IIR Seminar on benchmarking, Helsinki, Finland, 11.6.1992.

Pryor, L.S., "Benchmarking: A Self-Improvement Strategy", The Journal of Business Strategy, November/December, pp. 28-32, 1989.

Simon, H.A., Reason in Human Affairs, Stanford, Stanford University Press, 1983.

Simon, H.A., "On the Behavioral and Rational Foundations af Economic Dynamics", in Day, R.D., Eliasson, G., The Dynamics of Market Economies, Amsterdam, Elsevier Science Publishers, pp. 21-41, 1986.

Stolper, W.F., "The theoretical bases of economic policy: the Schumpeterian perspective", Journal of Evolutionary Economics, vol. 1, pp. 189-205, 1991.

9

FROM UNCERTAINTY TO CONTROLLABILITY, LEARNING AND INNOVATION

Jan Holmström and Karri Kosonen

Department of Industrial Management, Helsinki University of Technology

Otakaari 1, 02150 Espoo, Finland

ABSTRACT

To take the step from uncertainty to controllability and learning the industrial firm needs to recognize that both operational problems and problems of conflicting interests within the firm may be caused by its own planning and control procedures.

Capacity and delivery problems in the manufacturing industry are commonly a result of poor communication between members of the supply chain and of the chain members' internal practices to deal with variety. Pull control cannot always be applied to the extent necessary to solve the problem of distorted information. Thus, it is essential to recognize and accommodate the problems of distorted communication in the design of planning and control procedures. Based on a number of case studies from the Finnish manufacturing industry it is demonstrated that clarity and process overview are the cornerstones to reduce self induced control problems.

It is also argued that down-stream process transparency is not sufficient to achieve fast and fluent operations. To secure the cooperation of all actors in the value adding chain it is necessary to directly expose upstream operations to the final customer. A cultural benchmarking of employee attitudes and behavior in the Finnish metal working industry and the banking industry reveals the importance of direct customer exposure as a determinant of employee commitment.

Key words: operations development, self-induced operational uncertainty, customer exposure

1. INTRODUCTION

The difficulty of dealing formally with operational uncertainty is a primary reason for why operations management is necessary. However, uncertainty of the operational situation is often self-induced and amplified by bad communication in a changing environment. Simply put, it is possible for an industrial firm to acquire enough relevant knowledge about its operational situation to avoid much uncertainty and reduce the need for direct management control and decision making.

Senge (1990) very succinctly defines the complexity caused by the delay between action and outcome as dynamic complexity and observes that this type of complexity is extremely difficult to deal with or even acknowledge. However, ignoring the effects of slow communication has, as will be demonstrated in this paper, dire consequences on the firms ability to be competitive.

From the study of supply chains it has been demonstrated that small demand fluctuations are amplified to upstream operations (Houlihan, 1987; Towill, 1992).

This phenomenon was first described by Forrester (1958). In a supply chain a small and controllable ripple of the end customer demand is easily distorted to a substantial and strongly cyclical pattern of surging and falling demand for upstream companies. Because the time interval between demand surges may be several years the problem is easily associated with changes in the economic environment.

For a company far removed from the consumer it is very costly and difficult to deal with uncertainty of demand and to provide customers with reliable service. Unreliable delivery performance is the direct consequence of upstream companies having too little capacity to deal with unexpected demand surges. The response of down stream companies is to hold buffer inventory against unreliable supply. However, since communication of end customer demand is distorted by attempts to control buffer inventories, the companies of the supply chain are caught in a positive feed back trap where the efforts to deal with uncertain demand and delivery only increases the uncertainty of the situation.

There is, besides the length of the supply chain, another equally important source of demand distortion, which is variety. Product variety in combination with inflexible capacity encourages firms to extend response times in order to achieve some degree of scale benefits, which in turn results in demand distortion very similar to that of a long supply chain. Product variety is also a key disabling factor for improved profitability of industrial firms. Analysis of the extensive PIMS database indicates that customized products are the single most significant factor in explaining poor profitability in manufacturing industries (Buzzel & Gale, 1987). Both increased operational cycles and small batch production has a negative effect on efficiency, and thus also profitability.

2. REDUCING SELF-INDUCED OPERATIONAL UNCERTAINTY

The issue of variety implies that distortion of demand is not purely a problem of interfirm communication. The internal organization of many companies is not designed to master the dynamic complexity of an uncertain environment. The controllability of the value adding process is poor, i.e. the firm is not able to effectively pursue its market opportunities due to the constraints posed by long throughput times, inflexible capacity and inability to quickly procure the necessary materials. Buffers and attempts to achieve economies of scale slows down communication within the firm and misconstrues decision makers understanding of the value adding process.

When the perceptions of decision makers responsible for different aspects of the operation diverge the ability to adapt to changing circumstances weakens. Additionally, inadequate communication and control practices may create situations where it appears that the difficulties experienced are solely caused by the environment, even though substantial causes are to be found within the organization.

Burbidge (1989) observes that a major internal source of uncertainty are the load surges caused by multicycle control of production. Weekly loads fluctuate strongly in a manufacturing process where several different parts with different planning cycles are made. The solution to this self-inflicted problem of varying capacity requirements is to control the production of all parts within the same cycle. The quantity to be produced of each part varies from period to period. (see e.g. Burbidge 1994)

Unsynchronized control practices induces uncertainty by distorting communication in the value chain. An example of this relationship between operational uncertainty, controllability and communication is evident in the case of a small Finnish electronics subcontractor. In this case unsynchronized material control practices was found to contribute to operational uncertainty. In the company the use of MRP (Material Resource Planning) severed the link between production planning and purchasing. The result was that the company experienced severe difficulties to respond reliably to changes in demand.

The initial situation in controlling materials and the new practices devised by management to improve controllability and reduce operational uncertainty will be used to illustrate the point (Figure 1.).

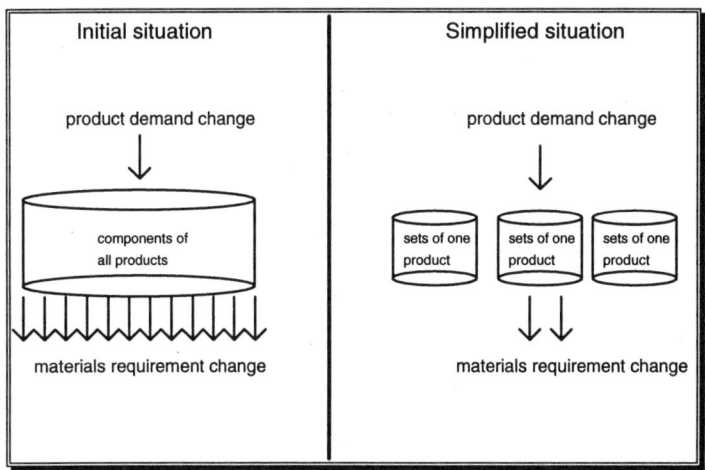

Figure 1: Improving controllability and reducing operational uncertainty through fast and accurate communication

Initially material requirements were controlled with a centralized information system. Components were controlled individually. This weakened the connection between product demand and materials requirements. A change in product demand caused an avalanche of purchase orders needing attention. Due to the number of materials in the individual products and the effort involved in adapting to changes it was difficult to keep up with whether enough materials were procured and whether unnecessary orders were canceled. This is a typical situation of self-induced operational uncertainty and low controllability.

The new practices devised simplifies the operational situation and improves communication by grouping materials in product specific sets. The communication of demand changes and their impact on materials requirements is direct and fast. There is little uncertainty of the operational situation and high controllability. This simplification also allows for reduced buffering, i.e. a speeding up of the operation. However, it must be pointed out, that to be effective the new practices requires that the firm is capable of defining sets and finding suppliers for the sets.

Controllability is the capability of the firm to pursue its goals, e.g. to seize market opportunities, reduce costs and improve profit, in a changing environment. The Just-in-Time model is commonly regarded as the simplest and clearest way to achieve high controllability and productivity in a manufacturing operation (see e.g. Schonberger, 1990). A Just-in-Time operation is demand driven with very little distortion of communication due to the direct feed back of pull control. Demand variations are equally visible in the whole value adding chain and the efficiency loss due to product variety is minimized by determined efforts to simplify material flows and reduce set up times.

The effect of introducing pull control in an operation that has previously been run in a batch production mode is dramatic. A Finnish supplier of engineered metal components for the electromechanical industry was able to cut throughput times and improve productivity by 50 per cent. This was the direct result of the elimination of waiting times and the material handling, production planning and control activities in batch production. A critical factor in the successful introduction of pull control was the ability and willingness of local suppliers to deliver Just-in-Time.

However, simplifying material flows and reducing set up times is often a difficult task. Very small production runs also make it difficult to achieve a consistent quality in many types of manufacturing processes. Long processing times in upstream value adding activities also complicate the situation. Therefore a certain degree of planning ahead, i.e. push control, is often necessary. The critical issue of planning is how to minimize internal sources of uncertainty while retaining an effective use of resources.

It must be emphasized that an effective plan is simple to understand and easy to monitor. If it is not possible to simplify planning and control by grouping a production process in independent parallel flows, then a fixed planning cycle should be considered for control of the sequential value adding stages. This simplifies the task of maintaining flow according to demand and coordinating dependent parallel activities.

The objective when defining which sequential activities should be controlled together is to have as few stages to control as possible and to group the activities in such a way that the total throughput time is as short as possible. This approach is practical especially when the process times of different activities vary greatly. Thus, by determining the planning cycle according to the most time consuming activity and grouping as many other activities as possible together, both simplicity of planning and fast throughput can be achieved.

The case of a Finnish clothing manufacturer illustrates the importance of simplicity and clarity of planning in a situation where pull control is not possible. The company produces a wide range of product varieties in an integrated process covering knitting, dyeing, printing, cutting and sewing. The process is currently loosely controlled in half a dozen overlapping process stages with varying control cycles. The consequence of this approach is an average throughput time of 10 to 11 weeks, and a prohibitively large number of dependent activities to consider and control. Production problems surface slowly and tend to accumulate in downstream operations at the end of each four week scheduling period. However, by reducing and separating the control stages and introducing a fixed relay schedule it is possible to project demand with minimal distortion throughout the process. The benefit of this is, besides a reduction of throughput time to three weeks, that production and sales can have a common understanding of delivery capabilities and that any problems in producing according to plan are visible within a week.

Controllability is the capability of the firm to pursue its goals in a changing environment. The ideal is to have pull control and organize production according to parallel product flows by using, for example, group technology techniques (see Burbidge, 1989). However, this is not always feasible. In that case much attention must be paid to avoid the undesirable surge effects of complex multi cycle control schemes. To achieve controllability it is critical that planning and monitoring of the operations is both simple and clear.

3. OVERCOMING CULTURAL OBSTACLES TO IMPROVEMENT

The spectacular competitive success of Japanese manufacturing industries is commonly ascribed to innovations in production and quality management. It has also been claimed that the introduction of these practices have been more successful in Japan than in the West due to cultural differences. This is a reasonable, but false, explanation. Differences in successful implementation can be explained by the fact that the practices address specific obstacles to improvement in many Japanese organizations. Implementation and adaptation is also more straightforward with a clear understanding of purpose.

Practices that address specific obstacles to improvement which are common in a large number of situations can be expected to have a dramatic impact. The impact of the quality control movement in Japan illustrates the effect of overcoming an obstacle to improvement that is rooted in the norms and values of management and employees. The respect for authority and seniority was identified as a very common inhibiting factor to tackle problems of bad quality in the Japanese industry (Ishikawa, 1985). Until the late 1940's quality efforts were guided by the perceptions of senior employees and not on the basis of analysis and facts. Even when a person performing the task knew the problem it was not possible for him to bring it to the surface so that something could be done about it. The "Quality-Story" format of problem solving with its emphasis on quantitative evidence was eventually adopted to overcome this cultural barrier to improvement (Lillrank, 1988; Ishikawa, 1985).

Leverage to achieve outstanding quality was found by breaking up the existing norms and values that prevented effective actions to improve quality. This development transformed the traditional work organization, where production work was divided into simple independent tasks and where planning and control responsibility was in the hands of specialists and supervisors. The emphasis on quality improvement and the introduction of statistical quality control techniques on the shop floor transferred control and problem solving responsibility from specialist functions to the value adding activities themselves (Drucker, 1990). In the wake of successful quality improvement, JIT production control could also be effectively introduced.

The Japanese quality movement illustrates an important point of successful operations development: There are specific actions that can have a pivotal effect on the norms and values of employees. Before copying "best practices" it is important to know the purpose of different courses of actions. This is especially the case with actions that affect cultural obstacles to improvement. In order to be effective "best practice" has to be "reinvented" from the organization's own situation and needs.

3.1 The Effect of Customer Exposure on Employee Behavior

The identification of obstacles to improvement and leveraged actions are the breaking points of successful operations development. The interesting question is: Could there be some norms and values that are obstacles to operations development and which are common to a large number of companies?

Cultural obstacles for successful operations development are clearly evident in a study comparing employee norms and values in the Finnish metal component manufacture industry. In an extensive, in-depth study, of three machine shops employing advanced manufacturing technology Kortteinen (1992) identified fundamental obstacles to operations development in both worker and management norms and values.

The starting point of the study was to explore whether the apparent agreement of employee and management interests in the Just in Time and Lean Manufacturing concepts have had an effect on worker attitudes and actions. Kortteinen identified an "ethos of survival" as the prevailing motivator of worker behavior in the case companies. This ethos can be shortly summarized as a quest for independence and self-esteem, which in the case companies took the form of determined attempts by individual workers to "out-wit" and seize control over the manufacturing system. Even though much effort had been spent in the case plants on reshaping and streamlining the organization, and developing training and technical competence, a high barrier remained between production workers and management.

The conflict of interests between management and workers seriously constrained the opportunities of the firms for a fast and fluent customer oriented operation. Generally, line management was well aware of this situation but was not able to do much in fear of the potentially disruptive actions of production workers. In a historical perspective this is very similar to the problem Taylor (1911) set out to solve by a rational and controllable work organization.

Kortteinen (ibid.) expanded the scope of his study from the male dominated manufacturing operations to female bank clerks. The objective was to explore the determinants of worker attitudes and actions in a completely different type of industry undergoing a similar process of dramatic reorganization and operations development. The unexpected result was to find the same basic "ethos of survival" among the bank clerks. However, the energies of workers to acquire independence and self-esteem were not directed against the organization and management, but towards building a circle of loyal customers preferring the personal service of the worker.

The different behavior driven by the same basic motivation prompts us to ask: Is very strong customer exposure the leveraged action to align employee and management interests? In that case a strategy for successful operations development in manufacturing industries would be to find ways to organize so that industrial employees are able to be as proud of their own customers as bank clerks are.

The significance of being exposed to the customer or not is essential. The problems with machining workers "out-witting" the work organization in the companies studied by Kortteinen (ibid.), can be seen as a result of the absence of customers, i.e. the operation did not have true customers with a choice. The machining operations were upstream operations servicing assembly cells or finishing operations. In one of the companies studied, a very substantial reorganization of assembly work has recently been concluded. Assembly workers manage the whole delivery process from order processing to installation support. The lead time from order entry to delivery for standard products has been reduced from three weeks to less than one day, including manufacturing of key parts.

The basic problem can now be defined. The "ethos of survival" may be directed either towards a destructive struggle for control and independence or towards a constructive effort to build self-esteem through servicing the customer. Critical determinants for employee behavior are the position in the value adding chain and the relation to the customer. Business process redesign can affect employee behavior by increasing process visibility and removing information barriers, i.e. to diminish differences between positions in the value adding chain. Pull control, i.e. building a chain of customers (Schonberger, 1990), is effective where ever it can be applied.

However, it is not enough to improve process visibility and information flow in one direction only. Visibility of the customer in upstream operations is only part of what is required. Upstream operations also need to experience that their performance is visible to the customer. It is necessary to expose upstream operations to the customer to induce employees to build self-esteem through servicing the customer. Nevertheless, it is not desirable that the customer has to deal with a great number of persons engaged in producing the services she requires. The ideal, from the perspective of customer exposure, is that one individual or group is simultaneously responsible for the whole value adding process and all customer contacts. When this is not possible, each organization must find its model to expose employees to the customer throughout the value adding chain without overwhelming the customer with communication requests and information.

4. CONCLUSIONS

Distorted communication is a major cause of operational uncertainty. By reducing response times throughout the supply chain demand distortion is reduced. However, the problem of demand distortion is not restricted only to long supply chains with long lead times and poor communication. Practices inside the firm also contribute to the distortion, especially if the complexity of demand is high. By introducing planning and control practices that preserve overview and clarity the distortion can be significantly reduced.

Efforts to improve the operational performance can only be successful if people's perceptions are affected and experiences are accumulated. Customer exposure appears to be a key factor to overcome problems of diverging interests within industrial firms. Regarded from the perspective of organizational innovation customer exposure may be an equally powerful enabling factor to improvement as quantitative quality control tools and formal procedures were in the ascent of the Japanese industry.

REFERENCES

Burbidge, J. L. (1989), *"Production Flow Analysis - For planning Group Technology"*, Clarendon Press, Oxford

Burbidge, J. L. (1994), *"The use of period batch control (PBC) in the implosive industries"*, Production Planning & Control, 1994, vol 5 , no 1, 97-102

Buzzell, R. D. & Gale, B.T. (1987), *"The PIMS Principles - Linking Strategy To Performance"*, The Free Press, New York

Drucker, P. (1990), "The Emerging Theory of Manufacturing", HBR May-June 1990, pp. 94-99

Forrester, J. (1958), *"Industrial Dynamics, a major breakthrough for decision makers"*, Harvard Business Review, July/August 1958, pp. 37-66

Holmström, J. B. (1993), *"Realizing the productivity potential of expeditious operations"*, Licentiate's Thesis, Helsinki University of Technology

Houlihan, J. B. (1987), *"International supply chain management"*, International Journal of Physical Distribution and Materials Management, 17 (2) 51-66

Ishikawa, K. (1985), *"What is Total Quality Control? - The Japanese Way"*, Prentice-Hall, Englewood Cliffs, N.J.

Kortteinen, M. (1992), *"Kunnian kenttä"*, Doctoral Thesis, Helsinki University (in Finnish)

Lillrank, P. (1988), *"Organization for Continuous Improvement - Quality Control Circle Activities in Japanese Industry"*, Doctoral Thesis, Helsinki University

Schonberger, R. (1990), *"Building a Chain of Customers"*, The Free Press, New York

Senge, P. (1990), "The Fifth Discipline -The Art & Practice of The Learning Organization", Doubleday/Currency, N.Y

Taylor, F. W. (1911), *"The Principles of Scientific Management"*, New York: W.W Norton, 1911, 1947

Towill D. R. (1992), *"Supply chain dynamics - the change engineering challenge of the mid 1990s"*, Proc Instn Mech Engnrs Vol 206, pp 233 - 245

PART TWO

Applications

10

Benchmarking ....to be the Best!

Bill Baker, Texas Instruments, Incorporated, 2501 S. Highway 121, Lewisville, Texas 75067, U.S.A.

Benchmarking has been written about in many books by Robert Camp[1], Greg Watson[2], Michael Spendolini[3], McNair and Liebfried[4] and others. Their writings have covered everything from the definition of benchmarking to the ethics involved and how to make a site visit. In order to gain a full appreciation of the process and use of the "tool" of benchmarking, all these aspects must be understood.

1. Culture Change

At Texas Instruments we've found that to attain world class status in all key processes requires a major cultural change. Most organizations that are truly successful attain world class performance in one key area but find it difficult to achieve that status in all key areas. The secret to the cultural revolution in companies is that each operating group has a clear focus on what it takes to successfully compete in that process...even though it may be an internal process not readily visible to outside customers. Benchmarking serves to be the "reality check" for internal processes. *Doing your best* is no longer good enough. Benchmarking can educate internal teams and mold the ingredients for this cultural revolution.

1.1 Deployment

At Texas Instruments, we have worldwide employment of 58,000 in over 30 countries. Deployment of any policy takes excellent planning and execution. One of the basic building blocks of this culture is the corporate philosophy toward continuous improvement and goal setting. Without these two factors and an aggressive competitive nature, a benchmarking program will not survive. Given these, it is relatively easy to follow the logic path to "How do we compare?" and finally "How do we create a competitive advantage?"

2. Management Support

Our CEO is an active supporter and he published a company wide policy statement requiring benchmarking to be an institutionalized process from the top to the bottom. This statement provides legitimacy across the organization and allows us to participate in any department and any decision making process.

3.   Training

To support benchmarking, we designed a comprehensive training program to help deploy process knowledge and provide guidance for benchmarking teams.

3.1   Overview Training

The overview class was designed around benchmarking from a historical perspective with help from The Quality Network, a consultancy established by former Xerox employees. This class serves as an orientation and motivational tool. The course runs for 3-4 hours and involves some fun classroom exercises that generate interest. In this particular class, we have instructed over 1700 employees and approximately 4500 customers and suppliers.

3.2   Business Process Management Training

The second level of training can be traced back to our policy. We want each benchmarking team to fully understand their own process before investigating other companies.   This is extremely important because our teams are usually cross functional and there can be a considerable learning experience just to arrive at a common understanding. This Business Process Management class takes one full day and the team produces a "product"...a detailed process map including inputs and outputs at each step and a clearer understanding of customer requirements. Reengineering can occur here as the team sees steps that can be modified, combined or eliminated prior to actual benchmarking.

3.3   Benchmarking Team Training

The third level of training is Benchmarking Team Training in which the team goes through the planning phase of benchmarking and is briefed on ethical and legal concerns as well as data research techniques. From this class, they emerge with a detailed plan for their benchmarking study.

4.   Benchmarking Champions Network

This culture change is further supported by our Benchmarking Champions Network that has members in every major group and site. The Champions are responsible to promote and facilitate teams and the process in their area. They identify core processes that need to be benchmarked, act as a resource for teams and follow-up to aid change implementations. We have 124 Champions in our worldwide operations. Communications, a key element in deploying best practices, is accomplished through an on-line database of completed studies and tools to aid facilitation.

5.   Benchmarking vs. Reengineering

In the current literature we can become confused about the interaction of some of the change processes. We often hear that reengineering efforts are underway in many companies and even governments. Our belief is that both reengineering and benchmarking are complementary and not in competition. Through reengineering, we use all of our internal brain power to eliminate waste and non-value added operations to increase process efficiency. In benchmarking, we use worldwide brain power to make step function improvements using techniques that have already been perfected. In fact, that's the goal, use everything at your disposal to speed up learning and constructive change!

6.   The Benchmarking Process

6.1   Management as a Team Member

One of the most important steps in cultural change is mid manager support for Benchmarking Team objectives and a willingness to implement changes recommended by the team. In order to achieve this support, we have implemented the use of an authorization form that binds management to benchmarking team goals. Gaining mutual agreement on expectations helps prevent downstream misunderstandings. Items covered on this agreement are:
        Team Members
        Scope
        Approximate Completion Date
        Expected Results

This simple contract validates the need for a team study, dedicates resources and creates the boundaries necessary for effective team operation. The team and manager can always review the scope to stay on track.

6.2   Benchmarking Process Steps

The four phases and ten process steps of benchmarking are necessary to obtain full benefit and leverage from a benchmarking study. Some of the steps can be done in parallel to reduce cycle time but they all must be done to gain maximum sustained benefit. (See figure 1)

Step #9 - "Implement specific actions" is the most important step, since, if this is not done, then no benefit will be derived from the study and simply will be reduced to "interesting data."

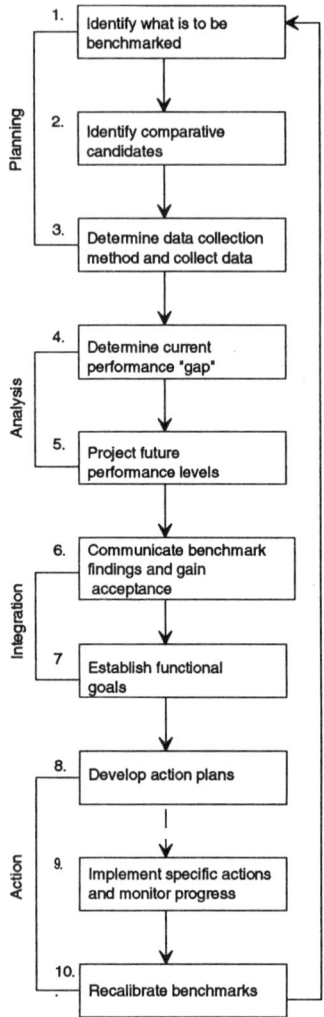

Figure 1.

7.   Benchmarking Application
      Opportunities

7.1   Production Operations

The benchmarking process is particularly useful in production operations that contain readily measurable indices such as:

-    Cycle Time
-    Defects
-    Equipment Capability
-    Factory Layout
-    Overhead Costs
-    Factory Management Levels
-    Set up Time

7.2   Administrative Operations

But of equal or greater importance are the administrative processes that lie at the heart of the company:

-    Order Entry Time
-    New Product Development
-    Design Quality
-    Accounts Receivable
-    On-Time Delivery
-    Customer Satisfaction
-    Training Programs
-    Teaming and Empowerment
-    Recognition and Reward

In fact, the white collar administrative practices can yield some of the most outstanding results because many of them have not been subjected to the same engineer - like scrutiny that production processes have historically undergone. Until you have benchmarked your core processes, how do you know if you are competitive?

8.   Success Stories

8.1   Printed Wiring Board Assembly

There are a couple of success stories I would like to share with you. The first is a study done in one of our Printed Wiring Board Assembly Shops that originally had been constructed to support one high volume product. As our continuous improvement philosophy became imbedded, the manager of this shop was concerned about the shop's viability when its sole customer product line became obsolete. He commissioned a benchmarking study that surveyed many PWB Assembly Shops in the USA. They looked at all aspects of operations from factory layout to defect levels to cycle time and even management structure. The benchmarking study concluded that to be competitive, all of these areas needed to change and change quickly. For instance, we set cycle time goals...that we thought unbelievable at the time because the change was so drastic. But during the past months, we have seen a 78% reduction as shown in fugure 2.

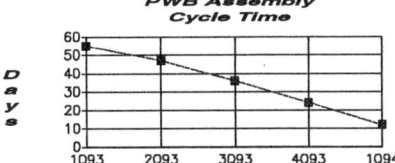

figure 2.

The benchmarking study caused us to take action and focus on gaining competitive advantage by reengineering the entire process..... cycle time, management organization, factory layout and defect levels.   In effect, the benchmarking study was our "wake-up call" and we saw radical change was needed and was possible. "If they can...we can too!" became our motto. A true cultural change occurred thanks to an enlightened management, a realistic benchmark and lots of hard work by everyone in the shop.

8.2   Machine Shop Coolant

A second study is one done by one of our fabrication machine shops. The study began when several machinists noticed a foul odor coming from their machines as they returned to work on Monday mornings.

They determined the odor came from the machining coolant sumps. In order to eliminate the odor, they formed a benchmarking team, studied their process and began making contact with other manufacturer's, the milling machine supplier and others. They discovered several best practices, including one coolant that did not create an odor and also was much less expensive!

During the year the coolant not only solved the odor problem but saved the company $153,000 in this shop alone. We passed this information on to our other machine shops and implemented change across the company.

9.   Recognition and Reward

Benchmarking Teams are like all other teams, they need to have rewards and recognition. At Texas Instruments, the teams present to Group and Site management both during their studies and at completion. They share best practices at the Quarterly Champions Reviews, are recognized in Site newspapers and are eligible for Annual Team Bonus Awards. The Benchmarking Champions are responsible for making sure their teams are recognized.

10.   Summary

Benchmarking is a powerful tool but it is *not* for the faint hearted. An organization that is committed to continuous improvement will find benchmarking to be a barrier buster. In many cases it leads to drastic change and management must be committed to support "management by fact," not intuition or we've always done it this way.

10.1   Lessons Learned from our Team Studies

Some lessons we have learned from conducting hundreds of benchmarking studies reveal some common mistakes that can be avoided.
- Always examine your own process first.
- Make team goals specific and not too broad.
- Maintain team member enthusiasm by having regular meetings and assignments for all members.
- Someone on the team must act as a leader.
- Do not take "industrial tourism" trips
- Remember metrics are not the answer...best practices are.
- Take Action!

References

1.   Camp, Robert C., Benchmarking, The Secret for Industry Best Practices that Lead to Superior Performance, ASQC Quality Press,1989

2.   Watson, Gregory H, Strategic Benchmarking, John Wiley and Sons, 1993

3.   Spendolini, Michael J, The Benchmarking Book, AMACOM, 1992

4.   McNair, C.J. and Leibfried, Kathleen, H.J., Benchmarking, a Tool for Continuous Improvement, ONMEO, 1992

11

Internal Benchmarking: Identifying Best Practices Within a Global Enterprise

Steve Crom[1]

BIBA (Bremen Institute of Industrial Science and Applied Work Science)
Hochschulring 20, D-28359 Bremen, Fax: + 49 421 218 55 10

INTRODUCTION

As technology, customer expectations and competitive responses continue to change at an increasing rate, companies that learn the fastest have a competitive advantage. Internal benchmarking can create a network through which information and ideas are exchanged. It is an effective way of identifying and disseminating best practices from one part of a global enterprise to others, with the ultimate objective of improving business processes to deliver greater value to customers.

What follows is a discussion of internal benchmarking and the opportunities it offers as illustrated by the case of Globalco, a Fortune 100 consumer products company. The case is divided into the following sections:

- The company's situation and why internal benchmarking was appropriate.

- Gaining corporate sponsorship and selecting the right people for the benchmarking team.

- Building the team and defining its charter.

- Collecting the right data, qualitative and quantitative.

- Creating the final product.

- The results.

- Lessons learned about the process of internal benchmarking.

- Conclusions.

The discussion is based on the experience of the author working directly with this Fortune 100 company from August 1992 to May 1993 on a world-wide internal benchmarking study.

---

[1] Thank you, Isabelle van Notten, for your editorial comments and contributions.

1. INTERNAL BENCHMARKING, THE OPPORTUNITIES IT OFFERS

In many large multi-national companies there are pockets of excellence, aspects of an operating unit that represent leading edge practices from which there is a great deal to be learned. However, most organizations look externally and benchmark competitors and/or organizations in unrelated industries that have world class reputations in a particular area of interest, rather than learn from themselves. The advantages that internal benchmarking offers are as follows. It allows you to:

- probe for details, including what went wrong as well as right

- educate those doing the benchmarking to work in new ways

- demonstrate successes within your own company's culture and business environment

- establish communication channels and a network for highlighting and disseminating improvements and innovations throughout the organization

- focus external benchmarking activities once you know your own company's strengths and weaknesses

2. GLOBALCO'S SITUATION AND WHY INTERNAL BENCHMARKING WAS APPROPRIATE

Globalco prides itself in its entrepreneurialism and the autonomy given to country General Managers to run their operations as they see fit. The company established itself in over 60 countries by picking people who thrived on the challenge of creating a local business in difficult circumstances, often in developing countries with very little infrastructure and basic services i.e., a reliable supply of electricity. At this time, seventeen of the company's largest subsidiaries account for 80% of sales and profits, each developing, marketing, selling, producing, distributing similar products. As the businesses mature and face stiffer competition the question arises: How can a General Manager in one country learn about what is working in another country, take the idea, modify it to fit his/her circumstances and significantly improve the business?

David Bonasit, Vice President of Global Continuous Improvement, decided that benchmarking was a tool for identifying best practices throughout the company. He looked toward outside organizations: competitors and Best-In-Class non-competitors. A proposal for doing external benchmarking was developed by David and his team. When the proposal was reviewed with other executives, they commented:

> *"The culture in Globalco is such that successes developed internally are more credible than those that come from the outside. We do not have an accurate picture of what it is we do well."*

They therefore suggested that Bonisat visit certain operating units instead of looking outside. These Globalco operating units each exhibited excellence in several of the strategically important business processes. In response David Bonisat and his team decided to leverage the opportunity and benchmark eight key Globalco operating units or subsidiaries. Afterwards an external benchmarking study would be commissioned focused by the internal findings.

The options David considered for conducting an internal benchmark study were to hire an outside consulting firm or develop an internal team. First, as a former General Manager, David concluded that if a team of Globalco managers did the research themselves, packaged and communicated the results, these would be better accepted than if they came from an outsider. Second, members of the internal benchmarking team would become experts in benchmarking and be a valuable resource to others wanting to do benchmarking. Third, by selecting a cross functional team, David wanted to create a nucleus of influential managers who would help sell the application of leading edge practices throughout the company (in their respective functions), because they would discover themselves what differentiates successful from average practices. In addition to being cross-functional, David assembled a team that represented the company's divisions in Europe, Asia, North and South America to help spread the learning geographically.

David thought it would help to have a neutral person in the role of facilitator of team meetings and site visits. A consultant was selected based on his facilitation skills as well as experience in benchmarking and organizational development. The next big step was to get corporate support and the commitment to free up the right managers internally to participate in this seven month project.

3. GAINING CORPORATE SPONSORSHIP AND SELECTING THE RIGHT PEOPLE FOR THE BENCHMARKING TEAM

David sold the President of Globalco on the idea of an internal benchmarking team saying that the more involvement there is from the regions and functions at this stage the more commitment there will be to implement best practices globally. "Benchmarking is only as good as the operating improvements and results it leads too," David asserted. "While it takes longer, an internal team is in the best position to help translate the findings into concrete operating improvements." The proposal for doing internal benchmarking with a dedicated team of Globalco managers was approved.

The higher the caliber of the people on the team the more stock General Managers worldwide would put into the findings. David selected the members of the team based on who had reputations for being:

- opinion leaders
- functional experts
- insightful
- able to work in groups

Each team member was expected to devote 25 to 30 days to the project.

Three senior managers from David's staff with extensive experience in leading improvement efforts were apart of the team. Harold Pinter, the Continuous Improvement Manager responsible for South America was chosen as the project manager given his experience in data collection and quality audit programs. Before joining Globalco he learned about quality and Total Quality working in the automobile industry. Harold had exactly the right personality for the job of project manager. With a sense of humor, he kept the group organized, educating them along the way about benchmarking and the art of data gathering.

David and Harold picked a consultant, Scott Beckman, who was experienced in benchmarking but would leave the ownership of the project to Globalco, acting as a facilitator for the process and team meetings. Having an outside facilitator meant that David and Harold were free to participate in the team's working sessions while Scott kept the team focused and made sure that there was balanced participation. Each team member's role was clarified during a week long team building and planning event that launched the project.

4. BUILDING THE TEAM AND DEFINING ITS CHARTER

The more team members knew about the state-of-the-art in manufacturing and knew about each other, the more they could contribute to the project. Accordingly, Harold arranged the first team sessions to be held in Austin Texas, immediately following an international conference at which companies were presenting best practices in supply chain management. The first three days of the week the team attended presentations, bringing back insights about what they had heard. Most importantly the team members had time over dinner to get to know one another and learn how they wanted to contribute to the project.

Immediately following the conference the team spent two days off-site in a hotel room working on the following:

Day 1 - Building The Team

- The team's mission
- Each individual's learning style
- The team's learning style
- A strategy for working effectively together
- The role of each team member

Day 2 - Defining The Team's Charter & Task

- The team's charter
- Categories of data to collect
- A protocol for gathering data
- The site visit schedule
- Pre-work requested of the host country before the visit
- Calendar of the project and team events

As team leader David made it clear at the outset of day 1 that the team's mission was to document best practices and not to make recommendations during our benchmarking visits about what should be improved or how. The team wrestled with the question of what value they would offer the host country if they did not make improvement recommendations. "Too often," David said, "corporate task teams descend on a subsidiary, get a partial picture of what is going on, and leave the GM with a list of things to do as if he didn't already have enough to do." Instead, the benchmarking team would be on site to document the work processes and/or managerial practices that the subsidiary feels contributed the most to excellent customer service. Documenting best practices, how they were achieved and the principles involved turned out to be a service much appreciated and valued by the host country. It motivated them to continue the work they were doing with a renewed focus on the critical success factors.

At the end of the two day off-site meeting, the scope of the team's inquiry was defined as the total supply chain, from ordering raw material to delivering finished goods to the customer. Included in this definition was sales, marketing and product development activities. The team agreed that understanding excellent customer service practices meant identifying enabling factors such as leadership, organizational climate, training and development that contributed to creating and maintaining outstanding business processes and results.

5. COLLECTING THE RIGHT DATA, QUALITATIVE AND QUANTITATIVE

Practices that contributed to excellent customer service, on-time and complete deliveries in particular, was what the team set out to understand and document in a way that would be useful to others. The sites to visit were selected based on their reputation for excellent customer service, and historical on-time/complete delivery performance. Subsidiaries were selected from each of the company's geographic regions representing markets in which the company had dominant market shares as well as those in which it did not.

David called the General Managers of each subsidiary nominated, explained why he and the President had undertaken the project and enlisted their support in sharing successes throughout the company. Because of David 's reputation as a results oriented General Manager the subsidiaries agreed to host the team's visit.

The General Manager of each subsidiary received a pre-work packet from the Best Practices Team before they arrived on site. The packet contained:

- a description of the project, the team's charter and members
- advice on nominating successes or best practices to be documented
- a fact sheet to be filled out on each nominated best practice
- an agenda for the team's visit
- an explanation of how the team would go about collecting data once on site

The prework package requested that the General Manager meet with his/her staff and identify the practices of which the subsidiary was most proud. They were asked to provide performance measures documenting improvement, plus a high level flow chart of the current process. The Best Practices Team reviewed the pre-work data before the visit, deciding which of the practices should be documented and by whom.

The site visits were scheduled to last three days. A typical visit started by meeting with the General Manager and his/her staff. In this initial meeting introductions were made, the team's purpose was reiterated as documenting the subsidiary's customer service successes. The General Manager made opening remarks about the subsidiary's performance and the context within which to place the subsidiary's results (e.g., social, political, economic, market and competitive conditions). Next, Scott facilitated a group discussion of the important historical developments that led to today's performance. In particular:

- What were things like before the current General Manager arrived?
- When was the current management team formed?
- Where there changes in organizational structure?
- What prompted the focus on customer service?
- How was support galvanized among employees to change?
- Were consultants involved? If so, how were they selected?
- How did the management team measure progress?
- What were the results over-time?
- What was the General Manager's role and the management team's role throughout?
- What would they do differently next time?

This opening conversation gave the Best Practices Team the same background information and allowed them to all see the General Manager and his/her staff in action together. In the final analysis the role of the General Manager as team leader and champion of customer service excellence surfaced as the single most important enabler of business process improvement.

At the close of the opening discussion the list of practices nominated by the subsidiary as their best were prioritized. Members of the management team formed small working groups with their functional counterparts on the Best Practices Team and another team member. The task of the working groups was to quantify the customer service benefit of specific practices, create a flow chart of the processes, show the magnitude of improvements made and document how the changes were made. Lessons learned and suggestions for how to get started were solicited during the interviews.

In the opening meeting and small work group meetings, improvement projects surfaced that merited documenting though they were not mentioned in the pre-work exercise. In fact, some of the better cases were not included in the pre-work. The team quickly learned that there was no substitute for face-to-face discussions, "walking the processes" and meeting with everyone involved including customers and suppliers.

The data collection protocol used during the site visits was the following:

Data Collection Protocol

- Clearly define the business process, improvement goal, measures and implementation time frame.
- Identify the motivation for improvement. How was a sense of urgency to change created? Was there a competitive threat? Was the improvement goal focused on the customer?
- Capture the resources (people, expenses, outside consultants) that were involved.
- Flow chart the process as it was and it is in its improved form.
- List the people involved in the project, including the sponsor, the team leader, team members and supporting members paying special attention to their selection criteria, qualifications and roles.
- Document the improvement methodused and the steps involved.
- Write down the story of what happened and how it happened, in as much detail as possible. What was unique about this project? Capture any anecdotal stories about the team and the project to communicate to others how they worked together.
- Probe for data and results beyond the performance measures submitted in the pre-work. Be sure you have the latest data.
- Document the reasons for any backsliding or reversals in performance.
- Document the team's plans for continuing to improve the process.
- Explore what they would have done differently. What advice would he/she give others embarking on a similar process improvement effort?

The keys to capturing useful information were:

- Before conducting the first interviews generate a prototype of what a finished write-up should look like so everyone has the same end product in mind.
- Have a standard protocol that each member of the team follows to collect qualitative as well as quantitative information.
- Quantify performance and bottom-line benefits.
- Follow the flow of the business process as if you were a component of the process i.e., walk the process.
- Talk to those doing the work to really understand how it works, not how it is supposed to work.

In daily debriefs, each member made a short presentation of what was documented and reported progress. Other team members were called on to explore a particular best practice that involved their functional expertise. Next steps were identified and the schedule adjusted accordingly.

The third day of the site visit, the team met to agree on which best practices and enablers would be written up. Scott listed them on flip chart to review with the management team. The agenda for the wrap-up session with the management team was:

- Review the list of best practices to be written up

- Review enablers of those best practices

- Ask how to package the findings in a way that would be most helpful to others

- Ask for feedback on the best practice team's approach and how it could have been improved

- Agree on a schedule for drafting the cases and getting feedback from the subsidiary before publication

- Summarize each team member's view of the best practices they documented

Validating the team's findings and correcting any misconceptions right away was the main purpose of the wrap-up. It was an opportunity to formally thank everyone involved. According to Harold, "in every visit the wrap-up was a smashing success because in every case the management teams had never really sat down and reviewed all the good things they had done over the past two or three years. It was a chance for their contributions to be recognized by a group of peers."

Before leaving the subsidiary, the team would document who was responsible for documenting each best practice and establish a documentation/review schedule. Upon returning to their home offices, each team member would prepare their reports and forward them to Harold Pinter according to the schedule. Some team members brought laptop computers to the site visited which allowed them to summarize their notes on the trip home. Harold would compile the draft reports and circulate them to the team and the subsidiaries. Both the team and the subsidiary personnel would review the cases and make corrections and additions as needed.

In summary, the documentation process was to:

1.   Send pre-work to the host to identify best practices ahead of time.

2.   Review the completed pre-work as a team before the visit.

3.   Conduct an introductory meeting with senior management to set the tone, get to know the style of the management team and clarify the best practices to document.

4.   Meet in small working groups to collect as much detailed information as possible.

5.   Debrief as a team daily to share impressions and adjust the data collection schedule.

6.   Follow-up on site to fill as many data gaps as possible.

7.   Conduct a wrap-up meeting on-site to validate the team's findings, thank the host and clarify next steps.

8.   Review the draft cases with the team and host following the visit so everyone is comfortable that the information being published is accurate.

From Harold and David's perspective, value was added and knowledge gained in every step. At the same time, everyone was amazed at the amount of work, focus and persistence necessary to get a detailed, accurate and useful story.

6. CREATING THE FINAL PRODUCT

When asked how to best package the team's findings there was consensus that a book of write-ups by itself would not be very helpful. First, the challenge was to help people see the importance of customer service, to create a sense of urgency to change; second, to capture people's imagination about what was possible; third, to provide useful examples of how others have solved common problems; and last, to provide support in developing a coherent process improvement strategy and implementing it.

Compiling the final handbook was a big job. The corrected drafts from the subsidiaries, were given to an editor. In the process of polishing the text, the editor (a neutral party reading the cases without any particular functional expertise) prepared a list of items that needed clarifying. "This was a critical step in making the cases accessible to the widest possible audience within Globalco," remembers Harold. The editor's comments were forwarded to both the authors of the reports and the subsidiaries for their input. At this point, the cases were given to a desktop graphics person to format the final reports and integrate the graphics with the text. There were a total of over 80 cases and 500 pages of text.

The team developed a method so that operating unit management and improvement teams, the primary audience of this effort, could identify and prioritize the over 80 cases to select those most useful to them. The team developed a one page self-assessment form in which management or improvement teams would evaluate their current Customer Service status. The evaluation attributes were cross-referenced with the various cases. The assessment form was piloted, reviewed with operating unit management teams that were visited by the team, and revised accordingly. David Bonisat made his staff available to work with the operating units to facilitate the assessment and develop their action plans based on the best practice cases.

Before the handbook was published, the Best Practice cases were reviewed with key stakeholders: Divisional Presidents and Vice-Presidents, and functional Vice-Presidents. The final approval to publish came from the functional Vice-Presidents of human resources, manufacturing, finance, technology and sales. After their final edits and recommendations the handbook went to press.

The plan is to add best practices to the handbook as they are discovered and/or developed, for it to be an archive that grows and is updated continuously.

7. THE RESULTS

The first operating units to benefit from this study, even before it was published, were those who participated in the study and those functions that provided key personnel to serve on the team. Ideas flowed freely between these constituents and many ideas were adapted and quickly instituted.   There were many ideas of how Customer Service and Customer Service Improvement should be approached, "this study provided some key parameters to a unified Globalco approach," says Harold. Benchmarking (both internal and external) led to the establishment of a re-engineering effort in one of Globalco's key operating units.

The goal now is to develop a world class Customer Service methodology that is supported by a full time staff from several geographic regions and all key functions: just like the internal benchmarking team. "Several of the team members have been promoted and are now in position to really leverage what they learned," Harold reports. "All the team members are very proud to have served on the team and say that it was one of the best experiences of their careers. The members know that they can call on each other for support at any time and they do just that."

The Best Practices Handbook will be useful for the smaller and emerging operating units. Bonisat's department is small and, in reality, can and should only focus on the larger 17-20 operating units.   "The book with the assessment form is a great reference for the rest of Globalco: 50 or so operating units," says David.   "This handbook provides them with a unique resource. The feedback has been very positive."

8. LESSONS LEARNED ABOUT THE PROCESS OF INTERNAL BENCHMARKING

Internal benchmarking is ideal for global enterprises that have pockets of excellence but a tradition of autonomy rather than collaboration. It is best done with a team of the company's own managers trained in the best method of collecting and analyzing data. Through their direct involvement the organization learns how to learn, a strategically important process or capability in itself. Comments from team members reveal how much they learned:

> *"Get high level corporate sponsorship. It is essential in order to free up the caliber of resources needed to command people's respect and attention. The best people are always the busiest people, already assigned to high priority activities. It is only at the top that high priority activities can be reallocated so the right people have the time to devote to the project."*

"Build a group of people that know one another and how to work effectively together. It is just as important as knowing what data to collect and how. When doing benchmarking the message is in the method. If you are gathering information about excellent performance it should be done in a highly professional, well organized way."

"Skilled facilitation accelerates the team's development and economizes the time the team spends together. That person can be an outsider or someone from within the company as long as they view their role as guide rather than team leader."

"The team leader has to be a senior manager of the company with the authority and resources to convert the benchmarking findings into action."

"The staff work required to keep the team's activities organized, educate the team about the benchmarking (find and distribute articles) and edit the best practice write-ups is substantial. Harold devoted 40% or his time to this project for six months."

"Be flexible in the site visits and take your direction from your hosts."

"Expect your host to be cautious about the purpose of your visit until you have a chance to discuss the team and it's charter face-to-face."

"Sending pre-work to your hosts in advance as a courtesy but do not expect it to be self-explanatory since internal benchmarking is not yet widely practiced and understood."

"Everyone on the team should collect information about the qualitative aspects of the company that contribute most to its success, the enablers."

"Devote several hours to interviewing the General Manager one-on-one since his/her philosophy, values and behavior is the most important enabler to outstanding corporate performance."

For those benchmarking for the first time internationally, be sensitive to culturally appropriate behavior. If team members are not experienced in traveling to a host company destination, give them material to read ahead of time so they avoid basic cultural faux pas. A special warning to Americans traveling to England, remember Mark Twain's admonition that "America and England are two countries divided by a common language." Don't assume that because you speak the same language (more or less) that the same behavior is appropriate in both countries. In probing for information, respect the protocol of asking senior managers for access to information before approaching middle level managers directly.

9. CONCLUSIONS

If learning to improve processes is what separates the best companies from the average performers, it is in creating a learning network and environment that internal benchmarking has the most to offer. Think of the end-product of internal benchmarking as the building of a network and dynamic archive of knowledge rather than a final report. Build a team of the best resources available, managers who are most likely to move on to influential positions within the company. Assign a full-time team leader/coordinator. A key function that person fulfills is to communicate, communicate, communicate; it is an essential ingredient for the success of a global team. Take the team off-site to develop the team's ability to work together and clarify its own charter. Allow plenty of time early in the project for team building, planning and educating the team about state-of-the-art of benchmarking. Coordinating the first team events with a benchmarking conference is ideal.

Chose any consultant you work with carefully. Look for strong facilitation skills, benchmarking expertise and ability to fit into your corporate culture. His or her goal should be to transfer skills to the team as well as ensure the integrity of the process and data. Watch out for consultants who are more interesting in selling follow-on work than in successfully completing the benchmark study.

Involve the key stakeholders throughout the project. Get their visible endorsement of the project and commitment to use benchmarking to drive process improvement. Ask for their input on who should be including on the benchmark team. If senior executives are involved in the team selection process, they can help identify influential managers and will be more amenable to freeing up their time. This applies as well to selecting the sites to visit. Regular briefings and a final review of the practices documented with stakeholders helps keep the team focused on the company's top priorities.

Dedicate the resources necessary to create a world class product. Engage a professional editor and desktop publishing resource. To be a useful reference the final product must include contacts names (phone and fax numbers) for each best practice documented. A system should be created whereby a best practices archive can be kept, updated and accessed electronically by interested parties world-wide.

For internal benchmarking to be useful it has to be coupled with the resources to implement process improvement. While the two activities must be integrated, it is most effective for an internal benchmarking team to document best practices and not make improvement recommendations to their benchmark hosts. A self-assessment guide helps the users sort out the information that is most relevant. Above all, a practical process improvement strategy/plan must be in place that is supported by the company's leadership.

12

Benchmarking in Norwegian Industry and Relationship Benchmarking

Bjørn Andersen, Msc.Eng./Ph.D.-student, Department of Production and Quality, Engineering, Norwegian Institute of Technology, N-7034 TRONDHEIM, Phone + 47 73 59 71 22, Fax + 47 73 59 71 17, e-mail bjorna@protek.unit.no

1. BENCHMARKING IN NORWEGIAN INDUSTRY

Benchmarking is probably at the moment one of the most popular management tool. Courses, conferences (as you can self witness today), books, papers, etc., are flooding the market. Benchmarking is hot and "in" right now, especially in the US. The interest in benchmarking is rapidly growing in Europe as well, although it has not reached the same extent as in the US. Still, the number of companies actually using benchmarking, either sporadically or on a regular basis, is fairly low. Some European branches of large, often multinational, corporations do benchmarking. A few other companies, sensitive to the latest trends, do as well. Otherwise, the propagation of active benchmarking is slow.

In Norway, the situation is probably even worse. Except a very few Norwegian subsidiaries or affiliates of larger corporations, there have been no reports of benchmarking being used. As a response to this situation, the TOPP Program, a Norwegian program aimed at improving the productivity of Norwegian technology industries, has started an effort to introduce benchmarking to Norwegian companies. The model TOPP has used for organizing and financing the promotion of benchmarking can be used to get benchmarking started in other countries, areas, industries, etc., and will be described in this paper.

1.1 The TOPP program

The TOPP Program is running from 1991 to 1995, with a budget of about NOK 200 millions, partly financed by the Norwegian Research Council and partly by the companies involved. So far, approximately fifty companies are "members" of the program. Research personnel from the Norwegian Institute of Technology, NTH, and SINTEF, the Foundation for Technical and Industrial Research, in cooperation with consultants within different areas, perform the different activities of the program.

These activities range from productivity analyses, research projects, education (10 Ph.D.'s and a master degree in technology management), networking between the companies, improvement projects, and projects aimed at developing or promoting different tools and techniques.

1.2 The benchmarking project

One project of the latter type is on benchmarking, mainly as a consequence of the situation described above, an immense lack of knowledge and competence about benchmarking, both in Norwegian companies as well as among consultants and academic institutions. The project was started in December 1993 and is supposed to run for one year, with a budget of NOK two millions.

The purpose of the project is twofold:

*   To upgrade the TOPP companies' knowledge about and abilities to perform benchmarking.
*   To upgrade the competence within benchmarking at NTH/SINTEF.

Three of the TOPP companies were invited to participate in the project, where two researchers from NTH/SINTEF, of whom I am one, represent TOPP. The two of us have read an extensive body of literature on benchmarking as well as taken courses arranged by the International Benchmarking Clearinghouse. TOPP covers all expenses of the two researchers, both hours and travels. The companies contribute with the direct man hours of those from the companies working on the project and the cost of their travels. In addition, we cooperate with Xerox Quality Solutions UK, for project planning and implementation support and practical guidance on specific problems. The Xerox support is also covered by TOPP.

All in all, the financial structure of the project is very beneficial for the companies, which receive an extensive amount of support in return for the work they put into the project.

1.3 Project approach

The project focus is to bring benchmarking into the companies by attacking a specific problem of current interest. This way, we avoid using benchmarking as a hammer just looking for a nail to punch. If benchmarking is perceived in such a way, chances are the companies would not give the project the necessary support. When benchmarking is used as a tool for solving problems identified as important by the companies themselves, the project is likely to get far higher priority.

Before starting the actual implementation of benchmarking, a foundation of basic knowledge had to be built. For this purpose, a Norwegian, very practical handbook in benchmarking was written, partly prior to and partly in parallel with the implementation activities. Based on the handbook, we gave short training sessions, which covered only the most basic principles of benchmarking. To avoid the feeling of being overdosed by a massive amount of information even before the project really started, further training was/is given throughout the project as the need arose/arises, i.e., on-the-job-training.

We use a six-step benchmarking process (Watson, 1992), covering the phases of planning, searching for partners, collecting information, analyzing, adapting best practice to our own organization, and recycling the study. To guide the three companies through a complete benchmarking study, we arrange quite frequent meetings, where we actively participate in the process as facilitators. Activities that need to be performed are split between us, TOPP, and the companies, so that we take on part of the workload.

At milestones, i.e., completion of phases, Xerox is brought in to review what has been done in the phase, and prepare the team for the next phase. At this point, one of the companies is in the partner visit stage of the project, having visited some partners and planning another visit. The others are still trying the complete the planning phase. By November this year, all three of them will have performed a complete benchmarking study.

1.4 Conclusions

Naturally, these three will have learned the method of benchmarking. However, to ensure that also the rest of the TOPP companies gain from the project, the handbook will be made available to them. Any learned experiences throughout the project, i.e., both positive elements and things to avoid, will be incorporated into the book. To further boost the interest, a seminar will be arranged where the project is described and representatives from the companies put forward their experiences and obtained results from benchmarking. Some financing for other companies will probably also be made available, but not to such an extent as in this project.

It is still too early to evaluate whether the project has given any results, but we see that the companies involved in the project has gained thorough insight into the concept of benchmarking. Furthermore, as other companies hear of the project, reactions are they would want to participate themselves. At the end of the year, I think we will see that a large number of the TOPP companies have started using benchmarking. How much of this can be attributed to this project and how much is a result of a general increase in attention, is hard to estimate. Still, I think this model is useful for promoting benchmarking.

2. RELATIONSHIP BENCHMARKING

2.1 Types of benchmarking

There are basically two criteria for defining different types of benchmarking - what is being benchmarked and who is being used as benchmarking partner. Dependent on what type of company is being used as benchmarking partner, three types of benchmarking are defined (Andersen and Pettersen, 1994):

- Internal benchmarking, where other units within the same corporation are used as partners.
- Competitive benchmarking, where direct competitors are used as partners.
- Generic benchmarking, where partners are chosen regardless of industry.

Each of these all displays some advantages and disadvantages. Both internal and generic benchmarking avoids the problems of sensitive information and the fear of upgrading a competitor that clouds competitive benchmarking. However, the incentives for companies asked to be generic benchmarking partners are not overwhelming. The partner will probably have the opportunity to learn something himself, but this has to be balanced against the time spent on the benchmarker. Although surprisingly many companies accept to be benchmarking partners, many benchmarkers still experience problems gaining access to partners. One answer to the lack of access to partners might be what we have called *relationship benchmarking*, i.e., using companies, that one already has relationships with, as benchmarking partners.

2.2 Benchmarking against suppliers

One of the three companies participating in the TOPP benchmarking project has chosen purchasing as the process to improve. Prior to the start of the benchmarking project, this company had already planned a visit to some of their suppliers to discuss improvements in the supplier-customer relationships. As this visit was already planned, we decided to include it in the benchmarking project, using the suppliers as benchmarking partners as well.

This way, the visit had two purposes; both to discuss and improve the interface between the two companies, e.g., more and better prognoses, EDI-ordering, etc., and to discuss their respective purchasing practices, in order to learn from each other.

Using suppliers, or any companies with which the benchmarker has already got an established relationship with, as benchmarking partners, does not directly fall within any of the earlier defined types of benchmarking. Viewing the whole supply chain as one super-organization, it has some similarities with internal benchmarking. As the other company might be in some totally different industry, it might be thought of as a variant of generic benchmarking. Still, none of these really match.

2.3 Relationship benchmarking

The first reaction of the TOPP company, when they realized some of the unique advantages and disadvantages of benchmarking their suppliers, was that this did not fit into any of the three types of benchmarking we had told them about at the start of the project. They saw it as a "new" type of benchmarking, which they liked to name *relationship benchmarking*. Relationship benchmarking can be defined as benchmarking against companies with which the benchmarker in advance of the benchmarking study already has established a relationship.

It is probably not in the interest of benchmarking, as a field, to explode the number of different types of benchmarking, nor the number of different available process models for performing benchmarking. One of the truly strong sides of benchmarking is its inherent simplicity. On the other side, the defined types of benchmarking should not prevent benchmarkers from considering using a particular group of companies as benchmarking partners.

Simply to make benchmarkers aware of the possibility of using companies they have already established relationships with as partners, I would like to describe some of the advantages and disadvantages of this approach.

Some advantages with relationship benchmarking are:

- Strategic partners, i.e., suppliers, customers, or horizontal partners, will usually benefit from improvements their partner makes. A supplier will benefit from having a more stable customer with higher sales, a customer will benefit from a more stable supplier with better performance, etc. Therefore, strategic partners will generally be willing to act as benchmarking partners, and will probably give the benchmarker high priority and easy access to information.
- As the two companies know each other in advance of the study, time is saved as general presentations are not necessary.
- As the companies are already involved, there will be no problems concerning sensitive information.
- During the benchmarking study, there are possibilities for directly improving the relationship between the two companies, as between the TOPP company and their suppliers.

- As a consequence of acting as benchmarking partner, the partner, through increased attention to the process in question, can also make improvements. This both makes it more attractive to act as partner and, in turn, is beneficial for the benchmarker, as he gets a higher performing partner.
- Again, through acting as benchmarking partner, the partner will probably learn how to perform benchmarking himself, which can lead to improvements.

Some disadvantages are:

- When benchmarking a supplier, there is an obvious danger that the supplier will use the opportunity to "sell himself", i.e., devote far more time to presenting products and processes that work well.
- There is also a possibility that any strategic partner will want to appear better than he actually is, which will give false input into the benchmarking study.
- There is no guarantee that the benchmarker has any strategic partners that are sufficiently good to justify spending resources on using as a benchmarking partner.
- If the benchmarking partner is a horizontal partner, e.g., a partner in a joint venture, it is likely that this partner is or will be a competitor, in which case benchmarking can contribute to upgrading the partner.
- If the two companies are engaged in a customer-supplier relationship, there is a danger of revealing information that can impact price or delivery condition negotiations.

A conclusion is that strategic partners, if sufficiently good, might be a good choice of benchmarking partners, as they will generally be willing partners, which will benefit from improvements the benchmarker makes. The benchmarker should though be aware of the risk of biased information and time wasted on presentation outside the scope of the study. Still, keep in mind that *strategic* partners might make well-suited *benchmarking* partners, as well.

REFERENCES

Andersen, B. and Pettersen, P.G. (1994) *The Basics of Benchmarking: What, When, Why, and How*, Proceedings from the 1994 Pacific Conference on Manufacturing, Djakarta, Indonesia

Watson, G.H. (1992) *The Benchmarking Workbook: Adapting Best Practices for Performance Improvement*, Productivity Press, Cambridge, MA

13

Technological Benchmarking : A Case Study in Product Development by a Small Manufacturer

W. P. Lewis and A. E. Samuel

Department of Manufacturing Engineering, University of Melbourne, Grattan Street, Melbourne, Australia, 3052

Benchmarking originated in large organizations with considerable resources for conducting internal reviews and special investigations. This paper is concerned with benchmarking by a small manufacturer endeavouring to survive and prosper in a global economy. The company has embarked on an exercise to benchmark its core technological activities. The results are reported as a case study to demonstrate the development of methods appropriate to small scale, discrete product manufacturing.

1. INTRODUCTION

This paper examines the development of benchmarking procedures for the important core technological activities of a small manufacturer engaged in batch production of discrete products for niche markets in Pacific rim and other countries. The focus is on small firms since in many industrial economies they form the backbone of manufacturing industry and their role has often been under-valued [1]. Benchmarking is conceived here as " the search for industry best practices that lead to superior performance " [2], and a case study is presented of the approach adopted by one such firm.

The Company concerned employs 100 people with an annual turnover of $ 50m (Australian currency). It is engaged in the batch production of motor and engine driven centrifugal pumps and related water handling and filtration equipment, sales of around 100,000 units per year, 75% for the domestic market, 25% exported. The company has manufacturing facilities for material removal, cold forming by pressing and deep drawing, diecasting, welding, manual and automated assembly, product and component testing; there is close liaison with plastic injection moulders. During 1993 the Company's management decided that benchmarking was an essential tool to ensure that their firm's core activities were in accordance with world best practice. Expert advice was received that the planning and execution of a traditional benchmarking study would require well over 1,000 hours of staff time [3] and thus be a relatively expensive operation, so the Company planned a five-stage strategy in which there would be a progressive commitment of resources as each stage was successfully completed.

The actions to be taken in the Company's five-stage strategy are set out below in the context of discrete product manufacturing considered as a business process in which there are inputs of energy, information and raw materials which are transformed into outputs of marketable products by the application of human and physical resources in an effective and efficient manner.

(1)   Compare costs of manufacturing inputs to those obtaining in industrialised nations in Europe and around the Pacific rim. If input costs found to be competitive proceed to stage (2), if not reassess future of company.

(2)   Identify performance drivers which are crucial determinants of long term company success, and the sub-processes with which they are associated.

(3)   Analyse the performance drivers, dealing separately with those which are amenable to objective evaluation and those which require benchmarking.

(4)   Select a typical product for a pilot study and evaluate the performance driver(s) carried forward from (3) for this product. Review the results of the evaluation, and proceed if the predicted benefits outweigh the expected costs.

(5)   Conduct a full scale benchmarking study of the performance driver(s) carried forward from stage (3).

This strategy has been applied in the investigation described below.

2. INVESTIGATION

2.1. Stage 1 - Manufacturing Inputs

The Company's manufacturing manager toured the world and collected data from Germany, Italy, U.K., U.S.A., Korea, Japan, Taiwan on local costs of (a) steel, aluminium, copper, (b) purchased items such as capacitors, seals, bearings, (c) skilled labour, (d) electrical power. His conclusion : all input costs for the Company were lower than those in Europe, U.S.A. and Japan; some costs were higher than in Korea, Taiwan, but overall costs in those countries were similar to those in Australia.

2.2 Stage 2 - Performance Drivers

A means/ends analysis of the operations of a manufacturing organization yields a hierarchy of goals and goal statements in which the statement at any level of the hierarchy constitutes the means for achieving the ends represented by the statement in the level immediately above it, as indicated in Fig. 1. The top level corresponds to the ultimate goals of the Company, namely profitability and growth. The hierarchy in Fig. 1 proceeds down to the level of new product development and shows the relationship of this activity to higher goals. What are the " performance drivers " [4] for the Company - the factors which have a crucial influence, direct or indirect, on its successful operation and future prosperity ? The company's management identified the following factors, both related to new product development.

(a)   *Product Range* - a range of models sufficiently all-embracing to reflect the markets' requirements for performances covering a wide range of fluid flows in different applications.

(b)   *Short Lead Time* for new product development to enable fast, flexible responses to changing conditions.

This information was carried forward to the next stage of the investigation.

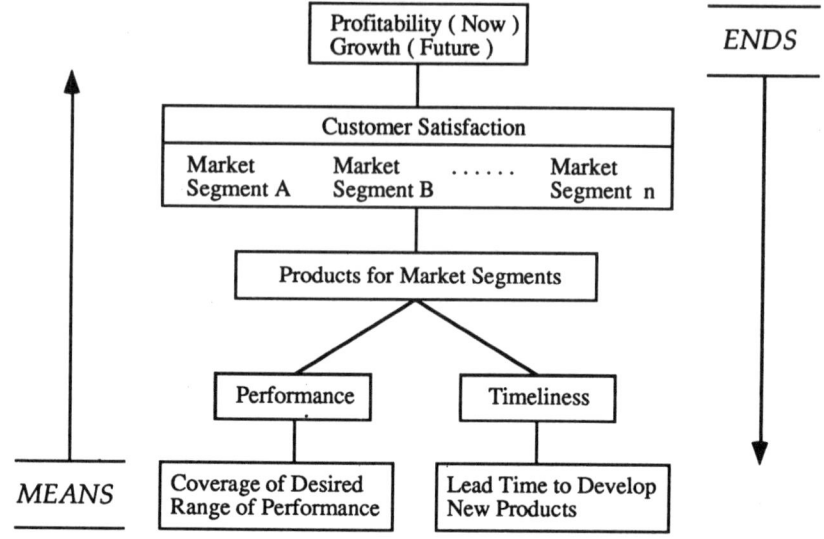

Fig. 1.   Means / Ends Analysis : Relation of New Product Development to Company Goals.

2.3 Stage 3 - Analysis

Product Range. The performance characteristics of the Company's products are continuous variables such as pump flow rates and efficiencies, motor powers, fluid storage capacities and so on. Products are discrete items and each model in a particular range will provide optimum performance for one discrete duty and less than optimum performance for other duties. The problem is then one of matching the number of models in a product range covering a spectrum of performances to the manufacturing resources available, when there is a discrete number of models and a continuously variable performance characteristic. For centrifugal pumps this problem has been investigated and effectively solved [4], and the methods developed can be extended to other products manufactured by the Company where the product range is also amenable to objective evaluation.

The picture changes if we move to a higher level of the hierarchy, say to "customer satisfaction". This is not a holistic concept because of the segmented nature of the market being served. One segment which has become prominent in recent years is concerned with solar energy applications, e.g. for heating swimming pools (with the 100% vision of hindsight this might well have been expected). The demand here is for low capacity pumps to circulate water through solar booster coils where there is a large pressure drop and a need for higher pressure pumps.

However, the Company was slow to recognize this demand and lost a significant market share as a result. It has now benchmarked its market research in terms of capacity to identify markets for products associated with or employing new or emerging technologies. Measures of performance on this criterion were obtained from comparisons with a strong local competitor and a large and well-established international manufacturer, using an ordinal scale with ratings of "high", "medium" and "low". As a result of this benchmarking exercise weaknesses in market research procedures were exposed and action initiated to remedy them.

Short Lead Times for New Product Development. The quest for short lead times in product development led the Company to consider benchmarking its product development processes. This presented problems some of which have been solved while the Company and the authors are still grappling with others. For example, the rate at which competitors introduced new products to the market was known, but their development times were not, although reasonable guesses could be made. In the event the company decided to embark on a self-assessment of its product development processes in depth and began to monitor this aspect of its operations by recording avoidable delays and calculating a performance index (P_i) for each product development programme as

$$P_i = \frac{(\text{Time to develop product}) - (\text{Time lost due to avoidable delays})}{(\text{Time to develop product})} \times 100.$$

2.4 Stage 4 - Pilot Study

The Company's management and the authors then agreed to select a typical product for investigation whose development could be studied as part of an on-going research programme into benchmarking manufacturing systems. The 6 litre tank assembly chosen is part of a domestic pressure system, a recent product added to an existing range by the manufacturer who already makes and sells systems having water storage capacities of 10, 15, 20 and 30 litres. It was developed in response to a competitor known to be about to market a similar product.

The design concept was obtained by scaling down from the 10 litre system, although it was realised that the sizes and shapes of some, possibly all, components must change if only because water and air pressures do not change. Fig. 2 shows a typical arrangement of components in a product of this type. An expandable diaphragm of nitrile rubber receives water under pressure from a small centrifugal pump and discharges it through a nozzle to provide pressurized water to a domestic system in a house or building without access to conventional water supply in a town or city. The diaphragm is mounted in an enclosing tank and kept in place by a keeper ring. The air in the tank surrounding the diaphragm is compressed both to cushion the expansion of the diaphragm and to assist the rapid egress of water from it. Other details of construction and operation can be inferred from Fig. 2.

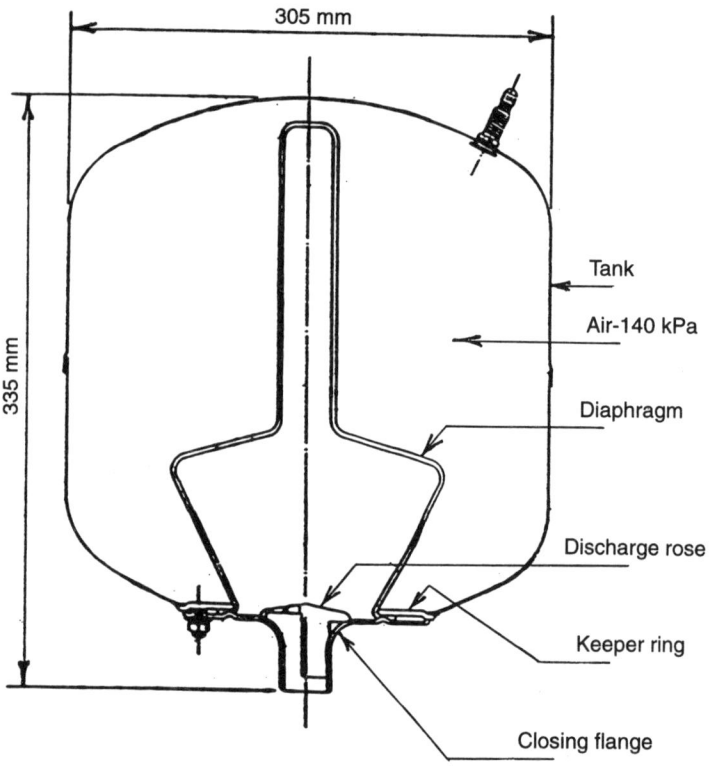

Fig. 2.   Sectional Arrangement of 6 Litre Tank Assembly.

Delays were experienced in the development of the 6 litre tank assembly in that it took longer than expected to create a new geometry of diaphragm to suit this application and to seek out and liaise with specialist suppliers of steel pressings to ensure that the cost of the tank was in line with the programme budget. There were also unforeseen manufacturing problems associated with the welding of the keeper ring to the outer casing and with forming the curved profile of the discharge nozzle. As a result of these delays the performance index P_i was 63%, a sufficiently low value to cause the company concern. As a result of this benchmarking exercise the Company has instituted a database for each product development programme comprising :

- short description of product
- graphical depiction of decision making via relevant DIN (see next page)
- checklist for deployment of design resources (see next page)
- performance index P_i, and if P_i < 100% the reasons for this.

In addition the Company has initiated research into information flows, decision making, and design resource allocation during product development. The results from the pilot study are summarized in the Appendices. Table A1 in Appendix 1 shows the Design Decision Table (DDT) for the 6 litre tank assembly, while Appendices 2 and 3 show the Design Information Network (DIN) and Design Resources Matrix (DRM) respectively.

In Table A1 the key design decisions taken during development of the product are listed in the left hand column and are identified by a two-letter code, e.g. DV for the decision on the magnitude of the diaphragm volume. The information input to this decision was the customer need (denoted CN in column 3) as defined by a cross-functional team working on quality function deployment (QFD) as noted in column 2. The outcome of this decision is shown in column 4 as the numerical value 6 litres. Similar explanations apply to the other entries in the Table. In the formulation of the project two other key design management decisions were taken re performance targets, namely (i) the diaphragm life was to be not less than 250,000 cycles of operation, a figure based on use of existing products by domestic comsumers, and (ii) the manufacturing cost (TC in Appendix 2) was not to exceed a value set by the selling prices of competitors. Other notation in Table A1 consists of EP, knowledge of existing products, and Δp, pressure loss through the discharge nozzle.

DIN's make visible and display the path through the array of key decisions by which a final outcome is reached, and exhibit this path to demonstrate the process of decision making in an objective manner. Traceability of and accountability for decisions are important themes in the ISO 9000 series of quality standards. In ISO 9001, for example, there are clauses dealing with "Verification that design outputs meet design input requirements", and with "Responsibility for initiation of action to prevent the occurrence of product nonconformity" - clearly requiring traceability of design decisions. In this case the Company's management considered benchmarking of product quality to be a matter for future investigation when the DIN in Appendix 2 would be an important analytical tool.

The Design Resources Matrix in Appendix 3 is an example of a tool adapted from Eppinger's work [6] by the authors to assist the Company commit resources to the key design decisions for which they are to be deployed. (In graph theoretic terms [7] a DIN is a directed graph and the DRM is the corresponding adjacency matrix.) "Resources" here is to be interpreted broadly to include intellectual design skills such as creativity and judgement as well as marketing knowledge about customer needs and manufacturing knowledge of the capacities of specialist suppliers. A study of the availability and timeliness of these resources and the use made of them in practice is under way at the time of writing this paper. Preliminary results suggest that the Company's current procedures are far from perfect and that designers are prepared to make guesses in order to meet tight deadlines.

The pro's and con's of proceeding to stage (5) are currently being weighed up by Company's management. The results from the first four stages of the benchmarking exercise have yielded valuable data on the strengths and weaknesses of existing operations and the Company's capacity for future growth and profitability.

3. CONCLUDING COMMENTS

Benchmarking has been considered from the point of view of a small manufacturer which is internationally competitive and intends to remain so. Small manufacturers are the backbone of many industrial economies, but the need for them to benchmark their practices has been insufficiently acknowledged, hence this paper.

All industrial companies but particularly those with a limited capital base have to marshal resources carefully to ensure profitable operation now and long term growth in the future. Benchmarking procedures for small manufacturers should therefore proceed in carefully designed stages, so that the value of the results obtained is proportionate to the effort expended in each stage e.g. via a series of hurdles and a pilot study as in this case study.

During the investigation reported here benchmarking was carried out in accordance with a carefully devised strategic plan, working through various levels in the means/ends hierarchy using subjective and objective rating scales as appropriate and relying on both internal and external performance data as bases for the comparisons made. On each occasion the Company gained new insights into the effectiveness and efficiency of its operations, thus demonstrating the flexible and wide-ranging nature of the benchmarking process. It is hoped that the paper will provide a useful guide to benchmarking proce-dures suited to the needs of small manufacturers and suitable for adoption by them.

REFERENCES

1. McKinsey and Company, Emerging Exporters : High Value-Added Manufacturing Exporters, Australian Manufacturing Council, Melbourne, 1993.
2. R.C. Camp, Benchmarking : The Search for Industry Best Practices, Quality Press, Milwaukee, 1989.
3. National Industry Extension Service, Benchmarking Self-Help Manual, Australian Government Publishing Service, Canberra, 1993.
4. C.J. McNair, Benchmarking : A Tool for Continuous Improvement, Harper Business, New York, 1992.
5. P. McAree, Planning an Range of Centrifugal Pumps, Proc. I. Mech. E., 186 (1972) 595-602.
6. S.D. Eppinger and K.R. McCord, Managing the Integration Problem in Concurrent Engineering, Working Paper No. 95, International Center for Research on the Management of Technology, Massachusetts Institute of Technology (1993).
7. N. Biggs, Algebraic Graph Theory, Cambridge University Press, 1974.

APPENDIX 1

TABLE A1.   Design Tasks and Key Decisions.

Nature of Decision - denoted (xx)	Decision Making Procedure	Information Input to Decision	Quantitative Result
Diaphragm			
Diaphragm volume (DV)	QFD team	Customer need (CN)	6 litres
Replaceable diaphragm (DR)	QFD team	Customer need (CN)	Adjustment to layout
Diaphragm shape (DS)	Personal creativity	DV, TS, KD	Special 3D shape
Water pressure (DP)	QFD team	CN, EP	~ 350 kPa
Tank			
Tank air pressure (TP)	QFD team	CN, EP	~ 200 kPa
Tank volume (TV)	Simulation - Prediction by calc'n	DV, DP, TP	18 litres
Tank shape - L/D (TS)	Judgement	CN, EP, TV	L/D ~ 0.75
Tank material (TM)	Judgement	EP	Mild steel
	(Informal analysis of cost-effectiveness of alternatives)		
Wall thickness (TW)	AS 2971	TS, TP, TM	1 mm
Internal surface (TI)	Design logic	DR, TM	0.1 mm thickness; anti-corrosion paint
Tank Outlet			
Keeper ring diameter (KD)	Adjustments to suit DS	DR, DS, TS	12 mm
Ring thickness (KT)	Experimental trial and error to ensure manufacturability	TW, KD	3 mm
Discharge nozzle profile (NP)	do.	CN, (Δp), KD	radius of curvature of profile

APPENDIX 2 - Design Information Network

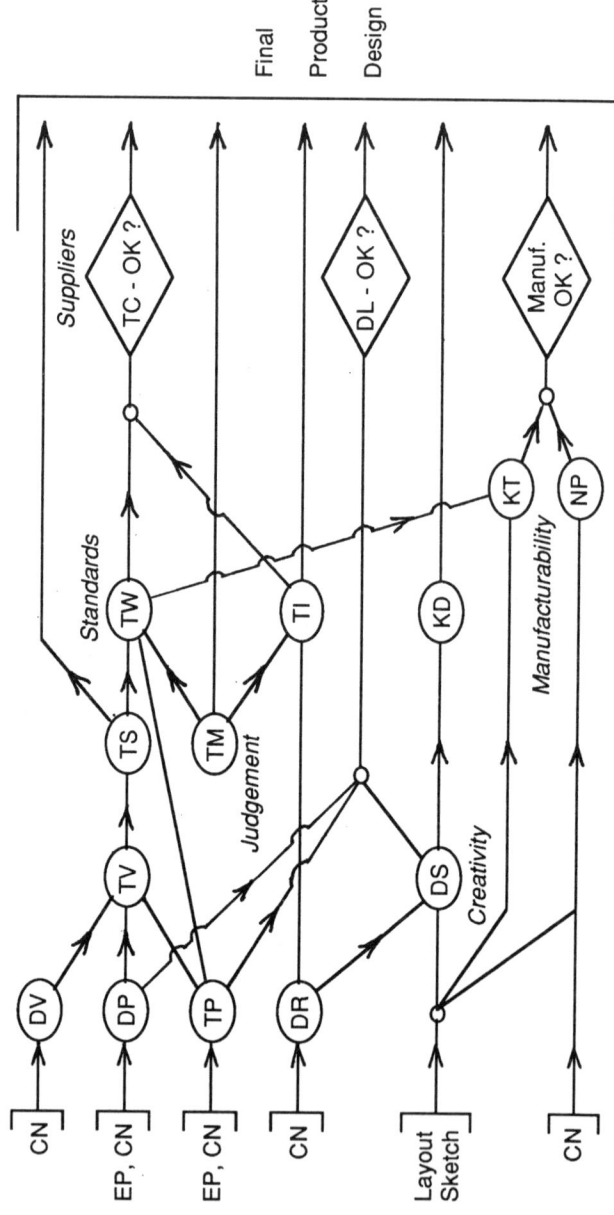

APPENDIX 3 - Design Resources Matrix

		Customer Needs	Existing Products	Creativity	Judgement	Design Logic	Standards	Product Integrity	Tests and Simulations	Manufacturability	Specialist Suppliers
Diaphragm Volume	DV	✓									
Diaphragm Air Pressure	DP	✓	✓								
Replaceable Diaphragm	DR	✓									
Diaphragm Shape	DS			✓							
Water Pressure in Tank	TP	✓	✓								
Tank Volume	TV	✓									
Tank Shape	TS		✓		✓						
Tank Material	TM		✓		✓						
Tank Wall Thickness	TW						✓	✓			
Internal Surface of Tank	TI					✓		✓			
Keeper Ring Diameter	KD					✓					
Keeper Ring Thickness	KT									✓	
Discharge Nozzle Profile	NP	✓								✓	
Diaphragm Life	DL	✓						✓	✓		
Cost of Tank Assembly	TC	✓									✓

A tick in a cell of the matrix identifies a Design Resource (column) which has to be allocated to the process of making a key Design Decision (row).

14

Benchmarking for implementing a new product strategy

P.G. Pettersen

The Norwegian Institute of Technology, Faculty of Economics and Industrial Management, N-7034 Trondheim, Norway

1. INTRODUCTION

Benchmarking has often been criticized for being a numerous game, comparing more or less comparable metrics between different companies. These comparisons have mostly been used to evaluate a company's performance by looking at what is possible and necessary to keep up with competitors. In many benchmarking studies, little information has been generated in respect to why there are gaps in performance between the company and other companies.

This paper is meant to give an example of how benchmarking can be used as a problem solving tool, more than merely a tool for comparison or evaluation. A case* is presented where a company is facing problems with implementing a chosen product strategy. Implementation requires process re-engineering as well as understanding and adoption of the new strategy throughout the organization. Benchmarking is, in this case, used as a tool for identifying and analyzing different ways for implementing the chosen strategy, based on learning from other organizations in similar situations.

Definitions and theories concerning the subjects of benchmarking and product strategies will also be briefly touched in this paper.

2. THE COMPANY AND IT'S CHALLENGE

The case company belongs in the engineering industry. It manufactures quite complex products, which requires high skills among both engineers and operators. The company has about 400 employees, and faces international competition as 70% of the turnover comes from export. The company is recognized for it's high product quality and ability to meet customers' requirements. Although the company offers a set of more or less standard products, the customers are often technically skilled and therefore define in detail the product they want and the set of capabilities they expect from the product. This leads to an extensive need for engineering for each customer order, but also gives the customer a very customized product.

Considering the three main performance indicators *(1)quality, (2) cost* and *(3) cycle time*, the company struggle with the latter two. Product quality is satisfactory, but as in most industries, customers' expectations to cycle time and cost reduction have to be considered. The degree of customization drives both cost and cycle times, due to the extent of engineering initiated by each customer order. The extensive engineering work often leads to problems with meeting the scheduled delivery time, and it also delays the purchasing orders and

---

*The referred case is a part of a benchmarking project in the TOPP program, which is a government supported program for productivity improvement in Norwegian engineering industry. In the project, researchers and consultants work as facilitators for the companies, to help them implement benchmarking as a improvement tool, and to solve a problem of current interest for the companies.

specifications to the company's suppliers so that purchased parts arrive too late to fit into the production and delivery plans. As a consequence, the company must either give up the delivery time, hold a considerable stock of components, or pay the suppliers more to make them shorten their delivery time to the company.

3. A BRIEF LOOK AT THE PRODUCT STRATEGY

It can be stated that the company is following a differentiation strategy, as it is offering higly customized and specialized products to it's customers. In a theoretical context, using Michael Porter's (1) model for categorization of product strategies, three generic product strategies are defined. These are:
- *Cost leader strategy*, which implies low cost products, often produced in a large scale.
- *Differentiation strategy*, which implies that the product contains attributes or features that make it different from other products. This can be made by using special brands, specific materials and capabilities etc.
- *Focus strategy*, which implies offering the product to a specific market segment, defined either by geographic or demographic characteristics.

In Norwegian engineering industry, the differentiation and/or the focusing strategies tend to be dominating. This can be explained by the relatively high production costs in Norway, and the fact that most of the mechanical industry delivers to the industrial market, which often requires customized products. This is also the situation for the case company.

The latest years, one has experienced a significant globalization of markets. Needs and desires converge and become homogenized all over the industrialized world, and at the same time, barriers of trade are removed. It will often be a mistake to believe that one can justify high costs (and price) by following a differentiation or focus strategy. This may be possible over a short period of time or under market conditions with high entry barriers or restrictions. However, given a strong international competition, someone will sooner or later attack your market segment, offering a better product at a lower price. A prime example of this is Mercedes Benz, who has lost a significant market share in the U.S. to Toyata's Lexus-model. At one-half the price of a Mercedes-Benz, one can have a virtually trouble-free, high performing Lexus with the same (or better) features as the Mercedes. This also shows that brand loyalty do not overcome the power of scarcity. It can therefore be argued that Porter's (1) model with the three generic strategies to a certain extent looses it's relevance for companies that compete on an international arena.

Andrews (2) states that the only long-term success strategy is to offer high value products (which means products that give the customer at least what he wants) to a low price. It is well known that mass production gives a potential for lowering costs. But from Henry Ford's mass production philosophy in the early part of the century ("everyone can have the color they want on their T-Ford - as long as it is black") and up till now, there has been a tremendous development in manufacturing systems. Today, there is a much higher degree of flexibility in the production line, and through new philosophies as for instance Just-In-Time-Production, a great variety of products can be produced in the same production line, and to a lower cost. This opens the possibility for mass production (which in general implies low costs), but also gives the customer a great variety of products to choose among. This is achieved by standardizing both parts of the product and parts of the production process.

The case company realizes that their customers must be given a high value product to a competitive price. Today, the customers are given a relatively high value product due to the heavy customization, but this results in high costs because of the extent of engineering work for each order. Therefore, there must be a trade-off between expensive customizing and cost reducing standardizing. The company has identified a solution to this dilemma, by distinguishing between basic parts of the product, and auxiliary parts. A strictly limited number of "basic" models are defined, and these are supplied with different combinations of auxiliaries to constitute the end product. The number of auxiliary parts being used will be limited and defined up-front, but can be combined in different ways to make the requested end product. In this way, the extent of engineering will be reduced, as well as the variety of parts or components being used in the production line. This will reduce the engineering costs, and simplify the handling of components. Still, the end product can be given a wide range of different features by combining the auxiliary parts in different ways. In this way, a great deal of the process from order to delivery can be standardized, without reducing flexibility and ability to meet customers' requirements.

However, the problem for the company is to implement this strategy, resulting in questions like: "How can this be done? What systems will be needed? What will the process look like? How will this affect the different functions in the company?" A lack of answers to these questions have been obstructing the implementation of the strategy. The main question is therefore if benchmarking can be a proper tool for helping the company find a way to implement the strategy.

4. DEFINING BENCHMARKING

There are lots of different definitions of benchmarking, and many of them tend to cover every aspect of benchmarking in one sentence. For all practical purposes, a simple definition of benchmarking, in business terms, can be as follows: *"Benchmarking is the process of comparing something or someone with best practice."*

This is a generic definition, and out of this, benchmarking can take place in several forms. Benchmarking is certainly about comparing, but can be defined in respect to *what to compare* and *whom to compare against*. Depending on what to compare, benchmarking can be divided into three main categoriess:
- Performance benchmarking.
- Process benchmarking.
- Strategic benchmarking.

Performance benchmarking is very much about comparing performance levels against "the best." The focus can be on the whole company or parts of it (e.g., functions, processes, products, departments etc.). Given that one find measures that are comparable, this can be an important contribution for motivating and convincing the organization that improvement is both possible and necessary. The problem is often that the metrics themselves do not say anything about *how* you can reach a better level of performance.

Process benchmarking goes one step further, as it tries to identify *why* someone is better. It focuses on other organizations' practices and methods that enable their high level of performance. Which companies to compare against, depends on who is perceived as the best. The best companies can be identified by doing performance benchmarking up front, and a typical benchmarking study will often be a mixture between performance and process benchmarking.

Strategic benchmarking is more about comparing strategic corporate decisions, on for example the allocation of resources, investments, selection of business ventures, technological evolution, or development of market segments. This can be an important information to the company's own strategic planning.

In respect to whom to compare against, benchmarking can take place in one of the following three categories:
- Internal benchmarking.
- Competitive benchmarking.
- Generic (or functional) benchmarking.

Internal benchmarking is comparison between units within one organization. This method is often used by large or worldwide corporations, and is often easy to undertake, as the information is quite easy to get and is often standardized between the different units. However, the potential for real performance breakthroughs is rather small, as you seek information within your own, already well-known environment.

Competitive benchmarking implies comparison with your competitors. This could give very interesting and useful information. However, the problem is that one seldom get any useful information at all, because nobody want to share sensitive information with their competitors. Besides, there are difficult legal and ethical issues that must be considered in this type of benchmarking. Therefore, competitive benchmarking will most often either take place as a superficial comparison of key performance indicators (metrics) based on public available information, or as a comparison of metrics between a group of companies that have given their information in an anonymous way.

Generic benchmarking is comparison against companies in different industries or at least non-competitors. It is often hard to use metrics because the companies can be very different, but a more qualitative approach can be taken, and there are always some common processes or functions that can be compared. The result of benchmarking across industry borders can often be the identification and adaption of practices and methods that have been totally unknown to your own industry. Examples are the bar coding system that has spread from industry to industry, or the "invention" of the Just-In-Time concept, which can be traced back to Taichi Ohno's observation of how the supermarkets handle their throughput.

A typical benchmarking study consists of a number of steps. Watson (3) presents an useful six-step model for benchmarking, consisting of:
1) A planning step (what to benchmark, documentation, measurement)
2) A searching step (search for benchmarking partners).
3) An observing step (visit benchmarking partners).
4) An analyzing step (analyze findings, find and explain gaps).
5) An adapting step (identify what is worth adapting to your own organization).
6) An improving step (implement changes, follow up).

Whatever way benchmarking is defined and whatever model is used, it is of most importance to stress that benchmarking is about learning. The focus for benchmarking has in fact chanced over the past years from being a tool for evaluation of current performance (and trend), to be an effective way of learning methods and practices from the best. Second, it is important to understand the difference between benchmarking and so-called "industrial tourism". Effective benchmarking requires a much higher degree of planning and systematic approach in general, than a plain observation.

5. BENCHMARKING IN THE CASE COMPANY

The company has been discussing how benchmarking can be used to solve some of the problems concerning cycle time and cost. Benchmarking is expected to contribute in two different, but still connected ways. A benchmarking study will be used both to analyze the critical process for the company, and to learn from others how a successful change in product strategy can be carried out.

A benchmarking team has been formed, consisting of four participants, representing the areas of marketing/sales, engineering, planning, and division management. Thus, roles as process owner, operators, internal suppliers and customers, and sponsor (management commitment) are all taken care of in the team. The team has defined the following areas as subjects to benchmarking:

1) The process from order entry to purchasing.
2) The way of changing product strategy from customized to more standardized products.

By focusing on the process from customer order to purchasing, the team is given a good understanding of the current situation, a basis for discussions with benchmarking partners and a possibility to streamline and optimize the process given it's constraints.

By focusing on the changing process, going from customized to more standardized products, it is possible to collect information concerning how this has been done in other companies, with specific focus on companies with successful experiences with such changes. The challenge is to decide the appropriate mixture between standardization and customization (given the customer requirements), to design new management systems, information systems, and technical systems that are needed to carry out the strategy, and to identify a way for implementing these systems.

The planning step in the project is crucial. According to experience from previous benchmarking studies, 50% of the total time should be used in the planning phase of the study. After choosing the area to benchmark, a lot of work must be done concerning documenting the chosen process and identifying measures. This is necessary to get knowledge of what the current situation is, and is done by going through written procedures and descriptions, and interviewing process participators. Based on relationship maps, flow-charts and verbal descriptions of the process, a discussion will take place among the team members. During this work, the team members become aware of what the entire process looks like, and where there seems to be needs and room for improvements. At the same time, the work results in a documentation that will be used later in the benchmarking study as an introductory explanation of the situation to benchmarking partners.

Compared to the definitions of benchmarking stated above, it is clear that the company wants to use benchmarking for learning. During the planning phase, the team members identify areas of improvements just by analyzing and discussing the process across the traditional borders between different functions inside the organization. In addition, by observing other successful companies that are or have been in the same situation, a lot of information and knowledge for improvements can be gained.

The type of benchmarking the company carries out, is a mixture between both strategic, process, and performance benchmarking. However, most effort is laid on process benchmarking. The first part of the benchmarking study will be focusing on the process itself, while the last part will be dealing with the enablers that make the company able to carry out the process as they would like to, and to implement the chosen strategy (i.e., the change from customization to standardization).

When it comes to whom to compare with, the company has chosen to look at other companies with similar challenges. This will neither be companies inside their own group (internal benchmarking), nor competitors. There are no appropriate partners with the same type of challenge within their own group, and a comparison against competitors is difficult because information concerning the chosen area for benchmarking is too sensitive and strategic important to be shared. Therefore, it has been decided to carry out a generic benchmarking study. As a start, the search for benchmarking partners is done inside the company's own industry, but not among competitors. A company with similar structure and technological level will be the most appropriate for comparison. For a future study, it is also possible to look companies in totally different industries. However, because this is the company's first benchmarking study ever, the ambitions has to be held within what is realistic. It may also be important to create fast and tangible results from this first study, to use it as motivation for future studies. Anyway, the team is very open minded to the thought of looking over the fence to a complete different industry in a future study.

6. CONCLUSION

Benchmarking has been looked upon as merely a method for comparing and evaluating performance among organizations. Except for the fact that benchmarking implies comparing against the best or best practice instead of the average, there has not really been anything that has differed benchmarking from other methods for comparing performance measures or metrics. In the process of comparing numbers, the output is never better than the input. Consequently, if one do not know what is behind the numbers, one can not interpret or make any conclusions from the benchmarking results. Benchmarking has therefore been criticized for being a "numerous game", where "apples-to-pears"-comparisons have been made, and where the findings seldom have given information concerning *why* someone has been better than others.

However, the last few years, the term "process benchmarking" has gained more attention. This has lead to more focus on *how* other (i.e., the best) do their work, instead of just what they achieve. Thus, benchmarking has become a method for learning and problem solving, as well as a tool for evaluation.

The purpose of this paper has been to argue that benchmarking can be used in a more active way for problem solving and attacking different challenges by learning from other organizations. It has also been argued that the best pay-off from a benchmarking project occurs when focusing on process benchmarking and comparing against other industries (i.e., generic benchmarking). Although the benchmarking project, referred to as a case in this paper, has not been completed yet, it shows how benchmarking can be used as a problem solving tool rather than merely a tool for evaluating and comparing performance.

LITERATURE

1. M. Porter, Competitive Advantage, Creating and Sustaining Superior Performance, New York, The Free Press, 1985.
2. B. Andrews, Standardizing Products and Services, Continuous Journey, No. 3 (1994) 22.
3. G. Watson, The Benchmarking Workbook, Cambridge Massachusetts; Productivity Press, 1992.

15

Benchmarking of Tandberg Data against EFQM's (The European Foundation for Quality Management) Assessment Model

Colin Jeneson*
* *Corporate Manager, Strategic Quality & Partnerships, Tandberg Data A/S, Post Box 134 Kjelsaas, 0411 Oslo, Norway*

The following article describes Tandberg Data's limited experience of Benchmarking and is related to Total Quality Management criteria. But first, it is necessary to put the activity into perspective by reflecting on some of the issues facing the company I represent and the business world at large.

1.0 GENERAL TRENDS IN EUROPE

In today's turbulent period of European politics, and of economic recession, despite encouraging signs of growth potential, the emphasis of many businesses has been placed on cost reduction. There has been, and will be, unrelenting pressure to downsize, to cut costs, to de-layer and to defend market positions.

In many cases this difficult environment has reduced the urge or the drive for quality; quality has been seen as a "low impact" investment or an investment which can be given a lower priority. I'm sure that to many industrialists this statement may seem fairly strange. After all, the emphasis on ISO 9000 and TQM has almost achieved proportions that could lead one to think that a new religion is in the process of being formed. But just how real has this focus been for many companies other than those with high export shares and demanding customers ?

The cry for ISO 9000 may often be all we hear - a desperate cry of necessity in order to enter markets through complying with European regulations.

Currently, energies are being directed in many consultative circles at slaughtering the TQM emblem in favour of "Business Process Re-engineering" instead of showing greater responsibility by marketing the necessity of and relationships between the sub-elements in both approaches. We get too easily hooked on terms in our eagerness to play with strategic theories divorced from the realities of our own business world and to blame the lack of results on the wrong choice of philosophy or inappropriateness of the acedemic model chosen.

Our company achieved ISO 9001 status in 1992. The procedures, audits and improvement activities that constitute our Quality Management SYSTEM are very important elements of our Total Quality process. We see them collectively as a vital discipline - a way of ensuring that we do not let standards slip when times are difficult and business pressures are significant. We do NOT however, regard them as the PRIME driver in our commitment and determination to delight our customers.

At this stage it is appropriate to call to mind the many companies who complain of ISO 9000's so-called total inadequacies. Such statements may often, at further glance, prove to reflect the different companies' own total inadequacies at deploying the standard in the way it was intended, through supplementing it with appropriate methods, tools and strategies for technical and administrative gains. The company's focus on its ISO 9001 quality system is often directed at procedures, papers and product traceability issues, without realizing that these issues, important though they are, will never be effective without an even stronger focus on the company's people, processes and policy deployment.

Many companies are obtaining ISO 9000 registration simply to get on a customer's list of suppliers. They comply with the word of the standard but not with the spirit of Total Quality, and are probably not gaining much benefit from it either. They lack the insight and willingness to be self-analytic and to be severe in their otherwise sincere appraisal of methods employed and results achieved. The fixation is still only on compliance with standards (quality assurance) instead of optimizing the whole company and its results for others.

We are living in a world which, because of good communication and technology, is a much smaller place. Word gets around fast. News of failure, of poor quality, of mismanagement can reverberate around the world and when it happens, customers go elsewhere, because with a shrinking world there is more choice.

But there is an upside. Companies also become recognized as world class, with a stamp synonymous with quality. IBM, Motorola and Digital are international examples. As supplier to all three, Tandberg Data here in Norway recently achieved its "stamp" - the IBM Supplier of the Year Award for Data Storage solutions.

2.0 SPECIFIC IMPLICATIONS FOR TANDBERG DATA

For many years this belief - that Quality is THE key factor for global business success - has echoed throughout the Tandberg Data organization, although at troublesome times maybe with varying sound-levels. Our preoccupation with quality started in fact in 1933 with the birth of our forerunner - Tandberg Radiofabrikk, known world-wide for its quality radios and tape recorders.

Tandberg Data defines quality as:
"A product's or a service's total ability to delight the customer through:
- standards
- cost-levels and price conditions
- user requirements, user-reliability and user-friendliness
- defined and latent requirements".

However, it would of course be futile for Tandberg Data to lull itself into thinking that because it practises TQM, and because it has recognized standards, awards and certification that it will automatically have customers that will remain with us for all eternity. This will depend on our ability to continuously identify, select and adopt the correct strategies and methods for best results at lowest possible use of resources - time included.

Nowadays, we have come to realise that total quality has to be integral to our corporate strategy. It is really quite natural that this is so - after all, as we all know, customers of electronics continue to demand a continually increasing degree of continuous improvements to the products and services companies provide in partnership with them and to the value for money also provided.

For Tandberg Data, these are now related to PC's, work-stations, terminals, systems for digital storage of data, sound and pictures, as well as LCD peripheral equipment for PC's and the audio-visual market. Global and even national competition in many of these areas is significant. There is an increasing awareness that in this fierce competition, business results related to customer satisfaction, employee satisfaction, environmental and social aspects are important pre-requisites to solid financial results, and that the pre-requisites can only be fulfilled through focusing on strategic management, policies, resource deployment issues and process management techniques and principles that are loyal to the concept of a comprehensive, company-wide Total Quality infrastructure.

Tandberg Data began the Total Quality journey in 1988. Even though we were also at that time highly regarded for our innovative thinking regarding the introduction of new technologies, products and processes, it was obvious that our focus need to be more all-embracing for own benefit, besides that of our shareholders and customers. We had always listened carefully to what our customers were telling us in surveys, in consumer groups, in customer contact programmes, through complaints, during product launches, and so on. They pointed out our shortcomings only too clearly. Although our product quality levels more than held their own in the face of comparable competition, two main things were happening:
1.   The quality of our competitors was steadily improving.
2.   The quality of their services was improving almost equally.

The clarity of our shortcomings was directly proportional to the interest we showed in revealing them ! We needed to bring our house in better order, to structure our approaches, our deployment of methods and our company-wide involvement to deliver more than satisfactory results for our customers, owners and fellow workers - suppliers included.

We began this process in the management team through increasing the knowledge and appreciation of Total Quality as expressed through our important strategic alliances with companies such as IBM, Digital Equipment Corporation and Motorola. Our desire to sustain these relationships was dependent on our ability to involve all parts of the organization in a joint, continuous effort to improve quality, productivity, responsiveness and flexibility in internal as well as external relationships.

First we had to involve all employees and functions in documenting what was expected of them and their services and how these should be performed for optimum results, then we had to provide a set of tools for them to continually improve the handling and results of these services. All workers on all levels had to do two things daily:
1. Perform work tasks according to accepted guidelines and principles.
2. Continually improve their work performances as well as the guidelines and principles these are based on.

The first sentence of the European Foundation for Quality Management's policy document reads: "Quality is a key factor for global business success". NOW, my company would go beyond that and say that "Quality is THE key factor for global business success".

3.0 OUR NEED FOR BENCHMARKING

What has all this got to do with Benchmarking ? Well, it must be said that adherence to quality system standards, receiving customer quality awards, and numerous company-wide ongoing improvement tasks, impressing though it may seem, quite simply isn't enough ! After all, what is the point of continuously improving a process that shouldn't be there in the first place ! We may not know that a process is superficial until we have studied how others receive similar results.

In addition, the company had to continually evaluate - through Benchmarking and Self Assessment - three main things:
1. The relevance of the strategies chosen.
2. The effectiveness of their translation to specific actions at different stages of the value creation process.
3. The actual results obtained for our customers, shareholders,personnel and society.

The more we began to work with these matters, based on knowledge and experience from our large international customers and suppliers, from research institutions at home and abroad, and more recently also from the European Foundation for Quality Management, the more we realized that Benchmarking and Self Assessment of the company's ability to practice TQM principles in its dealings with partners internally as well as externally, was the mechanism by which we could improve our whole organization and all processes in order to increase the value output for everyone; most important of all, to continue to improve the quality beyond that expected of our all customers, with steadily reduced use of resources.

The key to achieving this lay in our ability to recognize, implement and follow through appropriate activities. In short, our ability to do the right things and to do them correctly. To do this we needed to start looking at the way others did things and the way still others meant we ought to do things.

It was utterly inconceivable for Tandberg Data to divorce Quality from the day to day running of the business. It has to be seen relevant to and infuse the thousands of transactions that occur every day. There has to be a continuous focus on outcomes, a relentless avoidance of bureaucracy, and a visible recognition of actions, not just words. Total Quality must be in the corporate bloodstream.

In Tandberg Data's experience, the investment in Quality, and in Bench-marking & Self Assessment in particular is mandatory. It not only works, but provides substantial results. It is based on the belief that to continuously improve existing processes is essential but equally inadequate; you have to also look at other organizations' choice of processes to obtain similar results. These OTHERS may be competitors or companies in other industries. For Tandberg Data they also include - at company wide level - the European Foundation for Company Management's Assessment Model.

4.0 PREPARING OUR CASE FOR BENCHMARKING

Tandberg Data's first attempt at Benchmarking turned out in fact to be more of a Self Assessment activity. The company at this time - early 1993 - had not acquired sufficent knowledge of the terms deployed. We studied opportunities for benchmarking at different levels:
- Strategic (Management's Business Policies),
- Macro (Business Plan Implementation), and
- Micro (Operational at Dept./Functional level).

We looked at the approaches used by PIMS (Profit Impact of Market Strategy) and those of TOPP (Norwegian Productivity Improvement Programme) as well as those of our customers.

As mentioned earlier, we decided first to benchmark our strategies and activities aginst the European Foundation for Quality Management's Assessment Model, and then compare the results obtained through the use of our approaches with those of competitors and other businesses regarded as Best in Class.

We were spurred on by increasing demands for increasing productivity and quality levels, but also by a simple story.
The story came from a japanese company and is now well known. The Ford Motor Company was attempting to reduce costs in their accounts payable department which at the time employed some 400 staff. Through an arduous process they managed to reduce staff numbers down to 300, then to 200 but could NOT take it below that.

When they looked at Toyota they saw a similar department being run by just 15 staff! Toyota had distributed responsibility and had empowered staff to carry out more tasks at a relatively junior level. Specifically, they had given the receiving clerks the authority to make payments directly.

Consequently, the person at the back door of the plant was able to sign goods in, had the authority to organize distribution and make decisions like authorizing payments that previously were handled by other staff within the organization. The net result was a more efficient, less costly operation. Benchmarking provided a good example of using empowerment to speed up the decision making process. Quality IS time. Or to state the Spanish philosopher Baltasar Gracian who lived 300 years ago - "The wise do sooner what fools do later".

5.0 READY, GET SET, GO !

Returning to Tandberg Data's initial attempt at Benchmarking, I would first like to present to you the process initially adopted which consisted of the following 11 steps:

1. Gain insight into Benchmarking & Self Assessment
2. Develop Commitment
3. Plan the process
4. Communicate the process
5. Re-evaluate insight into Benchmarking & Self Assessment
6. Document Situational Report for all areas covered by EFQM Model
7. Compare with main criteria and sub-criteria in EFQM Model
8. Integrate results with findings from TOPP's pre-assessment
9. Interest and involve management teams in recommendations
10. Integrate in Strategic Planning Process for business units
11. Review progress

5.1 Gain insight into Benchmarking & Self Assessment
The following represented our main sources of information for this initial stage during which we focused our attention on learning the purpose and principles of Benchmarking:

*   Digital Equipment Corporation's Benchmarking Training Materials (Germany)

*   PIMS (**Profit Impact of Market Strategy**) 1993 Conference in Stockholm.

*   Self Assessment Workshops under the auspices of the Norwegian Technology Industry's Productivity Programme (TOPP).

*   Experience transfer with our largest customer IBM (Rochester) - the first company to win the prestigous Malcolm Baldrige Quality Award in the United States.

*   The Benchmarking concept of Bjelland, Dahl & Partners - entitled "Benchmarking with a capital B".

At this stage we focused particularly on examples of Benchmarking used in different functions, on different organizational levels, in different industries grappling with similar challenges and for different strategic purposes. This was very much a self-awareness phase where our minds were opened for the variety of opportunities to exploit Benchmarking for productivity, quality or business improvements.

We focused at this stage more on issues relating to micro, or if you like operational Benchmarking where the logical deliveries from an individual function - be it products or services - as well as our existing challenges, could be compared with functions obtaining good results in other organizations.

We came to the conclusion that our first attempt at benchmarking had to come from one of the following broadly defined areas:
• Issues most critical for customer satisfaction
• Expenditures & Incomes
• Human Resource Development
• Total Quality Management
• Our largest problem processes
• Choice of Technology
• Shipping
• Supplier Development & Partnerships
• Main cost components
• Measurements of performance

Many different urgent needs for benchmarking or assessment were identified and at the time the company was undergoing a significant restructuring programme. Our management approaches and human resource deployment were under the loop and we decided therefore to take a fresh look at the company as it was immediately prior to restructuring from another angle. As mentioned previously, this angle was provided by the EFQM's Assessment Model. We had earlier been studied by our major customer who certain aspects of the Malcom Baldrige Model to assess our worthiness as potential volume supplier of streamers to the IBM AS400 series worldwide. Now we wanted to take a COMPREHENSIVE look at the company by focusing on ALL aspects of the EFQM's model. Having decided this, we then had to develop commitment in the management team.

5.2 Develop Commitment
Commitment would not have been easy to develop, were it not for the fact that ...
1. we had very demanding but also supportive customers

2. the company had already been accepted as member of the EFQM based on its previous quality achievements

3. the company was at the time in the final stages of a pre-assessment activity conducted by the TOPP project.

We began by developing commitment among those contributing by offering local support for the activity. These were the quality managers of each newly formed subsidiary following the restructuring process. Though interest for the activity was already present among us, it was not always easy to maintain the level of interest. After all, having restructured the company, the new subsidiaries were anxious to show what they were good for without the interference of corporate staff - and rightly so. However, we were also very conscious of the fact that the information obtained from our benchmarking activity would provide us with invaluable insight into appropriate strengths, weaknesses, threats and opportunities that the individual subsidiaries had. This insight could then be infused into the strategic planning process and appropriate priorities made.

Having obtained commitment among ourselves - the quality professionals - we then embarked on selling the idea to the company's Quality Council, which consisted of the company's directors and quality managers. The Council readily approved the activity, although we did not feel at this stage that we had obtained REAL commitment. The exercise was still seen as something of an acedemic exercise rather than a business necessity. This was due to the fact that we had limited time for presentation and that the task in itself was very comprehensive. But at least we had the go ahead and even support in form of allocated time and priority.

In fact, a direct result of our benchmarking was that we disbanded our Quality Council and integrated its responsibilities into the normal management meetings. This was done in order to be loyal to the concept of Total Quality Management in our business practises, by dealing with issues cross-functionally as well as empowering and listening to all relevant functions in quality planning matters.

5.3 Plan the Process

Our chosen benchmarking activity could now get under way by first planning the process involved. We decided to pursue the following ten steps:

1.   Translate the EFQM model into Norwegian.

2.   Study the model's main elements.
        These were, ranked in order of % weighting:
        - Customer satisfaction (20%)
        - Business results (15%)
        - Processes (14%)
        - Leadership (10%)
        - People Management (9%)
        - Policy & Strategy (9%)
        - Resources (9%)
        - People satisfaction (9%)
        - Effect on society (6%)

2.   Study the main criteria and sub-criteria of each of the above-mentioned elements in the model.

3. Collect information from reports, meetings, personnel interviews and the TOPP pre-assessment report and relate the information to descriptions of each of the model's elements, main criteria and sub-criteria.

4. Compare (Benchmark) the company against the standards required by EFQM for the different elements and criteria.

5. Document strengths and areas for improvement in each area.

6. Communicate the findings to the Quality Council and subsequently through the management meetings.

7. Feed the information into the Strategic Planning Processes.

8. Review strategic and operative plans to see if they increasingly reflect or take into account EFQM criteria.

9. Identify possible links between results obtained and activities these were based on.

10. Spread information to appropriate process managers for synergy effects.

5.4 Communicate the process

At this stage we were concerned with communicating the process beyond that of directors and quality personnel to functions such as finance, personnel, logistics, design and manufacturing. We wanted to inform people of what we were doing and why it was important to make a critical, but structured assessment of what we had prioritized in different areas and how effective the results of these priorities were.

This communication stage was handled informally where we exploited our "open-door" culture to more or less chat about the activity to let people get the feeling that the company's quality drive was not just about fine mathematical measurements such as ppm or sigma, but also very much about how effectively we run our business based on total quality management approaches.

5.5 Re-evaluate insight into Benchmarking & Self Assessment

At this point we took a breather and reviewed our knowledge of benchmarking. It was at this point that we had a discussion about whether or not what we were doing was really benchmarking or maybe self-assessment. The conclusion was that we were adopting a benchmarking approach in order to assess ourselves ! Quite honestly we didn't care too much about issues of nomenclature; we were more keen on working in a disciplined way to obtain a useful tool for improvements, no matter what this discipline was really called.

5.6 Document Situational Report for all areas covered by EFQM Model

Here we took a "global" look at ourselves through the eyes of EFQM viewing each of the boxes or elements in the model. We asked ourselves what our own situation was with regard to each of the following five enablers and four results parameters:

ENABLERS:

* Leadership
  How the directors and managers inspire and drive Total Quality as the company's fundamental process for continuous improvement.

* Policy & Strategy
  How the company policy & strategy reflects the concept of Total Quality and how the principles of Total Quality are used in the determination, deployment, review and improvement of policy and strategy.

* People Management
  How the company releases the full potential of its people to improve the business continuously.

* Resources
  How the company's resources are effectively deployed in support of policy and strategy.

* Processes
  How processes are identified, reviewed and if necessary revised to ensure continuous improvement of the company's business.

* Policy & Strategy
  How the company policy & strategy reflects the concept of Total Quality and how the principles of Total Quality are used in the determination, deployment, review and improvement of policy and strategy.

RESULTS:

* Customer Satisfaction
  What the perception of our external customers is of the company and of its products and services.

* People Satisfaction
  What our people's feelings are about our company.

* Impact on Society
  What the perception of our company is among the community at large. This includes views of the company's approach to quality of life, the environment and to the preservation of global resources.

* Business Results
  What the company is achieving in relation to its planned business performance. The company's degree of continuing success in achieving its financial and non-financial targets and

objectives, and in satisfying the needs and expectations of everyone with a financial interest in the company.

5.7 Compare with main criteria and sub-criteria in EFQM Model

At this stage we looked beyond the five enablers and four results parameters mentioned above to study the more specifically defined demands which these include. To take an example, for People Management the specific demands were:

• Continuous improvement in People Management

• Preservation and development of the people's skills and capabilities through recruitment, training and career progression

• The setting of targets by individuals and teams and the continuous review of performance

• The promotion of everyone's involvement in continuous improvement and the empowermnent of people to take appropriate action.

• The achievement of effective top-down and bottom-up communication.

The other eight elements of the model possessed more or less equally detailed demands to be met.

5.8 Integrate results with findings from TOPP's pre-assessment.

The findings from TOPP's (Norwegian Technology Industry's Productivity Program) pre-assessment of Tandberg Data were now studied and related to each of the criteria and sub-criteria in the EFQM Assessment Model. Our own information was then supplemented and modified. It is important to note that modifications were only made to quantitative analyses, NOT to our own qualitative evaluations.

We quickly discovered that many of the observations and recommendations made by TOPP were in line with our own assessment. But in some areas - particularly matters relating to cash-flow, return on invstments, liquidity, etc. we obtained the added advantage of TOPP's more detailed analysis.

Our combined focus - broadly speaking productivity from TOPP and quality from Tandberg Data (if these aspects indeed are separable) enabled us to paint a clearer and more meaningful picture of our total situation. We also had the added advantage of drawing on the expertise from TOPP project personnel from universities and research institutes. These personnel either had the advantage of many years business experience at top level, or the equally important advantage of fresh knowledge of internationally accepted methodologies through their doctorate degree studies at the University of Engineering in Trondheim.

5.9 Interest and involve management teams in recommendations

After integrating and supplementing the findings from TOPP, we now had the total picture as seen by both of us. We could now further structure the findings in terms of strengths and areas for improvement. We were careful not to choose the term weaknesses at this stage (although it was mentioned earlier on in our exercise). We did not want to act as judges, more as facilitators or advisors to management and personnel.

Our final benchmarking report consisted of 22 pages of strengths and areas for improvement. We did not, on purpose, make recommendations in the form of priorities. This was a cross-functional task for management teams of which we were members on more or less equal terms with the other participants. Our aim at this stage was not to SELL anything but to PRESENT. The picture we presented was to be as clearly identifiable with our own activities, challenges, frustrations and ambitions as it was with the EFQM Assessment Model on which it was based. We were very concerned with avoiding the use of provocative statements but were equally concerned with not being too protective by hiding away sensitive but essentially important observations.

5.10 Integrate in Strategic Planning Process for business units

This task was given to two functions within each business unit:
1. The Director
2. The Quality Manager

The Director's task was to use the benchmarking report actively in all relevant stages of the unit's strategic planning process. For this to be effective it was necessary to debate priorities, opportunities, capital restraints, economic opportunities, competitor issues, and so on.

The task of the Quality Manager in each of the business units was to keep the process alive in spite of "other short-term, or even long-term urgencies". This was not an easy task. The electronics business does not exactly enjoy the best profit margins these days and time is not only quality as mentioned earlier, but also money.

This challenge for the Quality Manager or facilitator was probably the most demanding one in the whole process. It called for perseverence and determination sometimes in the face of management signals of quite another kind as immediate pressures were brought to bear either from owners, specific customers, dramatic market fluctuations or our own personnel.

The Quality Manager's task of keeping the process alive will continue to be challenging and increasingly call upon his personal and social attributes, his ability to command authority and respect in additon to possessing the technical insight also necessary for the quality assurance aspect of his job.

5.11 Review progress

Finally, it is the task of all management teams to review regularly:
- plans made
- improvement measures identified and accepted
- results obtained from use of recommendations
- results not obtained through lack of use of recommendations
- the appropriateness of the actual assessment model used
- the process of assessment itself

6.0 CONCLUSION

Most examples of benchmarking in industry are based on comparison of quantative data. The data is related to measurements of cost per unit, level of customer satisfaction, return on investments, quality performance, delivery times, time to market, time to customer, etc. Experience, has however shown that concentrating exclusively on QUANTITATIVE analyses, often leads to an incomplete analysis that is worth very little.

Quantitative analyses explain WHY the company has chosen to use specific measurement parameters. They explain differences in performance/quality/customer satisfaction/profitability, etc. between company A and company B. This difference does need quantifying and expressing in a way that shows the effect on the different operations involved. However, as a rule, QUALITATIVE analyses ought to come BEFORE QUANTITATIVE analyses because the first is a result of the second, and not the reverse. Hence Tandberg Data's choice of a QUALITATIVE benchmarking of its business situation against the EFQM Model for Asessment. This qualitative appraisal is however based in parts on quantitative comparisons administered by TOPP through their relations with other companies in Norway.

It is Tandberg Data's conclusion that Benchmarking activities ought to focus on first obtaining a clear understanding of the working methodologies used in specific areas targeted for improvement, before diving into the intersting and indeed important measuring of results.

May I conclude by recommending that you take a closer look at the European Foundation for Quality Management's (EFQM) activities, or the similar assessment activities of your own national quality organizations, most of whom have adopted similar models.

May I also take the liberty of recommending that you examine Tandberg Data's products and services - the final test of our company-wide and global quality effort to serve the market and our partners with their own stated and implicit requirements.

16

Benchmarking at Bang & Olufsen A/S

J. Bräuner

1. Bang & Olufsen A/S

Bang & Olufsen is a medium-size/large company in Denmark. Bang & Olufsen primary business area is development, production and sales of audio/video products. We sell our products world-wide, however besides our home market the rest of Europe - especially Germany, France and England - is our most important market area.

The company has been through big changes during recent years demanding big efficiencies rises all over.

2. Why Benchmarking

At the end of 1992 Bang & Olufsen decided on a productivity project, and at the same time it was decided that Bang & Olufsen should use benchmarking as a method of obtaining improvements jumps all over Bang & Olufsen.

Benchmarking as method was chosen as the method fits very well into the changes that the company is undertaking at the moment. The productivity project causes changes, which result in a much lower organization structure. This means that the individual manager obtains a larger 'span of control', and some of the manager levels are removed, which causes that the lower organization levels are raised. Besides this selfcontrolling production groups are introduced. These groups consist of employees who themselves plan their working day and share out the work. In this respect benchmarking is in a strong position as it is the process owners who take part in the benchmarking project, and it is the process owners who themselves must go out and experience how companies, which carry out performances in the 'Best Practice' class, operate and how they have obtained these performances.

3. How is benchmarking introduced at Bang & Olufsen

In the beginning of the autumn 1993 it was decided to establish a job (full-time) as benchmarking project manager at Bang & Olufsen. The purpose was to develop the Bang & Olufsen concept for benchmarking and to describe and introduce benchmarking as method all over at Bang & Olufsen. This includes preparing of descriptions and instructions which make it possible for other employees in the company to use benchmarking as a method and of course as a method adapted to the Bang & Olufsen culture and requirements.

In order of be able to develop the method into a Bang & Olufsen method it was decided to use a pilot project concerning benchmarking in order to obtain the necessary benchmarking experience.

4.　Choise of benchmarking model:

As mentioned in various benchmarking literature a benchmarking process course may consist of several steps. We have chosen to divide our first project into 7 phases or steps as follows:

1　Planning. (What should be benchmarked, how, choise of model, coarse planning, team-building)

2　Internal analysis of own process

3　Searching and choise of benchmarking partners

4　Selecting of data from the partners

5　Analysis and adaption of selected results

6　Implementation

7　Follow-up on introduced changes.

5.　Choice of a pilot project

The choice of a pilot project was dependent on more factors. The pilot project should be very limited, the goals of the project should be realistic and necessary for Bang & Olufsen as it would facilitate the carrying through of the project, and automatically it would create more motivation in the concerned area. Besides this the area in question should - if possible - be open to changes. This would ease the starting up of the project.

On basis of the above the Bang & Olufsen Automatical Component Assembling Department (ACAD) was chosen. The goal of the project in the area is a higher exploitation of the assembly of machinery and by this a higher capacity. This means that the focus area of the pilot project is the exploitation degree and as a result of this focus on the loss factors in the area (readjustment time, repair, maintenance, wait between part processes, etc.). One of the main requirements on the ACAD area besides the current requirements concerning quality and costs is the requirement on high flexibility causing smal-scale series.
The ACAD employes approx. 50 - 80 persons dependent on the season and consists of 5-6 partly separated process areas.

5.1 Choise of pilot project group

The signer is project manager of the project group with a co-operation project manager who organizing belongs to the ACAD area. The purpose of a co-operation project manager is partly to give the signer a sparing partner - a person who can review methods, and who in practice selects the members of the project group.

It is very important that the members of the project group are process owners or representative for these - this is of essential importance to the success of the benchmarking project.

In connection with the selection of the participants in the project group the first problems arose. The persons that we wanted to take part in the project were the persons who already had a huge working pressure on them, and therefore they were not able to spare time for participation in a benchmarking project in which the biggest problem was release of machine operators. After some internal discussions in the area internal resources were removed from another area to the ACAD area. This meant that sufficient resources could be released for the benchmarking project. It also meant that the benchmarking project received very good support from the top management. It is a very good motivation for the project members that they can see that the top management also finds the project important.

In order to get the processes of the whole area represented the project group has become rather big. The reason of this is that we want to have all the process owners directly represented in the project group.

In connection with the starting up of the project the project group has taken part in the selection of a working structure. The project group has been divided into two - a SMD group and a group for the leaded assembly, which then at suitable intervals coordinate their points of views. The division is very clearly in the first phases, and at the end in the implementation phase. The division has been chosen in order to obtain a higher process time. Crosswise of the two groups a smaller group exists, which takes part in the work of both 'main groups'. This also applies to the project management. The structure is shown in fig. 1.

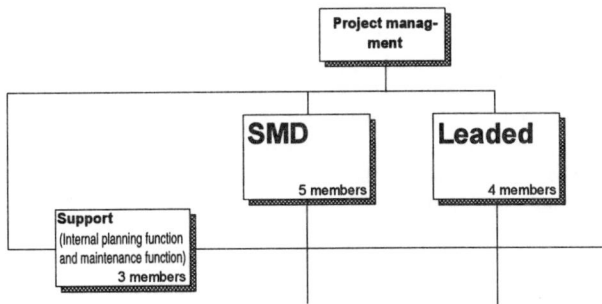

Fig. 1

6. The Benchmarking Course

Below the rough plan of the project including the planned terms and real terms is shown.

Phase:	Planned (end time)	Real (end time)
1: Planning phase with team-building	02/94	03/94
2: Internal analysis of own process	08/94	11/94
3: Searching and choice of partners	14/94	22/94
4: Selecting of data from the partners	22/94	34/94
5: Analysis and adaptation of selected resultss	27/94	37/94
6: Implementation	35/94	41/94
7: Follow-up	approx. 1/2 year	approx. 1/2 year

At present in the middle of August 1994 we are in the end of phase 4, "Collecting of data". Compared to the original rough plan we are therefore delayed some weeks due to various courses.

7. Experiences achieved via the pilot project

In the following I shall touch on some of the experiences that we/I have had until now in connection with our pilot project:

7.1 The project group

The composition of the participants of the project group was made from the point of view that all process owners should be represented if possible. However, this meant that we got a fairly big group. Due to the size of the group we thought it expedient to divide the work into two groups, who each of them dealt with two partly separated main process areas. As it is shown in fig. 1 some of the participants participated in both sub-project groups - e.g. the signer of course. The experiences from this sub-division and size is today of a doubtful character, but as in the main the following can be mentioned:

- Diffecult to coordinate efforts between the sub-groups
- Big strain on some of the participants - the participants who were in both groups started (more or less) in one of the sub-groups then in the other, which caused lack of teamwork - the group as a total became big and very difficult to work with.
- Many sub-process owners were participants in the project group.

Today I would choose a smaller firm group of max. 6-7 people and after than maybe include other ad hoc in the analysis phase.

On basis of this we decided to reduce the group from the 12 people to 8 people in total having in mind that those remaining should work a little harder. This group has now be working since the beginning of May, and I find it a better and more dynamic group.

7.2 Time schedule

As earlier mentioned the time schedule was only a rough outline, and I had no experience about how long time things take, and maybe therefore we are also delayed compaired to the original plan. A delay, which I do not find dramatical, remember it is the first time we are running a benchmarking project - and that we find that quality in benchmarking should have a higher priority than keeping the time schedule. However, there is one thing you should be aware of as project manager of benchmarking project - things which apply to all kinds of project work in which the participants of the project group are not full-time members - and it is very unusual that the members are full-time members especially at Bang & Olufsen. Often it will be the same persons, who take part in various projects/tasks at the same time - as it always will be the best employees who are chosen for these kind of tasks, and thereby it will often be the same employees.

7.3 Contact to external companies:

Because this was our first benchmarking project I decided that we not necessarily would look for "Best Practice" companies, but only for companies from which we could learn and from companies which were rather easily accessible - that is Danish companies. The final choice became two Danish and one Norwegian company.

Our way of choosing the companies was made by means of a gross list including potential partners. These were chosen from our internal network, suppliers, organizations and literature references. Among the listed potential companies (>50) approx. 15 were chosen. At these companies we would seak extra information, primarily by means of a questionnaire sent out to them.

To make a questionnare which gives you information about "How good is company X at...." is very difficult, especially when it is the first time. It took much longer than anticipated. We used at least 5 working weeks for making our questionnaire, a questionnare which is not at all the best questionnaire, however, it is a start, and I think it is a questionnaire which may form the basis of making questionnaires for the benchmarking projects at Bang & Olufsen. Together with the questionnaire I included a letter which told about benchmarking, the purpose of the application, and with information about us.

7.4 Benefits until now

What have we achieved until now:

- Establishing of a mutual measuring system and establishing of a number of measurings of the coefficient of utilization, on which we can make a follow-up.
- A number of "Entitlements", which we - compaired with results of the data collection at our partners - should have implemented.
- That the various levels/areas in the project work speak together much more than before (activation of employees).
- An almost finished method which, however, should be reviewed one more time.

7.5 What next

As mentioned earlier I am responsible for disseminating the method benchmarking all over at Bang & Olufsen. This is done by starting up projects in the individual areas of the company. When a project is started up a "cooperation project manager" is chosen as in the pilot project case. This person belongs to the area in which benchmarking should take place. This means that at least one person in the area after this should be able to carry out benchmarking projects. By choosing this method knowledge about and owner-ship of the method is obtained in the individual areas of the company. The owner-ship implies of course that the first project in the area is a success. Furthermore information to the whole organization must take place constantly concerning benchmarking and how the individual projects go and which results are achieved. Parallel with the starting up of the project a "benchmarking knowledge center" is built up consisting of the signer. This knowledge center should take care of:

- Building up of a data base with data about companies, which have processes of "Best practice" level.
- Collecting of data to the data base.
- Maintenance of the method.
- Follow-up and registration of results achieved at Bang & Olufsen.
- Take care of forwarding benchmarking newsletters in the company.
- Facilitator of new projects and project groups.
- Consultant for other project groups.

17

Strategic Performance of the Production System in a Machine-Tool Building Company

J.M. Goenaga[a] and C. Goikoetxea[a]
[a]Design and Management of Production Systems Department, Ikerlan S. Coop.,
2 J.M. Arizmendiarrieta, 20500 Mondragon, Spain

This position paper presents part of the work carried out in an application case-study[*]. The global project falls within the framework of what is known as Business Processes Redesign. To be exact, a strategic analysis of the production system of a machine tool building company has been performed.

1. THE COMPETITIVE THREESOME

The starting point of the work was *the competitive threesome* composed of the Company's Strategy, The Market and the Production System. The Corporate Strategy represents how the company wants to be the best in. The Production System represents how the strategy has been implemented. The third component, Market-Customer, represents how the customer perceives the company's way of being the best.

Competitive threesome

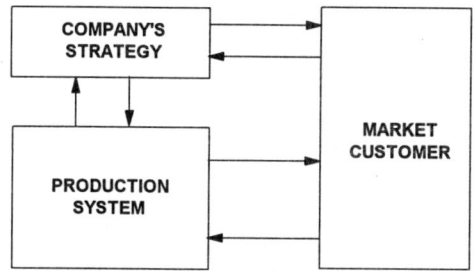

Figure 1. The competitive threesome

[*] We would like to thank Danobat S.Coop. for its active participation as test bed and Profs. Josep Riverola and Beatriz Muñoz-Seca at I.E.S.E. (Madrid) for its collaboration in the project. We also give gratitude to the Unidad de Estrategia Tecnológica (S.P.R.I. - Basque Government) for its financial support of the project.

The three components are related to each other by how the customer perceives the product-service offered by the company. The General Management of the company must define a long term strategy and objectives that have first, to be understood and assumed by the Production System in order to ensure that the product-service gets from the customer the perception it was expecting.

2. DEFINING THE STRATEGY

This customer perception game allows the company to select the desired way of competing. The game starts from the definition of the long term strategy of the company. The strategy must be defined by using production system related terms.

Several generic ways of competition such as cost, flexibility, innovation... have been defined by different authors. But their generic being makes it difficult to match them to the specific nature of each company because:

- A company usually competes in different market segments.

- Each market segment imposes its specific benchmark.

- These benchmarks may not be synergetic.

For this reasons the approach taken by the project has been to define objective criteria that allow the definition of what we call the *competitiveness profile* of the company. In order to define the competitive criteria the following must be considered:

*   The long term strategy and objectives of the company.
*   The parameters by which the customer perceives how the company wants to be the best.

2.1. The Five Criteria
The criteria defined in the project are:

CONSISTENCY: Reflects the degree of compliance between what was agreed about the production of each order and what has really happened. Two dimensions have been selected to be measured: Price/Cost (variability between the estimated and real price/cost) and Time (variability between the estimated and real time consumed from the approval of the order and the delivery of the machine).

TECHNICAL CAPACITY - MARKET SCOPE: This criteria aims to measure the wideness of the company's offer range. Two aspects defining and measuring this criteria in this particular application have been: the percentage of the objective market that the company's offer copes with and the level of parameterisation of machine component parts.

INNOVATION: The aim is to measure the implementation of new technologies (non-existent technologies in any machine tool built by the company before). The criteria measures the number of innovations introduced per year.

PRICE - COST: Cost measures cannot be compared with competitors' figures, so the approach has been to measure actual cost performance and to establish cost reduction objectives.

DELIVERY TIME: The time needed to produce a machine tool. The time between the order arrival and machine delivery is measured.

2.2. Stablishing the competitive profiles

To this extent, we have defined the competitiveness criteria. Now, it is time to determine the competitive profiles of the company. To do this task, a series of interviews with company's staff and customers were carried out. As a result, a range of values for each criteria was defined, with the highest level being that of the leading competitor while the lowest level is the minimum acceptable level (or viceversa if the low level is the optimum). So, it is a relative evaluation approach.

COMPETITIVE PROFILES

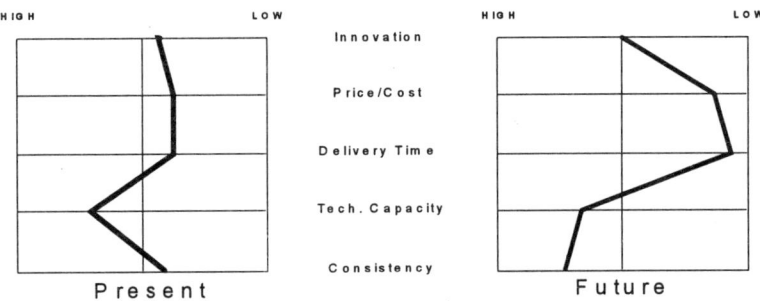

Figure 2. Competitive profiles

The difference between the shapes of both the present and future profiles indicates the direction of improvements, and the criteria identify the nature of the actions to be carried out. The next step is to identify the critical activities of the business and to determine the specific actions to be carried out in order to obtain the desired competitive profile, but this is a very detailed task, the explanation of which this position paper does not aim to go into.

3. MONITORING IMPROVEMENTS

Once the critical activities have been identified and improvement actions defined and implemented, monitoring of improvements is required. It is vital to check how our actions contribute to improve our competitiveness. This monitoring task requires the definition of real parameters related to the production system that show the system performance in accordance with strategic goals. The relationship between the parameters and the strategy is built by means of the competitive criteria. Some of the parameters defined in this application are shown in table 1.

PARAMETER	DESCRIPTION
P1	Percentage of parts in the objective market we can offer
P2	Number of parametric programs in operation
P3	Number of design hours worked in each order
P4	Mean deviation of parameter P3 by each machine family
P5	Mean of P3 by each machine family
P6	Design costs per month and order
P7	Purchasing costs per month and order
P8	Final Assembly costs per month and order
P9	Machine setup costs per month and order
P10	Machining tests costs per month and order
P11	Total cost = P6+P7+P8+P9+P10
P12	Mean deviation of the difference between real and estimated cost by each machine family
P13	Mean deviation of machines costs by each machine family
P14	Mean machine cost by each family
P15	Time worked per phase and order
P16	Date of first jobs per phase and order
P17	Date of last jobs per phase and order
P18	P15/(P17-P16)
P19	Mean deviation of P15 per phase and machine family
P20	Mean of P15 per phase and machine family
P21	Delivery date - order arrival date
P22	Delivery date - order launch date
P23	Mean deviation of P21 per machine family
P24	Mean of P21 per machine family
P25	Mean deviation of P22 per machine family
P26	Mean of P22 per machine family

Table 1. Monitoring parameters

4. CONCLUSIONS

To conclude let's point out the main aspects to be taken into account:

1. A globally balanced multicriteria approach has been applied. A company may compete in several market segments and each one influences the criteria with different weights. A balanced profile must be stablished.

2. The strategy as well as the customer perception must be stated in terms of production system parameters or variables.

3. We must fit the customer perception to our strategy, we must produce goods or services that are perceived by our customers as we want them to be. We therefore need to match the improvement actions to the differences between the market perception and our strategy.

4. These parameters must be easily measured to allow monitoring

18

Increasing Dynamic Adaptability through the Application of Benchmarking in Operational Navigation

Dipl.-Ing. Mathias Kirchhoff, Dipl.-Ing. Markus Weber and Bettina Layer (cand. rer. pol.)

Fraunhofer Institute for Manufacturing Engineering and Automation (IPA), Silberburgstr. 119 a, D-70176 Stuttgart, Germany.

1. CHALLENGES TO MANUFACTURING PLANTS

The world economy is in a state of radical change. Changes due to increasing globalization of acquisition and selling markets are having sweeping effects on the companies. Through increasing international competition the pressure on manufacturing plants is growing. The case of the German machine-tool industry shows that presence on international dynamically growing markets is necessary for survival if the national market is severely declining and if you do not wish to leave the field to international competitors.

Companies which are set up according to conventional principles of Taylorism, with sharply defined work content and correspondingly obsolete information flow policies, are no longer in a position to come up with the necessary inner dynamism and evolutionary speed. As a reaction to fundamental changes in the environment, however, strong self-dynamism and the ability for further development (evolution) are being required from the companies. Responsive company structures which assure dynamic adaptability must be developed.

Progress in information and communication technology has allowed almost any amount of information about markets and competitors to become available. This information is meant to be used and to be compared with a company's own performance parameters in order to achieve an increased dynamic adaptability by means of benchmarking.

2. DYNAMIC ADAPTABILITY

Growth processes of companies are essentially sustained by innovations in products, processes and structures. They are among the central success elements in the marketplace and in competition. "It are not always large things, however, which disturb the work process, but hundreds of little ones."[1] Innovations in successful companies, which as a rule occur intermittently and by leaps and bounds, are complemented by a variety of smaller improvements (continuous improvement process) introduced deliberately. In order to achieve longterm success, companies must possess the ability for dynamic adaptation, both by leaps and bounds through innovations as through continuous improvement processes (see figure 1).

A fundamental change how to view things is required from the companies. Considerations must turn from static to dynamic systems, from mechanistic to organic representation models. A manufacturing company must learn to understand its processes and structures as a system in its entirety. It does not develop in a linear and accurately predictable way, and its boundaries - seen from within and from without - are fuzzy and permeable. This in mind, the model of the "fractal factory" is being presented.

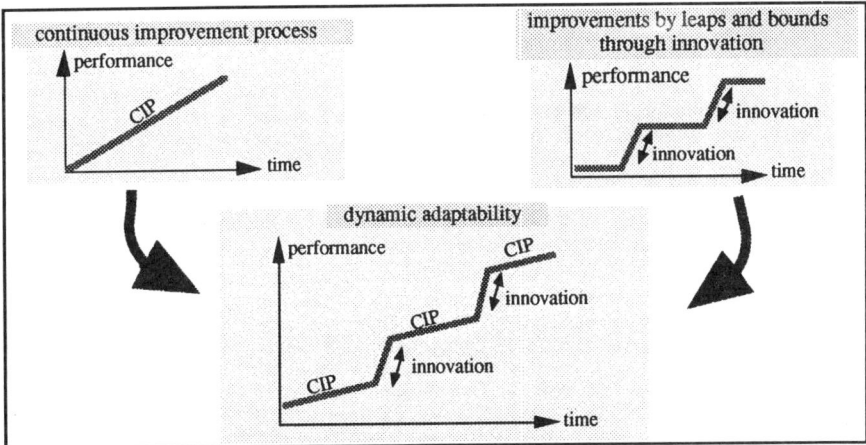

Figure 1:   Dynamic adaptability consists of improvements by leaps and bounds through innovation and continuous improvement processes.

3.   FRACTAL FACTORY AND OPERATIONAL NAVIGATION

The fractal factory gives a wholistic approach for solutions in order to initiate, strengthen and guide the self-dynamism of partially autonomous structures which are called fractals. A fractal is an independently acting unit within the company, with goals and performance to be clearly described. The description of fractal structures can be broken down into six horizontal levels: on cultural, strategic, socio-psychological, economic-financial, information and process level[2].

	vitality features:
cultural level:	• company culture • value system, guiding principles • "mission" of the company
strategic level:	• company strategy • value system • company structure • products
socio-psychological level:	• staff relation structure • awareness, ability and behaviour of the staff • structure organization in teams and work groups • information and communication among staff
economic-financial level:	• economy • profitability • resource consumption • value creation
information level:	• information flow • information systems
process level:	• processes • procedures/ methods/ instruments • technical systems

Figure 2:   Vital companies distinguish themselves through dynamic adaptability on 6 horizontally arranged levels.

Fractals can be further characterized with the following features:
- Fractals are self-similar with regard to the six levels.
- Fractals practise self-organization.
- The goal system emerging from the goals of the fractals is non-contradictory and must serve the attainment of company goals.
- Fractals are connected to an efficient information and communication system. They themselves determine type and extent of their access to data.
- The performance of the fractals are constantly measured and evaluated.

Thus fractals become the central formative elements in a company. Crucial factor is the increased transfer of management and control functions to the employees. This applies to employees on the strategic as well as on the operational level equally.

In order to guide the "movements" of the fractals in the respective appropriate direction of the entire process, a function is required which was given the nautical metaphor of navigation. Fractals navigate in a constant review of their position in the target zone and make reports and possible corrections. Applied to an industrial company, these functions have up to now been connected with the term "controlling".[3] Operational navigation could be characterized as decentralized control for partially autonomous work groups. Differences in concept and content, however, make a differentiation from the current term of 'controlling' necessary. Operational navigation aims at improving initiation, control and coordination of the continuous improvement process as entirety to improve the outcome, particularly with regard to the lack of orientation in the market (see figure 2). The emloyees within partially autonomous units of an organization are those agents of the continuous improvement process.

The improvement process runs continuously and responsively when high intra-departmental dynamism exists. Small and quick feedback loops are prerequisites for a high intra-departmental dynamism within the company. They allow prompt indication of a changed actual situation and the recognition of deviations between the desired goal and the actual position at the moment.

Derived from these objectives are the following essential features of the operational navigation:

- Employee orientation and participation: all employees work together in the team.
- Goal orientation: all team members work towards common goals.
- Reporting results: The achieved work results are made known to everyone in the team.
- Decentralism: Within the given goals, the team determine their respective use of resources independently.
- Process orientation through customer-supplier relations: The team regards the following work groups as internal customers.

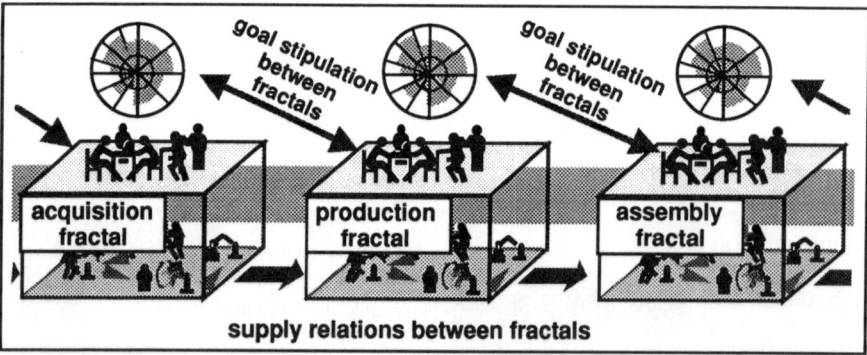

Figure 3:   Operational Navigation: process orientation through customer/ supplier relations

In detail this means the company goals must become transparent and be defined on all company levels. A constant feedback regarding the degree of goal attainment, especially in the value creation process, must guarantee - as up-to-date and process related as possible - a direct investigation of the causes of possible deviations.

Coordination and control of decentralized, partially autonomous units of an organization take place in a 5-stage-navigation process:

1) define and provide goals
2) report actual data
3) analyze deviations and make evaluations
4) formulate plans of action and alternative procedures
5) carry out measures

Special significance should be attached to the setting and definition of goals in operational navigation. It can only provide satisfactory results if proper goals are set. In many companies however, customer requirements and competition factors are not systematically analyzed. Goals are predominantly set by the company itself, according to the subjective estimation of its management.

However, in order to arrive at correct goals it is necessary for an enterprise to direct its view outwards. To analyze the performance parameters within one's own company systematically and to compare them with others known as 'best in class' became famous in the USA under the term of "benchmarking".[4]

4.   OPERATIONAL NAVIGATION AND BENCHMARKING

Benchmarking is a process of continously comparing and measuring an organization with other leaders anywhere in the world to gain information which help to take action for the improvement of its performance.

Operational navigation controls and coordinates processes. In order to establish a relation to benchmarking, emphasis should here be placed on process benchmarking. This is to benchmark discrete processes against organizations with performance leadership in those processes. It involves seeking the best procedure to conduct a particular business process.[5] Process benchmarking is a form of analysis distinct from other types of benchmarking, e.g. strategic, performance and competitive.

Here a representative variety of known benchmarking processes[6] is presented in seven steps:

1) Establishment of the object of benchmarking.
2) Judging own performance.
3) Selection of benchmarking partners.
4) Analysis and comparison of data.
5) Derivation of goals and measures.
6) Implementation.
7) Control and repetition of benchmarking.

Benchmarking and operational navigation are two parallel processes which correlate closely through a constant exchange of information, where the benchmarking process is meant to run on the strategic and the navigation process on the operational level.

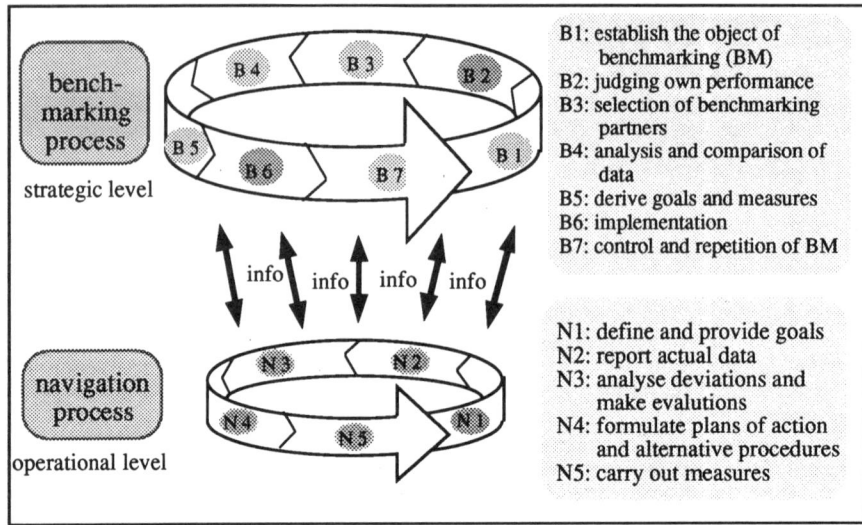

Figure 4:   Constant exchange of information allows benchmarking and navigation processes
to closely correlate.

Both processes are to operate continuously and information is to be exchanged after each step in order to achieve a permanent feedback effect. A loop cycle should not be expected from the benchmarking process in order to transmit extensive data to navigation. Instead continuous transmission of benchmarking information should take place. Then even fragmentary and superficial pieces of information can be useful for the navigation process. Reasons for a continuous exchange of information are:

• Employees are responsible for the transformation of the goal standards in operational navigation. Each piece of information passed on from the benchmarking process (e.g. by the selection of competitors) leads to its being tied into the entire process. Along with it comes a considerable motivational effect. In addition, the shock effect for the employees, caused by a confrontation with the often hard realities of external comparison, fails to appear.

• Significant indications for benchmarking, for example regarding critical processes, could be drawn from the navigation process. This can be dynamically adapted to the situation through constant feedback from the benchmarking process.

The linking with the navigation process is presented using the example of the benchmarking steps 2 (judging own performance) and 6 (implementation).

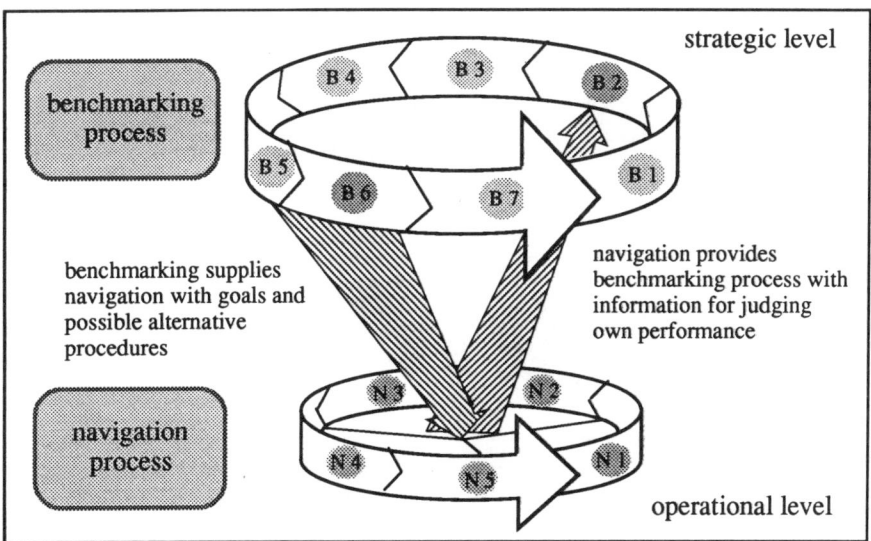

Figure 5:   Examples of relations between benchmarking and navigation

Connecting Point 1: Judging Own Performance

In order to carry out an internal or external comparison, the own performance must first be measured and analyzed. Many companies still lack the provision of such data and indicators about their own efficiency. Operational navigation comes into effect here. Navigation provides a variety of information about operational processes. On the one hand, this information can give impulses to a comparison using benchmarking. On the other, these data provide the starting base for a benchmarking process which is to be carried out.

Connecting Point 2: Implementation

In running through the benchmarking stages, goals and plans are formulated in step 5 and must be put into action in step 6. So far the implementation of goals and plans did not receive the necessary attention in discussions. It is the application of measures which seems to cause the greatest difficulties for industrial companies. Benchmarking provides strategic goals which lead to a continuous improvement of partial processes in the value creation chain.

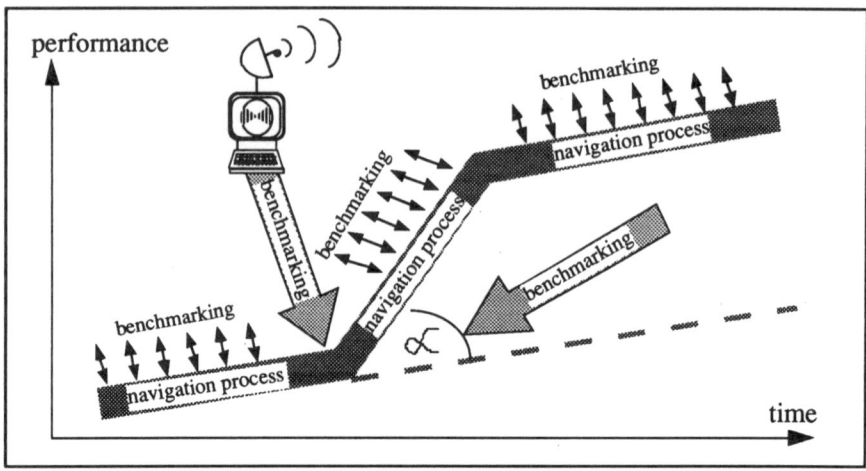

Figure 6:   Benchmarking: the strategic radar in the company.

Both processes therefore have a fruitful relationship with each other. With navigation, employees in the value creation process get a method which allows them to carry out an intended, continuous improvement. Benchmarking has three functions here:

- Benchmarking is a continuous motivating force for the employees through its constant provision of information.
- Benchmarking sets free creative potentials through external comparison and thereby provides important impulses for innovative leaps.
- Benchmarking describes the extent and speed of necessary performance increase.

To sum up, benchmarking can be considered as a strategic radar, as it sets the goal direction for dynamic adaptability. Operational navigation eventually leads the company to improved efficiency through the operational transformation of strategic goals (see figure 6).

5.   CASE STUDY

The connection of benchmarking with operational navigation was realized as part of an industrial project at the Fraunhofer Institute for Manufacturing Engineering and Automation (IPA) in Stuttgart. It was applied at a leading manufacturing concern in the building supply industry in southern Germany. The business employs 70 workers in the production of house doors. The production program can be divided into three varieties of products. The organizational structure in production includes six work groups which carry out the following work steps: lathe processing, preliminary assembly, final assembly, final inspection and delivery.

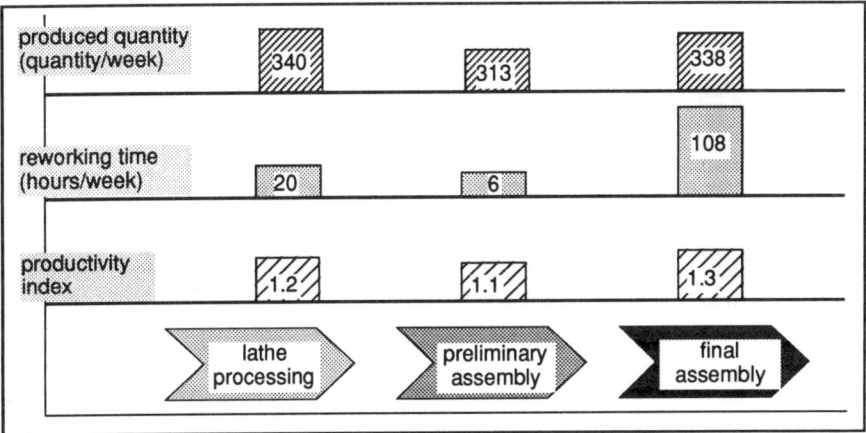

Figure 7:   Comparison of production work groups within the company.

The project revealed rationalization potentials in technology and in organizational structures of production, little self-dynamism and weaknesses in the ability of dynamic adaptability. In order to break the rigid organization and process structures, a navigation process was gradually introduced, thereby creating suitable conditions for benchmarking.

First the composition of the work groups, tasks to be carried out, methods of measuring performance, and goals were defined. The efficiency of the respective group was measured by means of structured indicators, processed and visually available on a display board. Each employee received reports on the achievements.

In the next step a continuous improvement team was formed to increase the dynamic adaptability. Current performance indicators, problems in attaining performance goals and approaches to improvements were discussed in teams at regular intervals. The team was put together from employees of all production work groups and IPA-consultants. Representatives of the work groups had the task of passing the findings discussed by the team on to all group members, to apply them and at the same time to bring improvement suggestions from the work groups to the team. This procedure included on the one hand as many employees as possible in the dynamic adaptation process, and on the other increased the in-plant dynamics through small, responsive teams.

An internal comparison of work groups was carried out as first step towards benchmarking (see figure 7). A simple efficiency comparison was possible because the same indicators for all work groups - such as manufactured quantity, reworking times, quality, production costs and productivity - were defined. The findings of the comparison were discussed by both the work groups as well as by the continuous improvement team and numerous causes for the differences were immediately determined. It was possible to achieve an efficiency increase in the work groups through the exchange of internal "best practices" and common problem solving.

A comparison with national indicators within the branch showed that the company fared well against the industry average. Compared with international benchmarking partners, however, there was a definite performance gap.

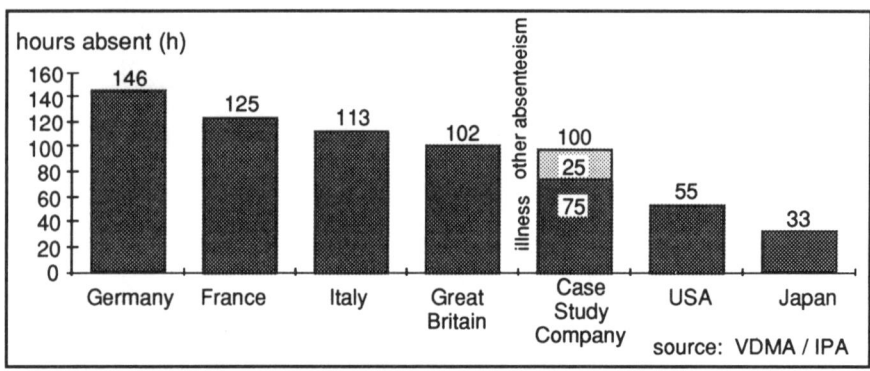

Figure 8:   International comparison of absentee rates (illness, accident, etc.) per employee per
             year in hours.

The comparison was carried out with indicators from operational navigation and data from
industry associations. This procedure represents an important step in the direction of more
comprehensive and detailed benchmarking.

The practice confirms the advantages of connecting operational navigation and benchmarking.
The project was able to establish awareness of the actual situation on all company levels. It
increased open-mindedness for changes and accelerated targeted improvements.

6.   BENEFITS AND OUTLOOK

The benefits of utilizing benchmarking in operational navigation can be divided into quantitative
and non-quantitative components. In the application example of a medium-sized building
supplier the following measurable gains in form of increase rates can be expected:

time reduction   > 40%
product cost improvement   > 25%
quality improvement   + 20%

The non-quantifiable benefits are reflected in a higher motivation of employees. They promote
creative ability as well as capacity for analysis, judgment and communication. Because they
change from people concerned into active participants they become more committed to the
company. This may result in a reduction of absenteeism and an increase in the amount of
improvement suggestions.

In order to operate profitably more knowledge must flow into theory and practice and be
successfully applied in the company. Benchmarking, the "strategic radar", which confronts the
employees with outside realities, together with the operational navigation can initiate and guide
this process.

As conclusion a piece of advice: "Begin benchmarking immediately at a vital, solvable problem
with courage, a small team and the active support of the employees."[7]

References

1 Warnecke, H.J. (1993): Revolution der Unternehmenskultur - Die Fraktale Fabrik, Berlin 1993.

2 Kühnle, H. / Spengler, G.(1993): Wege zur "Fraktalen Fabrik" in: io Management Zeitschrift 62 (1993) Nr.4, S. 66-71.

3 Horváth, P. (1991): Controlling, 4.Auflage, München 1991.

4 Camp, Robert C. (1989): Benchmarking: The Search for Industry Best Practices that Lead to Superior Performance, Milwaukee 1989.

5 Watson, G. (1992): The Benchmarking Workbook, Cambridge (Mass.) 1992.

6 Compare the 10 step process in Camp, Robert C. (1989)

7 Burckhardt, W. (1992):Benchmarking, VDI Berichte, (1992) Nr. 1014, S.89-116.

19

Benchmarking of Bid Preparation for Capital Goods

B. Hirsch, M. Krömker, K.-D. Thoben, A. Wickner

BIBA (Bremen Institute of Industrial Technology and Applied Work Science)
Hochschulring 20, D-28359 Bremen, E-mail: kro@biba.uni-bremen,de

Abstract. This paper describes an application example (case study) for the use of *benchmarking* strategies. The ESPRIT project 7131 focusses on the overall improvement of the bid preparation process. Three multinational companies representing industries with products engineered-to-order benchmarked their bid preparation process against each other.

The method used for enterprise modelling was IDEF0, a technique based on the SADT approach (hierarchical decomposing of processes). Applying this technology, the "as-is" situation was captured and analyzed. Both bottlenecks as well as best practices were determined. A generally applicable "to-be" situation was defined and implemented.

Keywords: Benchmarking, Bid Preparation, Enterprise Modelling, IDEF0, Software Design.

1. INTRODUCTION

Every day companies producing *capital goods*[1] face a serious dilemma — a potentially interesting request is presented, but to be successful a convincing, attractive, and reliable *bid*[2] must be prepared — and this causes costs in terms of time as well as other resources. The task may become enormously complex when all EC member countries are fused into one "home market" and — in a broader perspective — when international competition turns into a global encounter. Especially small and medium-sized companies will face difficulties in their attempts to exploit the market potential. Companies must undergo tremendous change if they are to survive, grow, and prosper in this environment.

Figure 1 gives an overview of the bid-related activities. Beginning with the reception of an inquiry, a technical concept meeting customer demands must be developed. Next, the costs for realisation have to be calculated and a realistic delivery date has to be estimated. Commercial conditions including the salesprice, transport, warranty, etc. have to be fixed. Finally, all relevant information is compiled into a document - the bid - which is submitted to

---

[1] Typical examples for *capital goods* are machines, robots or assembly lines - goods, companies have to invest in in order to facilitate production.

[2] Synonyms, and possibly more common expressions for the word *bid*, are offer, quotation or tender.

the customer. In case an order is placed (dotted line), the information generated has to be transferred to the relevant departments executing the order.

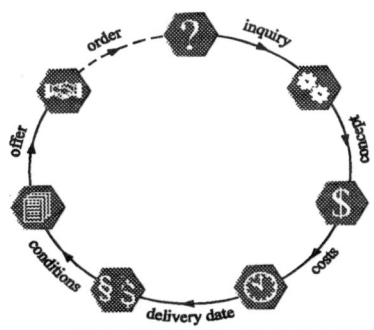

Fig. 1   The Bid Preparation Process

The situation in the capital goods industry can be described as a "competitive arena". Bids, based on poor or incomplete information, have to be prepared under high time pressure due to tight deadlines set by the potential customer. The concept developed within the bid often has to undergo "last minute modifications" - the customer changes his mind quickly. Generally, costs effected by bid preparation are not reimbursed directly but covered by the realized orders. Therefore, resources have to be minimized. The capital goods industry is characterized by strong competition. As a result, the winning chances are low (on an average 10% or less).

The Commission of the European Communities (CEC) is backing a project entitled "An Integrated System for Simultaneous Bid Preparation" (BIDPREP) which aims at developing a computerized system capable of supporting the bid preparation process by applying the concurrent engineering concept. The BIDPREP project runs from July 1992 to June 1995 and the human resource investment will total 23 man-years. The project is funded by the CEC under the ESPRIT programme (Project No. 7131) as well as by participating partners from Norway, Denmark, Italy and Germany.

One major focus of the BIDPREP project is the implementation and use of benchmarking strategies for developing a general applicable methodology for the bid preparation phase and creating new paths of supporting this process by software tools. In this context, benchmarking was understood as the process of identifying the best methods and practices and adopting or adapting the good features and implementing them to improve business processes.

2. THE INITIAL SITUATION

Within BIDPREP, three multinational companies, ABB, Krüger Engineering and Guehring Automation, formed a consortium together with three research institutes and a software developer in order to develop both an efficient methodology as well as a computer-based system supporting the preparation of bids.

Significant is the fact that - although the ABB subdivision in Strømmen, Norway, produces rolling stock for railways; Krüger Engineering, Copenhagen, is engaged in environmental protection; and Guehring Automation in Frohnstetten, Germany, manufactures grinding machines - the anticipated bottlenecks are similar and thereby independent of the product. The strategy for optimizing the process is to analyze the state-of-the-art proceeding, identifying bottlenecks as well as proven methods and thereby forming a generalized methodology which will be implemented in the final phase.

3. ANALYSIS OF THE "AS-IS" SITUATIONS

In order to compare the work flow of bid preparation at the users' sites, an appropriate modelling technique had to be agreed upon. The consortium decided on IDEF0, a technique that is based on the SADT approach. IDEF0 [2] allows the hierarchical decomposing of business processes down to the required level of detailing. It identifies activities, related *in-* and *outputs* as well as steering information, so-called *controls*. Moreover, *mechanisms* to support the activities can be identified. Figure 2 displays the graphical method.

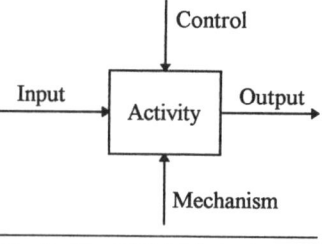

Fig. 2: The IDEF0 Method

As a first step, IDEF0 models for each company's bid preparation process were prepared. Figures 3 and 4 show how the preparation of bids is performed at Krüger Engineering and Guehring Automation. The analyses describe the "as-is" situations.

3.1. Bid Preparation at Krüger Engineering

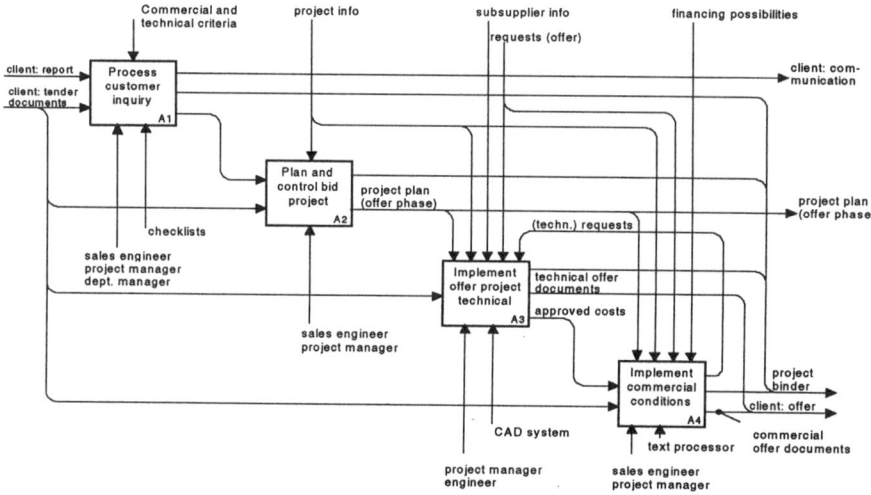

Fig. 3 Bid Preparation at Krüger Engineering

Krüger Engineering, a company within the Krüger Group, engages in activities of environmental protection, wastewater treatment, and energy supply in more than 30 countries worldwide. The bid preparation for a plant is handled as a project, managed by a project manager for the technical aspects and a sales engineer for the commercial aspects. The analysis revealed that the bid preparation is a very complex task involving various process, civil, mechanical, electrical and control engineers. Furthermore, an effective reuse of former information generated is very important. The reuse, however, is limited by the knowledge and experience of the individual employees. Currently, the process is supported by some PC-based applications (i.e. a text processor and a CAD system).

A major objective of Krüger Engineering is to increase the cooperation and teamwork through simultaneous and transparent access to the needed and relevant information during the preparation of an offer. Actually, no support for their subsidiaries is given. Best practices identified at Krüger's sites are the overall management of the bid preparation process including periodical meetings for all employees involved as well as well-defined checklists allowing risk analysis and an evaluation of all incoming inquiries.

3.2. Bid Preparation at Guehring Automation

Guehring Automation designs and produces high-speed grinding machine tools. The analysis (see Figure 4) shows that many decisions have to be discussed internally, justified, approved and communicated during the bid preparation phase. The number of employees involved depends on the complexity of the machine tool requested, i.e. whether an existing concept, a partly new concept, or a completely new concept has to be applied. The process is performed manually, supported by both an administrative and an MRP system.

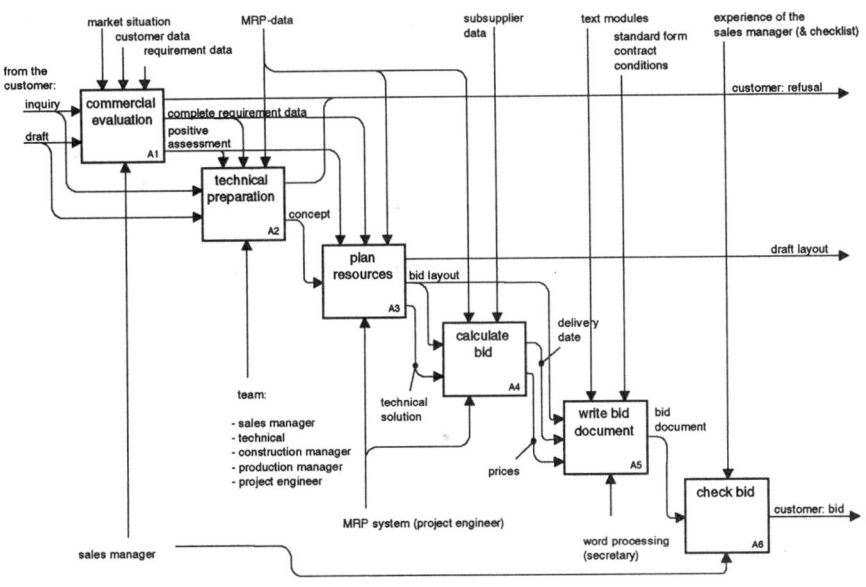

Fig. 4: Bid Preparation at Guehring Automation

A positive aspect of the bid preparation process at Guehring is the fact that consequences of a possible order can be simulated with the MRP system. A resource plan is prepared, for each bid. This minimizes the risk of not being able to keep delivery dates in case an order is placed. Currently, there is only poor support for the specification of the technical bid preparation and the document preparation. The bid layout is prepared manually and the text processor for the writing of the bid document is an isolated solution.

3.3. Bid Preparation at ABB Strømmen

ABB Strømmen produces railway cars and complete train sets for railway or tramway. The analysis revealed that bid preparation is a process involving many departments and resources. The types of bids vary from simple spare-part offers to complex bids for completely new design for train sets. Most of the bid preparation activities are based on manual routines, supported by a proprietary Design Support System, a production planning & control system, spreadsheets and text processors. Problems occurring have included structuring data for reuse in further bids, cost estimation, long turnaround times and communication with subsuppliers.

The strong point of the bid preparation process at ABB Strømmen is the product model concept which was developed to ease the technical bid specification. A Design Support System, with integration to an MRP system, was implemented using this model.

4. A REFERENCE MODEL FOR BID PREPARATION

Based on the analysis of the "as-is" situation, both best practices as well as additional requirements regarding the "to-be" situation of bid preparation at the industrial partner's sites were collected and classified. They were sorted into six groups:

- inquiry assessment,
- bid project management,
- product design,

- cost estimation,
- product scheduling and
- bid document compilation.

Each group can be seen as a core activity of the overall bid preparation phase; together they form the top level of a bid preparation reference model (Fig. 5). It describes a framework showing the roles, activities and interdependencies of the main characters during bid preparation within the context of the whole enterprise. It could probably be valid for most companies with complex products engineered- or made-to-order.

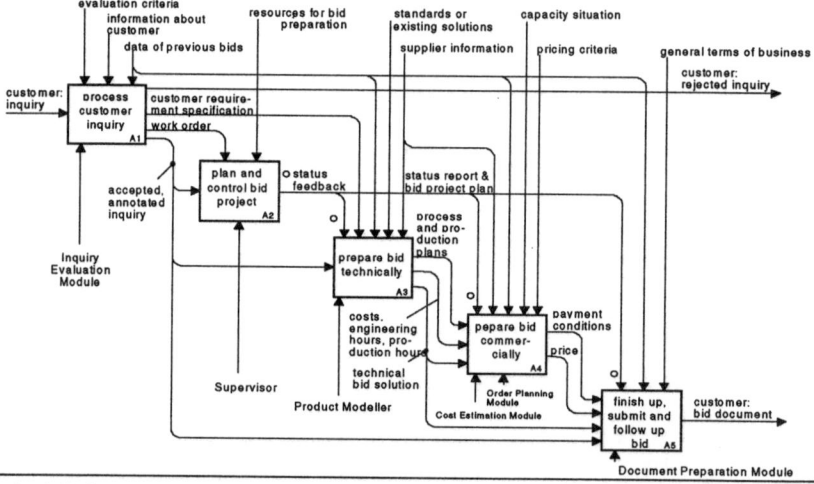

Fig. 5   A Reference Model for Bid Preparation

Within the BIDPREP project, the reference model serves as a basis for the design of an integrated bid preparation system. The system comprises dedicated modules for the individual pre-sales activities. The modules interact via an integration mechanism, the *communication bus* [3]. Figure 6 describes the overall system architecture.

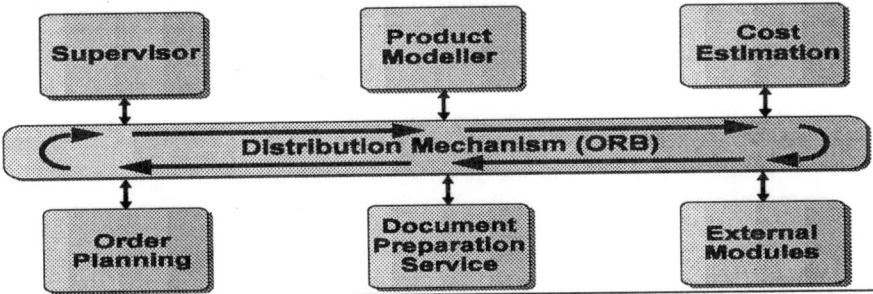

Fig. 6   The Overall Architecture of the BIDPREP System

5. CONCLUSION

In the BIDPREP project, benchmarking was applied in order to improve the bid preparation process by performing the following steps:

    1) Definition of the objectives (what should be improved),
    2) Selection of benchmarking partners (companies, research institutes),
    3) Forming of benchmarking teams (national subgroups),
    4) Capturing and analysis of current processes at the companies' sites,
    5) Definition and implementation of improvements.

This case example proves the theory that production processes of different companies can be compared with each other even if the product and/or the branch differs. Benchmarking strategies were used to analyze the execution of the core activities at the industrial partners sites. Both bottlenecks as well as "best practices" - like the product model concept from ABB, the management concept from Krüger - formed the basis for the specification of an overall bid preparation system capable of supporting all activities related to the preparation of bids.

The prototype system will by customized to the individual needs of the involved industrial partners. Expected results comprise improved chances of winning contracts, savings in time and money, wider support of bid-related activities, quicker average response times and more precise cost calculations and correct responses to inquiries.

REFERENCES

[1]   K.-D. Thoben, T. Kuhlmann, C. Lischke, R. Oehlmann: *"Concurrent Engineering in der Unikatfertigung"*, in *"CIM Management 2/1993"*.

[2]   G. J. Colquhoun, R. W. Baines *"A generic IDEF0 model of process planning"* in *"Int. J. Prod. Res., 1991, Vol. 29, No. 11"*.

[3]   Object Management Group: *"The Common Object Request Broker: Architecture and Specification"*, Object Management Group, San Diego, CA, USA, 1992.

20

Benchmarking in Software Development[*]

Nelly Maneva, Maya Daneva[a], Valia Petrova[b]

[a]Institute of Mathematics, Bulgarian Academy of Sciences
Acad. Bontchev str., bl.8, 1113 Sofia, Bulgaria
[b]Faculty of Mathematics and Informatics, University of Sofia
5, J.Baucher bul., 1126 Sofia, Bulgaria

1. INTRODUCTION

The Software Engineering solution of the so-called "software crisis" was proposed two decades ago. But there are still some alarming symptoms. A lot of software products are delivered with delay. They are of poor quality and with excessive costs (especially for maintenance). Though many methodological recommendations have been applied, our experience shows that pure theory itself is not enough for implementing the engineering approach to software development. So we propose a feasible approach based on common sense and pragmatism rather than on deep and complex theoretical results. In our opinion, people involved in software constructing are practically minded and such an utilitarian approach would be more comprehensible and useful.

2. A PROCEDURE FOR CONTINUOUS BENCHMARKING

The main goal of each software project is to ensure the efficient development of a high quality product. In order to achieve this goal we shall consider software project management as a sequence of crucial decision making which answers to the questions: What is the problem? What is the best alternative in given circumstances? What to do?

Usually decision making is done ad hoc and the motives of the resulting actions are vague. Next we are going to determine a unifying and systematic approach to supporting optimal decision making. This approach comprises two key ideas. First, the current objective must be posed (as Gilb write in [1]: "Actions without clear goal will not achieve their goals clearly"). Second, according to the objective stated, the kinds of objects and the set of their measurable attributes are chosen so as to benchmark them.

We suggest the following procedure for continuous improvement through benchmarking:

---

[*]This research is partially supported under the contract I-24 with the Bulgarian Ministry of Education and Science.

1. State the goal.
2. Define the relevant objects and their attributes.
3. Define the set of competitive objects and compare them.
4. Plan and accomplish the appropriate course of actions.
5. Measure results against goal.
6. Re-evaluate and continue.

The problems of applying this procedure to a software development and use are:

a) to submit a scheme of object description, a method of object comparison and a software tool supporting that method.

b) to identify the situations at selected moments of the Software Life Cycle which the procedure is used for.

Next we are going to propose a solution of the problems just mentioned.

3. OBJECT DESCRIPTION AND COMPARISON

Further on we shall stick to the following definition given in [2]: "*An abstraction of an object is a characterization of the object by a subset of its attributes. If the attribute subset captures "essential" attributes of the object, then the users need not be concerned with the object itself but only with the abstract attributes*".

A scheme of an object description and comparison based on the above definition will be given below.

Let S be a set of homogeneous objects $S = \{S_1, S_2, .., S_n\}$ and $H = \{H_1, H_2, .., H_m\}$ - a set of m attributes which describe the objects of S.

Each element S_i of S can be presented by an ordered m-tuple:

$$S_i = (E_1^i, ..., E_k^i, ..., E_m^i),$$

where E_k^i is the evaluation rating of the i-th object S_i with respect to the k-th attribute H_k.

The decision what elements should be included in sets S and H depends on the predefined objective. The latter can be related to a certain decision making problem. Each attribute must be weighted in accordance to its relative importance.

The heuristics algorithm [3] accomplished over descriptions is suggested to be used for transforming the set S into the set S', where the set S' is a completely ordered list of examined objects. This algorithm ranks the objects with respect to their capability of supporting the defined objective .

The object description and comparison can be performed by means of the software system SSS [4]. The latter was primarily designed as a tool for software product selection, but now being slightly modified it can be used for arbitrary objects. SSS comprises the following components:

The *Object Manipulating component* presents objects through tables where rows correspond to objects and columns - to attributes. The component ensures table creation, table deletion, data input in a table and a table structure modification through adding or deleting objects/attributes.

The *Weights Defining component* supports different modes of assigning the weights to the set of attributes.

The *Ranking and Result Presentation component* ensures the establishment of some parameters of the heuristic algorithm and its accomplishment. Next the component presents the ranked objects in different forms - textual or graphical.

4. HOW TO CONTROL SOFTWARE DEVELOPMENT THROUGH BENCHMARKING

The proposed procedure for continuous improvement can be applied in different situations. Each situation may be described as follows:

TO *<activity>*
FROM THE VIEWPOINT OF *<kind of software personnel>*
TARGET OBJECT *<object to be studied>*
SO AS TO *<objective>*

The item *<activity>* can be estimate, predict, choose, assess, describe, evaluate, etc. But all of them can be grouped in two main benchmarking activities:

a) *ANALYSIS* implies the comparison of a target object against the preliminary established model. The model is an abstract representation of the object with artificially constructed attribute values which must be achieved or avoided. So the model describes the success or failure and can be used for predicting the effect of some actions on the product or on the process.

This activity involves the describing of a "standard" object which will serve as a benchmark. Usually this requires a thoughtful study of the object and the use of some predicting procedures for determining unknown parameters and interpreting the results.

b) *ASSESSMENT* implies the study of a set of existing objects and their ranking so as to obtain the information needed for decision making, i.e. to select the most appropriate alternative for solving the problem under consideration, to see the position of a particular object among its competitors, etc.

The benchmarking can support the decisions made by each participant in software development and use, e.g. the item *<kind of software personnel>* can be managers, marketing staff, software or process engineers, programmers, vendors, users, etc. Their different points of view determine the diversity of objectives and the variety of target objects.

Modifying the classification given in [5], we can group the *<object to be studied>* as follows:

- Products - any artefacts, deliverables or documents which are created during the software development (specification, designs, programs, test data, reports, etc.);

- Processes - any software related activities (design, programming, testing, reviewing, auditing, etc.);

- Resources - i.e. personnel, teams, hardware, software, offices;

- Others - service policies, training programs, etc.

The *<objective>* stated depends on the problem under consideration.

There are no restrictions for the studied objects. If one can construct a set of measurable attributes and evaluate them, then the proposed procedure will work. But sometimes choosing the right target object and its attributes is not a simple task [5]. It requires joint efforts of experts, who keep track of evaluation practice and can advise when, what and how benchmarking should be carried out.

In case of complex target object a hierarchical characteristics structure can be used so as to define the measurable attributes. For example, the development process is such an object. It comprises a set of activities performed during the software development, the scheduling of these activities and the manipulating of the product. The development process is presented as shown below:

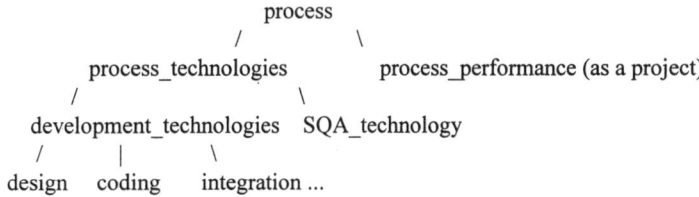

A development technology specifies the methods and procedures used during a particular phase of development. A SQA technology determines planning and control procedures referring to a certain software product and process. Each SQA technology includes all aspects of the process discipline, i.e. documentation, standards and organization rules which have to be followed.

The above described stepwise decomposition determines when the process decision making takes place. The SQA technology must be chosen at starting the software project while the development technologies depend on the current Software Life Cycle phase and they are determined in the process. The two types of technologies should be modified according to environment changes. At such moments the proposed benchmarking procedure will support the decision maker and it will help him to choose the optimal solution.

5. SOME EXAMPLES AND EXPERIMENTAL RESULTS

We have briefly described some specific situations of decision making through a Software Life Cycle.

5.1. Analysis phase
CASE 1
>   **TO** assess
>   **FROM THE VIEWPOINT OF** users
>   **TARGET OBJECT** software product of a chosen (given) type
>   **SO AS TO** get some information about users' attitudes towards such products

Attributes: Some users' defined quality characteristics.
Set of Objects: All software products available at the market.
EXAMPLE: The class of Illustration packages is studied by using some data from [6]. The attributes together with the user defined weights are given in Table 1.

The SSS system has been applied twice: with equal attribute weights and with the weights, presented in [6]. The results of ranking are shown in Table 2.

Table 1

Attributes:	Weights:	
a1 - Quality of Output	4.35	
a2 - Ease of Use,	4.25	
a3 - Firm's Reputation	4.12	
a4 - Value	4.12	
a5 - Charting Capabilities	3.89	
a6 - Presentation Features	3.88	
a7 - Drawing Tools	3.63	
a8 - Service and Support	3.47	
a9 - Spreadsheet links	3.32	
a10 - Price	3.27	

Table 2

Software Products	Ranking I	Ranking II
Harvard Graphics	2	1
Freelance Plus	3	2
PowerPoint MS	1	3

5.2. Feasibility phase
CASE 2
TO select
FROM THE VIEWPOINT OF managing staff
TARGET OBJECT virtual project
SO AS TO select the optimal virtual project on the basis of its economic,
            technical and market feasibility.
Attributes: Potential   Sales   Volume,   Level of Competition, Compatibility   with
            Marketing,   Compatibility   with Production, Patent Protection, Similarity to
            Existing Products, Environmental Compatibility.
Set of Objects: All virtual projects presented.

CASE 3

   TO analyze
   FROM THE VIEWPOINT OF project leader or process engineer
   TARGET OBJECT technology
   SO AS TO benchmark it

A typical task at this phase is to evaluate the production environment in a certain software firm. The process engineer has to establish the model of the desired project technology and has to evaluate it against the current technology used in the firm.
Attributes: The measurable attributes can be defined after the decomposing the following criteria:

   a) Functional criteria which represent the quality of a technology: modularity, integrity, clarity of the methods used, precision, effectiveness, level of complexity;

   b) Performance criteria which represent the quality of a current technology in use: adaptability, flexibility, level of automated support, level of standardization, reliability, efficiency, productivity;

   c) Organizational criteria which represent the quality of the technology discipline: the abilities of being controlled and coordinated, level of communication complexity (among groups and within groups), management complexity.

5.3. Design phase
CASE 4

   TO select
   FROM THE VIEWPOINT OF user or project leader
   TARGET OBJECT information technology
   SO AS TO determine the most appropriate information technologies for the software
          development.
Attributes: Ease of learn, Power, Efficiency, Program Volume, Structure, Portability.
Set of Objects: All information technologies available.
EXAMPLE: Let us compare a number of Programming Languages. The attributes used and their expert defined weights are given in Table 3.

Table 3

Attributes:	Weights:	Attributes:	Weights:
a1 - Ease of Learn	3	a4 - Programs Volume	6
a2 - Power	9	a5 - Structure	7
a3 - Efficiency	5	a6 - Portability	8

Two cases have been studied - with equal weights and with weights, defined by means of the SSS system.
The attribute values and the ranking related with the studied cases are given in Table 4.

Table 4

Programming Languages:	Ranking I:	Ranking II:
COBOL	4	5
FORTRAN	6	3
BASIC	2	2
PL/1	7	6
FOC	5	7
RPG II.	3	4
ADA	1	1

CASE 5

TO compare

FROM THE VIEWPOINT OF project leader or process engineer

TARGET OBJECT corrective actions

SO AS TO improve the process of designing

Attributes: process parameters improved by the performed corrective actions i.e. design quality, project cost and resources, degree of process control, etc.

Set of objects: Possible corrective actions for concerning the design process, e.g.:
- checking the guidelines followed;
- clarifying the design guidelines;
- re-organizing the design group;
- adopting the design standards.

5.4. Programming phase

CASE 6

TO compare

FROM THE VIEWPOINT OF quality engineer

TARGET OBJECT project state

SO AS TO control the quality during the development of a new software product

Attributes: Reliability, Authorization, File Integrity, Audit Trail, Continuity of Processing, Service Level, Access Control, Methodology, Correctness, Ease of Use, Maintainability, Portability, Coupling, Performance, Ease of Operation.

Set of Objects: Consequence of project states.

CASE 7

TO select
FROM THE VIEWPOINT OF designer
TARGET OBJECT integration strategy
SO AS TO find out the best integration strategy to be applied for constructing the
software system out of the program units.

Attributes: Partial Integration, Time Needed to Construct a Working Program Version,
Use of Drivers, Use of Dummy Section, Parallelism, Ability to Test, Program
Paths, Ability for Controlled Testing, Inefficiency.

Set of Objects: All available strategies.

5.5. Evaluation phase

CASE 8

TO assess
FROM THE VIEWPOINT OF project leader or user
TARGET OBJECT program documentation
SO AS TO rank the software product documentation according to its quality

Attributes: Style, Correctness, Completeness, Structureness, Clarity, Compliance with
Standards, Useful Examples, On-line Help.

Set of objects: Documentation of the competitive software products.

CASE 9

TO assess
FROM THE VIEWPOINT OF project leader
TARGET OBJECT participants in a software project
SO AS TO establish each participant's contribution to the software project progress.

Attributes: Productivity, Planned Participation in the Work on the Project, Real Participation
in the Work, Balance Based on Planned and Real Results, Quality of Results.

Set of objects: Participants can be divided into three groups: management staff (project
leader), specialists (designers, programmers) and administrative/service staff
(technicians).

5.6. Use.

CASE 10

TO evaluate
FROM THE VIEWPOINT OF user
TARGET OBJECT software product
SO AS TO establish the position of the new software product among the products at
the market.

Attributes: Correctness, Reliability, Efficiency, Integrity, Usability, Portability, Reusability,
Interoperatability, Testability, Flexibility, Maintainability.

Set of Objects: All products from a given class.

CASE 11

TO evaluate
FROM THE VIEWPOINT OF user
TARGET OBJECT software service
SO AS TO establish the quality of service provided by the competitors with
respect to a certain software type.

Attributes: Context-sensitive help, Unlimited free support, Toll-free support, Daily Support, Weekend Support, BBS Support, Fax Support, Other Extra-cost Training or Support.

Set of objects: Software products from a given type.

EXAMPLE: Using data from [6] the set of service policies provided for a number of Graphical Packages is studied. The attributes with the mentioned weights are given in Table 5.

The SSS system has been applied twice - with equal attribute weights and with the weights, presented in [6]. The attribute values and the results of ranking are shown in Table 5 and Table 6 corresponding.

Table 5

Attributes:	Weights
a1 - Context-sensitive help	8
a2 - Unlimited free support	5
a3 - Toll-free support	4
a4 - Daily Support	9
a5 - Weekend Support	3
a6 - BBS Support	6
a7 - Fax Support	4
a8 - Other Extra-cost Training or Support	3

Table 6

Objects:	Ranking I	Ranking II
Aldus Persuasion 2.1.	5	5
Charisma 2.1	1	1
Freelance Graphics	3	4
Harvard Graphics	2	2
Hollywood 1.0v2	6	6
PowerPoint 3.0 MS	4	3

CASE 12.

TO evaluate

FROM THE VIEWPOINT OF user

TARGET OBJECT training programme

SO AS TO compare the quality of training offered by the competitors and to decide which programme aspects have to be modified in order to ensure more efficient training.

Attributes: Number of computers used for Training, Training Time, Number of Participants, Number of Training Units, Place of Training (a Training Center or Firm's office).

Set of objects: Training programmes available.

6. CONCLUSIONS

The paper describes a feasible approach to benchmarking for software development and use.

Our further research will be focused on:

- defining or precising (if chosen) a set of all reasonable attributes for some objects often used in software development. Next the person dealing with benchmarking is supposed to select an appropriate subset of the attributes thus defined;

- designing and implementing a prototype of an intelligent system facilitating object description and ranking on the basis of different methods.

REFERENCES

1. T. Gilb, Principles of Software Engineering Management, Addison Wesley, 1987.
2. P. Wegner, Programming languages- the first 25 years, IEEE Trans. on Computers, C-25, 12 (1976).
3. E. Anderson, A Heuristic for Software Evaluation and Selection, Software Practice and Experience, 8 (1989) 707.
4. M.J. Daneva and N.M. Maneva, A Software Selection System - Description and Applications, Proc. of ACMBUL Conference "Computer Applications", Bulgaria, Oct. 4-10, (1992) 26-1.
5. N. Fenton, Software Metrics: Rigorous Approach, Chapman&Hall, 1991.
6. C. White, Harvard Still Tops in Graphics, PC WORLD, 11 (1992) 228.

PART THREE

Modeling

21

How to improve company performances from outside: a benchmarking model

Mario Lucertini[a], Fernando Nicolò[b], Daniela Telmon[c]

[a]Centro Volterra, Università di Roma Tor Vergata,via della Ricerca Scientifica,00133-Roma (I)

[b]Dipartimento di Meccanica e Automatica,Università di Roma III,via Segre 2, 00153-Roma (I)

[c]Tradeoff, servizi e consulenza per le aziende, vicolo del Cedro 3B, 00153-Roma (Italy)

Abstract. Benchmarking is an approach used for evaluating and improving company performances, by comparing them to the best performing companies. Benchmarking first studies the process to be improved, evaluate the performances, finds a best practice process in order to try to match two parts of the processes which have analogies, and then tries to change or modify the interconnections, structures or behaviour of the part to be improved using the analogy with the best trasformation process, to transform performance evaluation into improvement decisions. In the paper, we try to define benchmarking from a modelling point of view, on the ground of a suitable representation of the production system.

1. INTRODUCTION

Many authors conducted extensive investigations on managerial behaviour, in particular on managers who make decisions with incomplete information. Most non programmed decisions involve too many variables for a thorough examination of each and managers rarely consider all possible alternatives for the solution of a problem. Instead of attempting to maximize, the modern manager satisfices; he examines the five or six most likely alternatives and makes a choice from among them, rather than investing the time necessary to examine thoroughly all possible alternatives.

In fact, many of the so called non programmed decisions contain too many variables to be examined one by one. Thus, in a bounded rationality context, there exists, in every problem situation, a series of boundaries or limits that necessarily restrict the manager's picture of the world. Such boundaries include individual limits to any manager's knowledge of all the alternatives as well as such elements as policies, costs and technology that cannot be changed by the decision maker. As a result, the manager seldom seeks the optimum solution but realistically attempts to reach a satisfactory solution to the problem at hand.

From a modelling point of view, the boundaries include both decision variables and constraints: the decisions are taken on a subspace of the real decision space and on a subset of the complete set of the feasible decisions in the subspace. The main element conditioning decisions are resources.

By definition, anything that can be used to help solving a problem is a resource. Resources include time, money, personnel, expertise, energy, equipment, raw materials, and information.

Contraints are factors that impede problem solving or limit managers in their efforts to solve a problem. Lack of adeguate resources might prove to be a significant constraint.

Other element such as worker attitudes or government programs may prove to be a resource, a constraint, or both.

In decision making studies, many reserchers have concentrated on the analysis of alternatives with given constraints.

Limited attention has up to now received constraints analysis.

On the other hand, constraints seem to be the real object of benchmarking and the most important constraints that can be used as drivers of an improvement process depend on the embedding of a set of interconnected activities or operations in a set of interdependent organizational units (*ou*).

Different authors have developed typologies of organization structures, with boundaries and constraints. There are, first of all, constraints on resources. These may (at a certain cost) be or may not be removed. Non removable contraints are usually physical, technological or environmental constraints. There are contraints on the flows of materials, that can be in most cases modified only with regard to constraints linked to wastes. There are logical constraints, such as precedence and concurrency, that in many cases can be removed by a different assignment of operations to *ou*, although with an additional cost. In fact, an important subset of logical constraints are organizational contraints, that can be often modified by suitably modifying the links among *ou*, e.g. company procedures.

Benchmarking is a way to face the problem of modifying the organization structure and the operations assignment in a rational way. It is often defined as:

> *continuing search, measurement and comparison of products, processes, services, procedures, ways to operate, best practices that other companies have developed to obtain an output and global performances, with the aim of improving the company performances.*

This definition can be endorsed, but, to be more specific, it is important to focus the attention on structural changes in the organizations.

In fact, the performance optimization on the ground of a given structure and/or structural changes obtained with a sequence of small local changes, are classical way to operate, coherent with the incremental culture of engineering and management.

In a rapid changing environment such changes do not fit, in many cases, the competitiveness goal; we need more effective tools.

The problem is that major modifications, even if theoretically sound, could prove to be inpractical. To convince management that relevant structural changes could be done, we must give evidence that the new organization could work. A widely recognized good proof is that others companies have adopted with success the new organization.

This paper, although a simple work in progress without numerical applications, tries to formalize a benchmarking model, which could lead to an improvement metodology.

The performance indicators are introduced in this framework, not only for comparaison purposes, but also to find guidelines for improvement.

2. THE PRODUCTION SYSTEM: THE ACTIVITIES, THE ORGANIZATION AND THE DECISION PROCESS

The behavior of a production system is driven by three main elements: activities, organization, and assignment of activities to organizational units (*ou*).

An activity is any action aimed at trasformation of materials and/or information, including an activity aimed at the production of a decision. The activities are linked one to the other and they determine the flows of materials and/or informations throughout the system. Following a standard classification of activities, we can divide them in three main categories: production, support and management. Production activities concern products, i.e. the flow of materials/informations/money, which are the primary goal of the production system. Support activities concern all the facilities and informations needed for production and management. Management activities concern the decisions on the production flows, support and interconnection functions, all necessary for the system to function. Following this classification, a subset of the resources needed for the production belongs to the production flows, the remaining part belongs to the support flows. A transformation process is a set of interconnected activities.

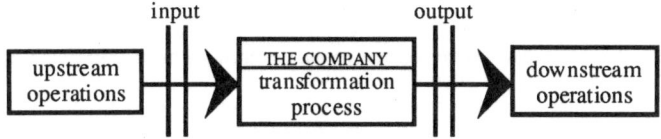

The organization concerns the choice, in a given set of feasible possibilities, of the set of resources the company is willing to use for production, the clustering of the resources into *ou*, and the way these *ou* interact one with the other, exchanging informations and materials, through pratices and procedures. Notice that resources are considered both in the flow representation of the production system and in the organization. In fact, a resource, e.g. a machine, belongs to a *ou* and is managed there, but its use becomes part of the production operations and therefore is included in the flows of activities. The ways to use a machine for a given set of activities are part of the activity management introduced above, the decisions concerning new machines or relevant modifications of the machine configuration or interconnection with other machines are part of the organization.

The assignment of activities to *ou* concerns the embedding of a production in an organization structure, so as to obtain a feasible system and a good performance. This is a standard task of the company top and middle management: who is in charge of what has to be done in each time interval and who is performing the job.

Following this representation,there are apparently two ways of looking at the system. The first begins with the existing organization and finds the best fit of activities onto the given organization, eventually with minor ex post adjustments. The second way, is to begin with the activities to be performed, and try to build around them the best suitable organization. This often drives the need of new activities to implement the new organization. These activities are difficult to foresee a priori and, in particular, it is difficult to determine before hand the quantity and quality of resources needed. This has serious effects on the performances and produces often significant delays in time.
One of the contributions of the the total quality mouvement has been that of understanding that the best solution can be obtained by looking at the system in an integrated way, together with a continuous redefinition of the company's resources involved in the organization, in a continuous improvement approach.

In this new approach, a process, traditionally seen as a set of activities and flows, is now seen as a set of activities and flows embedded in an evolving organization structure.

Most of the performance indicators concerning money, materials and informations (and the whole control activity) have been traditionally based only on the production flows. Even the total quality movement still has difficulties in creating indicators for support and managerial flows dimension on one side, and on the organizational dimension on the other side.

The field measures are taken in particular points of the system, suitably related to the transformation process considered (see the above flow chart describing the production flow), and are in terms of quantities (flows, levels and time). On the other hand, performance indicators are taken at different times or time intervals, and adequately elaborated. Typical performance criteria are: effectiveness, efficiency, productivity, profitability (or budgetability), quality.

On the other hand, examples of integration indicators, taking into account the organization structure and the embedding of the activities, are indicators of flexibility (range, uniformity, mobility,...) and complexity (connectiveness, procedures path lenght, number of *ou* interconnections,...).

In order to determine what and where to measure, and which are the right performance indicators and how they relate to measurements, we must follow two basic directions: internal relevance and external consistency.

As far as internal relevance is considered, it is important to define company goals and constraints. To put together goals, measures and performances, you need a conceptual model of the transformation process, that can be used to transform performance evaluation into improvement decisions.

As far as external consistency is concerned, it is important that all indicators will be the same in all the companies, and therefore they will be comparable.

The decision making process, its link with the value of a set of performance indicators, suitably depicting the company's behaviour, and the organization supporting the process, are a cornerstone of the benchmarking building.

In practice, company decisions lie on different levels. From the viewpoint of this paper we can introduce three levels of decisions: strategic, tactical and operational. Using decision models' language, we may characterize the three levels as follows:

Operational level
Given: environment, operational conditions and procedures, information flows, operational constraints (different types of technological and organizational contraints), a univocally defined objective function,...;
find: the value of decision variables directly connected to the process;
such that: the performance will be optimized (throughput maximization, lead time minimization, ecc.).

Tactical level
Given: environment, structural constraints difficult to modify, the company goals, a set of performance indicators;
find: operational constraints, information flows, operational procedures and the value of decision variables;
such that: the organization performances will be optimized (flexibility, complexity, ecc.).

Strategic level
Given: environment, global structural constraints and resources, set of interconnected decision centers;
find: company goals and performance indicators;
such that: the profitability of investments will be maximized.

Benchmarking focuses only on certain types of intermediate decisions, that we have called of tactical level.

These decisions do not concern, typically, basic company strategies, such as market selection, process selection, joint ventures. In the following, to focus the attention on the intermediate level, we will suppose that strategic decisions have been already taken and cannot be modified.

In the same way, benchmarking decisions do not concern operational decisions, such as material routing and operations scheduling. We will suppose that, given the set of constraints, internal efficiency (i.e. a good solution) is always achieved by the decision makers at the operational level. All actions at this level can therefore be considered completely determined by the upstream decisions and the environment. For instance, we may suppose that, in the constrained optimization of the operational level, where all actions are performed on the ground of the choices made at the tactical level, the decision process can be represented in an optimization format with a unique decision maker and complete information, where the output of the tactical level produces the constraints of the optimization problem.

To simplify the notation, in the next section the optimization model of the operational level is assumed to be in a linear programming format.

Benchmarking decisions focus on a tactical level, where you can modify organizational constraints, procedures and practices. We have had examples of this in physical material handling, distribution systems, assembly lines, production layout, make or buy, precedence or concurrency constraints (a usual way to represent the relations introduced by the information flow through the organizational units of the system).

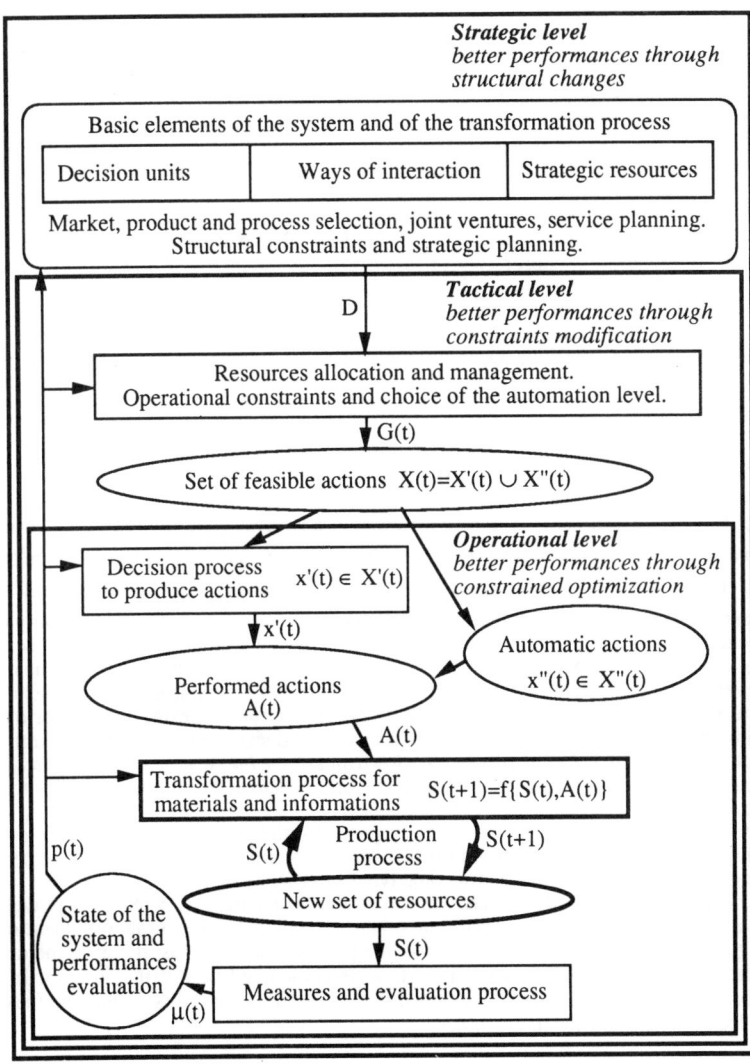

Traditionally, in managing a production system, you generally start by defining the activities of all the organizational units, then the interconnection network that links them, taking only implicitly into account the support services' need. Starting from this breakdown into units' activities, you aggregate the units on the basis of the interaction network in order to define aggregated processes and their coordination needs. But in this way the breakdown becomes an organizational constraint, and the performance optimization process is typically done in the framework of a given organization structure.

On the other side, organization structure is often the main obstacle to improvement, and the optimization of such a structure is a problem much too wide and undefined.

Benchmarking, as it will be better explained in the next sections, is a way to handle this problem. The activity breakdown is performed on the ground of a set of different organization structures and for each of the resulting systems, the performances can be evaluated. Instead of looking among all possible organization structures, we limit our search to existing ones. To make this effective, you must be able to compare not only similar companies, but also very different ones, and therefore we must work on subsystems that have the same input-output functional model. The subsystems to be compared are found bottom up, aggregating the unit activities on the ground of the relationship among units. The breakdown is performed on the process implemented by the two subsystems.

We may therefore say that we compare two companies where the same function is performed in two different ways, corresponding to two different breakdowns (i.e. two different organizations and/or two different assignments). These breakdowns are, in our opinion, one of the main aspects of benchmarking.

In order to make an activity breakdown, it is important to describe, the functional relations among input, transformation activities and output, the resources consumed and the resources available for the activities, the aggregation of resources into ou, the links among ou and set of ou. Given the resource/activity connections and the relations amongst activities, we can define the resource allocation process and plan our activities in time.

3. THE OPERATIONS BREAKDOWN TREE

Let P be a given process which transforms, in a suitable time interval, a given input in a given output. P is composed by a set of interdependent operations. Let U and U^p be the sets of ou in charge of performing the process P in company C and in a partner company C^p. In fact, for benchmarking, we assume the point of view of company C and we will use company C^p as the benchmarking partner; as already said, benchmarking is not a redesign from scratch of a process, but a redesign based on the transfer of processes existing in other companies.

Several structured metodologies and tools developed for organizational analysis may be of help to us for finding technological coefficients for resource constraints and for expressing performance indicators, usually defined in terms of products and final outputs, as a function of the activities in which the process breaks down and of the ou in which the company resources are partitioned. To do this we need to associate products to activities and activities to resource consumption.

The standard breakdown of activities required by the benchmarking analysis produces a decomposition useful for comparing processes in different companies and may be used for finding standards of performance for some types of processes.

In developing the activity breakdown, some subsets of activities corresponds to well defined ou, with formalized interactions with the other ou and with the external organizations, other activities cannot be properly embedded in the process' ou and must therefore be considered separately. In the following, these activities are considered part of the support. In general, support activities are only partially devoted to the process we are considering, but they are also included in different external processes.

The hierarchical breakdown produces a breakdown tree. Each breakdown of a node of the tree produces a set of nodes corresponding to a partition of the activities of the transformation process and, in case, two additional nodes: one for all support needs not included in the transformation process considered above (e.g. maintenance, purchasing, delivery, administration,...), another for the interconnection among set of activities belonging to different elements of the partiton and the interconnection between them and the outside environment (e.g. transportation systems, information network, rules and procedures,...). In the following, we will indicate the nodes of the tree as process, support and network nodes. Process nodes correspond to all the different phases of the transformation process at different levels of aggregation. Support nodes enable process nodes to operate effectively. Network nodes correspond to connection and integration procedures among all its brothers; more precisely, each network node induces an interaction graph among a set of process nodes (e.g. precedence constraints, information flows,...). Support and network nodes have no sons and represent part of the organization structure.

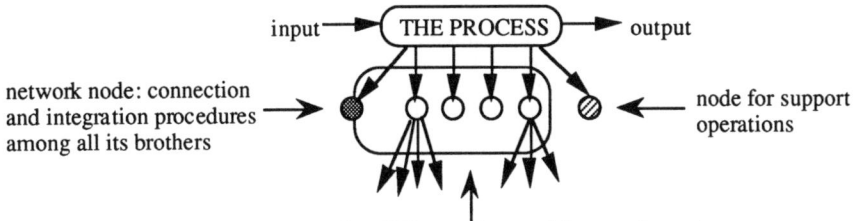

process nodes: different phases of the transformation
process at different levels of aggregation

The breakdown's first level of decomposition

Each leaf of the tree corresponds to a set of activities which are assigned to a unique *ou*; for process nodes, each *ou* is typically in charge of only one leaf. This is often the case when the organization has been designed ad hoc for the process.

Let T and T^p be two operations breakdown trees (*obt*) of P.

In general, T and T^p will have several subsets of equivalent nodes. By equivalent we mean that input and output (therefore the transformation process) are the same, although the transformation may be obtained with a different set of operations, different decision variables and using a different set of resources.

Notice that, if we consider a transformation process of company C^p (node of T^p) and we try to transfer this process to company C substituting it to the corresponding node, we must anyway redesign the process and adapt it to the caracteristics and features of company C. When we redesign the process, we may either find that the transformation is the same as the original one in company C, or that it is different. In the first case the two nodes are called identical. In the second case an opportunity for benckmarking becomes possible. Two equivalent nodes can be both leaves, both internal nodes, one leave and one internal node.

If a node of a tree has no equivalent node in the other tree, then it may belong to a subtree which have a corresponding alternative subtree in the other tree; the roots of the two subtrees are equivalent nodes, although not identical. In this case a benckmarking opportunity becomes possible, by replacing the whole subtree.

Every time that we find the opportunity of substituting a node or a subtree of C with a node or a subtree of C^p, we introduce an *or* node representing this choice.

Let T^{ao} be a tree with *and-or* branches, obtained as the union of T and T^p (always from the point of view of company C). The set of *or* nodes represents the set of choices available for the benckmarking. Each *or* is the root of two subtrees (that may also be a single node, the left subtree is a subtree of T, the right subtree is a subtree of T^p) which perform the same input-output functions (although in a different way, with a different set of decision variables and using a different set of resources).

and-or branches are obtained by integrating in the breakdown tree the benchmarking partner's processes, ways to operate and best practices

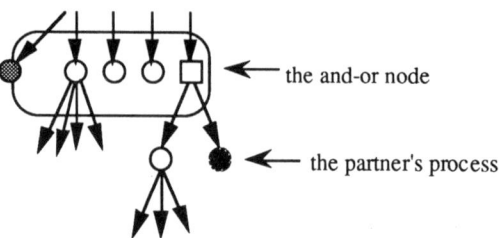

the and-or node

the partner's process

A detail of the and-or tree

Even though, in principle, all combinations are possible (with a different set of constraints), to better clarify the benchmarking process, we may consider separately meanigful aspects.

The most important aspect is *organizational benchmarking*, which studies the possibility of substituting only subtrees corresponding to different breakdowns of subactivities. In this case we suppose that company C has designed at best the elementary tranformation processes, their interconnection and the support; the only constraint in obtaining better performances is a not optimal breakdown.

A second aspect is *integration benchmarking*, which studies the possibility of substituting only network nodes. This means either a different set of interconnections, or a different way to manage them.

A third aspect is *implementation benchmarking*, which studies the possibility of substituting only single process and support nodes (supposing therefore that the interconnections and the nodes are the same in the two trees). In this case we suppose that the design of some single units may be improved.

A fourth aspect is *goal benchmarking*, which studies the possibility, for two trees with all identical nodes and a benchmark expressed in terms of a set of values of performance indicators, that the actions of company C does not fit the benchmark.

4. A BENCHMARKING MODEL

The decisions associated to the operation breakdown tree T introduced in the last paragraph, produce actions given in term of a vector of variables $x \in X$ (where X is the decision space of company C for process P). Each entry of x belongs to a unique leaf of T. It may be a continuous variable (tipically associated to process leaf) or a binary variable (tipically associated to a network node, where it represents a link). Variables associated to identical nodes belong to the same decision subspace.

The constraints of company C (in a linear programming format) can be written as $Ax \le b$, where A is a generalized technology matrix and b is a generalized resource vector. The value of the entries of A results from the design of the units and the network. In our model, we assume that the value b cannot be modified by benchmarking; this means, in particular, that b must support both the activities of T and the activities of T^p redesigned for C.

The objective function of company C (to be maximized) can be written as f(x) and can often be expressed as a linear combination of performance indicators. We can suppose that performance indicators can be expressed as function of the decision variables that drive the process. In fact, such variables, together with the initial state of the system, determine the state in all the following time frames (because of the assumption of a deterministic system), the state of the system determines the output and the output determines the performances. When, in order to find better performances, we incorporate some processes of another company into our company, the decision variables can change, but the performance indicators remain the same (even though, hopefully, with a better numerical value).

Let us indicate as $I(x)$ the performance indicators vector function of a given decision vector x and I^* the best performance vector, obtained examining the performances of the best companies (competitors or companies performing that function better then the others).

The whole operational level decision model can therefore be written as:

$$\begin{array}{c} \max f(x) \\ x \in X \\ Ax \leq b \end{array}$$

(OL)

In the following we will present a model based on organizational and goal benchmarking. The analysis can be easily extended to the other two aspects mentioned above. We therefore assume that in T^{ao} there are no *or* nodes between single nodes (i.e. all the single leaves are designed at best) and k *or* nodes between two subtrees with different breakdowns (i.e. for each path from the root to a leaf there is at most one *or* node).

The colums of matrix A and vector x are divided into blocks: one block for all the variables belonging to the leaves corresponding to identical nodes in T and T^p; one block for each *or* node, corresponding to the variables belonging to the leaves of the left subtree of the *or* node (the subtree belonging to T).

$$A=[A_0|A_1|A_2|\ldots|A_k]$$
$$x^T=[x_0^T|x_1^T|x_2^T|\ldots|x_k^T]$$

The bloks from 1 to k can be replaced by a suitable redesign of the corresponding processes of the company C^p. Let us indicate such blocks and the correponding variables (assumed to belong to a given decision space Y) as:

$$R=[R_1|R_2|\ldots|R_k]$$
$$y^T=[y_1^T|y_2^T|\ldots|y_k^T]$$

Company C verifies that its performance is lower then the market needs, decides that the performance benchmark in given by I^* and tries to reach such performance by introducing a suitable subset of the k alternative processes of C^p. Let ß be a 0-1 k-vector, each value of ß corresponds to one of the 2^k subsets of activities belonging to C^p ($ß_i=0$ if the T activities is taken, $ß_i=1$ if the T^p activities is taken). In practice k is relatively small and all possible combinations can be easily analized.

Let z be a decision variables vector obtained by chosing the decision variables x or y depending on the value of ß and Z the correspondent decision space:

$$z^T=[z_0^T|z_1^T|z_2^T|\ldots|z_k^T]=[x_0^T|x_1^T or_1 \ y_1^T|x_2^T or_2 \ y_2^T|\ldots|x_k^T or_k \ y_k^T]$$

where or_i chose x_i if $ß_i=0$, chose y_i if $ß_i=1$. In the same way we can obtain a new matrix D.

$$D=[D_0|D_1|D_2|\ldots|D_k]=[A_0|A_1 or_1 \ R_1|A_2 or_2 \ R_2|\ldots|A_k or_k \ R_k]$$

Notice that the number of entries of z is equal to the number of columns of D.

As the performance indicators are function of the decision variables, when we introduce activities of company C^p into company C, they become functions of the new decision variables.

We will indicate as I(z) the indicators vector obtained by chosing the decision variables x or y depending on the value of ß.

The organizational benchmarking model can be written as:

Model BM(ß)

Given:	A,R,I*	
Find:	z	(i.e. x,y)
Such that:	$D(ß)z \leq b$	(i.e. $A_0 x_0 + \sum_{i=1,\ldots,k} \{(1-ß_i) A_i x_i + ß_i R_i y_i\} \leq b$)
	$z \in Z(ß)$	(i.e. $x \in X$ and $y \in Y$)
	$\partial[I^*-I(z)]^+$ is minimum	

where $\partial[\cdot]^+$ is a weighted sum of the positive values of the argument entries.

In practice BM(ß) is a goal programming problem.

Notice that, if I(z) can be approximated by a linear function and there are no binary variables, we can apply the results of Data Envelopment Analysis [ccgss] in order to find the optimal value of ß.

If ß=0 we have an example of what we have indicated in the former section as goal benchmarking. In fact, if BM(0) has a solution with $\partial=0$ (i.e. $I(z) \geq I^*$), then C has the possibility of reaching the benchmark I* without structural changes, but the company goal f(x) pushes the company in a wrong direction. We must therefore change the company objectives, and consequently change the activities, to reach the goal.

5. CONCLUSIONS

The paper presents a modelling approach to benchmarking. The paper tries to produce a structured methodology for benchmarking: four types of benchmarking are put into evidence.

One type of benchmarking is *goal benchmarking*, which studies the possibility, based on the improvement of performance indicators, of trying to get the values of the benchmark.

Even more important, according to our opinion, is *organizational benchmarking*, as we have called it, which studies the possibility of substituting sets of activities of the whole process with other sets of the breakdown activities of the same process in the best practice company.

Another aspect is *integration benchmarking*, which studies the possibility of changing the interconnection pattern for the same activity breakdown (this case is fairly rare on its own, because, in general, only few interconnections are possible).

The last aspect is *implementation benchmarking*, which studies the possibility of redesigning process or support units. In this case we suppose that the design of some single units may be improved.

An optimization model, based on a linear programming format, has been introduced, to exemplify the approach, combining the goal and the organizational benchmarking.

From an algoritmic point of view, the solution of the benchmarking problem formulated here is related, for linear indicators, to data envelopment analysis.

The paper tries to give a frame for benchmarking analysis; a future aim is to validate the approach in a few meaningful cases. This validation activity is now in progress in two different fields: information systems for manufacturing and information systems for university services.

REFERENCES

[bon]   C.Bonini, *Simulation of information and decision systems in the firm*. Prentice-Hall, 1963.

[cam]   Robert C.Camp, *Benchmarking. the search for industry best practices that lead to superior performance*. ASQC quality pres, 1989.

[ccgss] A.Charnes, W.W.Cooper, B.Golany, L.Seiford, J.Stutz, *Foundation of Data Envelopment Analysis for Pareto-Koopmans efficient empirical production functions*, J. of Econometrics, vol.30, n.1/2, 1985, pp.91-107.

[hc] Micheal Hammer, James Champy, *Reengineering the corporation*, Hamper business, 1993.

[hub] G.Huber, *A theory of the effects of advanced information technologies on organizational design, intelligence, and decision making.* Academy of Management Review, 15, 1, pp.47-71, 1990.

[jk] H.Thomas Johnson, R.S.Kaplan, *Relevance Lost: the rise and fall of management accounting*, Boston: Harvard Business School Press, 1987.

[kap] Robert S.Kaplan (ed.), *Measures for manufacturing excellence,* Harvard Business School Series in Accounting and Control, 1990.

[kp] A.Kumar, P.S.Ow, M.J.Prietula, *Organizational simulation and information systems design: an operations level example.* Management Science, 39, 2, pp.218-240, 1993.

[lc1] Byron C.Lewis, Albert E.Crews, *The Evolution of Benchmarking as a Computer Performance Evaluation Technique*, MIS Quaterly, 9, n°1 (march 1985), pagg.8-16.

[lc2] Richard L.Lynch, Kelvin F.Cross, *Measure up: yardsticks for continuous improvement.* Basil Blackwell inc., 1991.

[ln] Kathleen H.J.Liebfried, C.J.McNair, *Benchmarking: a tool for continuous improvement,* Harper Business, 1992.

[lnc] Mario Lucertini, Fernando Nicolò, Daniela Telmon, *Integration of benchmarking and benchmarking of integration*, Nato Advanced School on Integration, Il Ciocco, Italy, 1993 (to appear on Int.J.on Production Economics).

[mas] Brian Maskell, *Performance Measurement for World Class Manufacturing: a Model for american companies,* Productivity Press, 1992.

[ms] T.Malone, S.Smith, *Modelling the performance of organizational structures.* Operations Research, 36, 3, pp.421-436, 1988.

[mt] Kathleen Malette, Joanne Tomlinson, *Benchmarking: focus on world class practices,* AT&T Bell Laboratories Technical publication Center, 1992.

[saa] Thomas L.Saaty, *Decision making for leaders: the analytical herarchy process for decisions in a complex world,* RWS Publications, 1988.

[spe] Micheal J.Spendolini, *The benchmarking book,* AMACOM, 1992.

[tzc] Fancis G.Tucker, Seymour M.Zivan e R.C.Camp, *How to measure Yourself against the best,* Harvard Business Review, jan.feb. 1987, pagg.2-4.

[wat] Gregory H.Watson, *The benchmarking workbook: adapting best practices for performance improvement,* Productivity Press, 1992.

[you] Edward Yourdon, *Modern structured analysis*, Prentice-Hall, 1989.

22

The use of process modelling in benchmarking

S J Childe & P A Smart

School of Computing, University of Plymouth,
Plymouth PL4 8AA, United Kingdom

The increasing interest in benchmarking as a tool for achieving radical improvements in a business' competitive performance has encouraged many companies to attempt a comparison of their performance to that of others. This has been attempted in two main ways. Companies who compare their performance to that of their competitors restrict their potential operating improvements to a position of equality with the competitor, who may by then have moved on. Those companies attempting to look into other industries to gain real originality may fail if they do not focus on activities which are directly comparable. This paper looks at the need to identify the correct activities to study and proposes a modelling technique which can help the company to establish a baseline for comparison.

1. WHAT IS BENCHMARKING?

The term "benchmarking" has become the vogue term in the management arena in the nineties. One may suppose its origin lies in other professions such as carpenters or drapers who may have made marks upon their benches to allow standard measurements to be taken. The term has also been defined [Chambers 1988] with reference to the discipline of surveying:

"A surveyors mark .... indicating a point of reference .... anything taken or used as a point of reference or comparison, a standard, criterion etc"

A useful definition of benchmarking is as the development and use of reference points or standards against which business performance can be judged.

Bob Camp [1989] describes Benchmarking as "the search for those best practices that will lead to superior performance for a company". His text presents a structured approach for searching for those industry best practices and implementing them into a business environment. A number of philosophical steps which are fundamental to the success of benchmarking are identified:

* Know your operation
* Know the industry leaders or competitors

*   Incorporate the best
*   Gain superiority

The ability to gain superiority is dependent upon a detailed understanding of the company's own operations and those of others and the ability to incorporate these to develop performance improvements. It is the authors' opinion that many companies which attempt benchmarking go ahead without the detailed knowledge of their own business which is vital to allow the appreciation and assimilation of the best practice exemplified in other companies.

The process of searching for industry best practices should be seen as a learning activity. Companies learn how they operate, they try to learn how their competitors operate, and aim to learn how industry leaders operate, so that they can apply the knowledge gained to their own businesses. Paradoxically, while benchmarking tends to focus upon learning about others, its success may depend upon learning about the company's own processes to set the agenda for the study and to allow the findings to be used.

External comparison depends upon being able to identify the leaders whose practice is the best. This knowledge may not be available within the company and assistance from external knowledge sources may be required. These sources might be consultants, academics or institutions such as benchmarking groups. As the company performs successive benchmarking exercises, this knowledge base can be expected to increase.

"Best practice" may be hard to find. It would be impossible for a company to conduct an exhaustive search of all the potential best practices across all industry sectors. This means that it is likely that a company would not identify the overall best exemplar. The company must conduct their search for best practice given the limitations imposed on them by their available knowledge base, that is to say, they may only search for those industry best practices which they feel to be most appropriate given their knowledge of the industrial sector. This base should grow and provide more pertinent pointers to the industry best as the number of iterations of the benchmarking process increases.

If companies are to emulate the practices of other firms it would appear that the best possible result would be to match their performance, since it appears to be impossible to surpass the standard upon which the exercise is based. However, this would be to overlook the benefits from formulating a hybrid "better" process comprising the best ideas from the studies conducted. To excel in this area the company must use the knowledge gained as a potential source for creativity. Without this sort of innovative process a company can never aspire to be a leader itself. Benchmarking must not be seen as a strategy to imitate other companies but as a mission to use the experience of others as a source for novel ideas. This approach is discussed by Smith et al [1992].

2. IDENTIFICATION OF ACTIVITIES FOR BENCHMARKING

The concepts of best practice and benchmarking sometimes appear to trivialise the problem by addressing a level of performance which can be described as "best" or as "best in class". This overlooks the simple question of how to measure performance. The company which performs a particular activity in the best way to suit its own particular business strategies - possibly quite well known as the most successful company in the business - may not address the same aims as the company who wishes to learn from them.

Simple performance measures can be used to illustrate this point. Manufacturing companies often base their strategies upon a particular balance between the measures of cost, quality, functionality, delivery lead time and delivery reliability etc. If a company has a strategy of operating a particular activity at the lowest cost, it may make compromises to the service it offers to its customers, for example. A company in the upper end of the market might compete primarily upon the basis of service quality. For these two companies, the best way of achieving the activity may mean two different things. They may be incompatible for benchmarking purposes.

The activities which are to be compared must therefore be identified not only on the similarity of the task but also on the basis of the competitive business objectives which govern the task. If they do not agree, there may be scope for learning from an alternative approach, as long as the differences are understood.

Similarly, the business environment must be considered when selecting a benchmarking partner. For example, if a company was benchmarking its procurement activities, it would not necessarily be wise for a company in the business of one of a kind production to benchmark itself against a high volume low variety producer, because the competitive requirements of one company would not be satisfied by the activities found to be good at addressing the requirements of the other. The way a company needs to manage its supply arrangements in a high volume business where "clout" is available is likely to be quite different to the arrangements which would enable the purchase of small numbers of widely different items from a wide range of suppliers.   However, we may also argue that to achieve radical performance gains, looking at activities in radically different companies might be a valuable exercise. A good example of this is provided by the well known case study illustrating the benchmarking exercise between Xerox and L L Bean [Camp 1989].

Following the implementation of a planning system in the inventory control area, Xerox identified the picking operation as the greatest bottleneck.

The benchmarking effort resulted in L L Bean - an outdoor sporting goods retailer and mail order house - being used as the main benchmarking partner. Whilst the two companies would appear to be dissimilar, the processes were comparable:

> "... L L Bean products may bear no resemblance to Xerox parts and supplies. To the distribution professional, however, the analogy was striking: both companies had to develop warehousing and distribution systems to handle products diverse in size, shape and weight" [Camp 1989]

If it is accepted that a similar activity with similar competitive priorities can be found in a potential benchmarking partner, one may assume that there would then be a reasonable foundation upon which to copy the superior performing company's practices. This reassurance would increase as a function of the degree of consistency in the nature of the task - the inputs and outputs agreeing - and the degree of consistency in the performance measures. Unfortunately the simple act of copying from another company may not be a simple as it seems.

It was pointed out by Juran in his address to the winners of the Malcolm Baldridge National Quality Award in 1989 "to learn from experience requires a transfer of knowledge. Such a transfer of know-how should not be done by mimicking what the winners did. Mimicking is risky because of the differences in respective cultures. A reliable transfer of know-how requires thinking through what are the lessons learned .... it is the universals which are transferable from one culture to another."

Hayes and Pisano observe [1994] "Two companies may adopt similar strategies and production processes, but one can end up being far more successful". A strategy shift is needed to that of "learning" from the experience of others rather than copying. This shift is from imitating other companies to using them as a source for novel ideas [Smith et al 1992].

The use of benchmarking as a source for novel ideas provides the basis for the potential creation of innovative activities. Viewing benchmarking in this fashion also provides the potential to surpass the industry best. Copying implies only matching the performance of a particular activity, whereas the creation of hybrid solutions based on the industry best has the potential to itself be the best, since it creates something original and new.

Benchmarking, viewed as the basis for corporate learning, therefore requires a technique to allow:

* the identification of the key business activities;

* detailed activity definition to allow comparison with other companies;

* the characterisation of the relevant performance measures.

This should allow benchmarking to operate across industry boundaries between companies who do not compete with each other.

3. DESCRIPTION OF BUSINESS ACTIVITIES

In order to ensure that the benchmarking exercise compares like with like, and to ensure the performance measures and other conditions are similar, a means for describing activities and business processes is required which allows the appropriate level of detail to be drawn out, and which allows the performance criteria to be compared. This can be achieved by the use of a hierarchical method which allows activities to be decomposed into various levels of detail, whilst showing the business context which provides some information about performance requirements. Some light is shed on this problem by the research work currently being undertaken in the field of Business Process Re-engineering. Companies are beginning to explore the questions of "how can I identify processes in my business?" and "what is a business process?"

According to Davenport and Short [1990], a business process is "the logical organisation of people, materials, energy, equipment and procedures into work activities designed to produce a specified end result". Davenport and Short also state that processes have two important characteristics. Firstly, they have customers and secondly, they cross organisational boundaries and are generally independent of formal organisational structure.

Similarly, Hickman [1993] defines a business process as "a logical series of dependent activities which use the resources of the organisation to create, or result in, an observable or measurable outcome, such as a product or service". The authors would add that a business process must be initiated by and must provide results to a customer, who may be internal or external to the company.

A useful structure established by the CIM-OSA standards committee [1989] sub-divides processes into three main areas: Manage, Operate and Support. The CIM-OSA framework regards *manage processes* as those which are concerned with strategy and direction setting as well as with business planning and control. *Operate processes* are viewed as those which are directly related to satisfying the requirements of the external customer, for example the logistics supply chain from order to delivery. *Support processes* typically act in support of the Manage and Operate processes. They include the financial, personnel, facilities management and Information Systems provision (IS) activities. These definitions serve as a framework which the company may use to focus its benchmarking efforts. Further focus comes from the identification of the processes within these groupings.

4. DEFINITION OF BUSINESS PROCESSES

In the authors' view a business process operates in a manner analogous to the operation of an industrial or chemical process in as much as it comprises "a series of continuous actions or operations ...." [Hawkins 1984] which are performed upon a commodity. It may also be regarded as a conduit along which a commodity flows. In this context, a commodity might be conceptual or material. Such commodities pass along their respective process conduits and are transformed, at different stages in their progress, as various operations are performed upon them. An activity for benchmarking can therefore be defined by the process of which it is a part.

4.1 Manage processes

Direction Setting
    This process includes all high level strategic planning activities. It acts as an overall managing activity which takes ideas about direction based upon business and environmental information, including customer feedback, and transforms these into a set of strategies, operational goals and performance measures.

4.2 Operate processes

Order Flow (Products)
    The Order Flow process takes the customer order and transforms it into a finished product. The commodity which flows through this process is the customer's specific product requirement. This initially takes the form of an order and is transformed into a product which embodies the customer's requirement. As the order flows in one direction, money flows in the other: thus the process ends only when the product is accepted and paid for by the customer.

Activities within this category may include raw material purchasing, product assembly, the production of the product, obtaining orders, delivery and installation of the product, invoicing and money receipt.

Service

This process takes the customer's requirement for a service and satisfies it by providing that service. For example the requirement could be the need to keep machines operating reliably, transformed by the service into an assurance of trouble-free performance. Activities include the management of customer enquiries and the provision and management of the necessary technical support to satisfy the customer.

4.3 Support processes

There are a considerable number of activities which are required to support the key business processes. These relate the company to its business environment, which can be thought of as a series of markets within which the company operates. These have been identified by Fine and Hax [1984] as capital markets, labour markets, technology markets, factor markets and product markets. Each of these markets is addressed by the company through a business process.

Capital markets

The process attracts investment into the firm and provides benefits (typically shareholder dividends) thus maintaining the company's position in the capital market.

Labour markets

The process of recruiting, training, remunerating, motivating, appraising and retiring employees. By processing employees, the company maintains its human resources and its position with respect to the labour market.

Technology markets

The assessment and development of available technology, and the selection, installation, maintenance and disposal of plant and equipment.

Factor markets

The establishment and development of relationships with suppliers, supplier development and liaison, and the termination of relationships with suppliers no longer required. This process may also be concerned with the make-or-buy decision.

Product markets (and the market for services)

The company retains its competitive position in the market place by a process which maintains the awareness of its potential customers. This "marketing" activity may be seen as part of the operating process since it involves obtaining orders and providing service, and since the company's position in the product marketplace must ultimately depend on the way in which orders are satisfied.

5. PROCESS MODELLING

Having identified the process which is to be considered for benchmarking, the process must be presented in a way which allows communication, understanding and analysis. Various types of process modelling tools fulfil these requirements. One of the most popular tools is $IDEF_0$.

$IDEF_0$ comprises:

* A set of methods that assist in understanding a complex subject;
* A graphical language for communicating that understanding;
* A set of management and human-factor considerations for guiding and controlling the use of the methodology.

$IDEF_0$ uses top-down decomposition to break-up complex topics into small pieces which can be more readily understood. The diagrams are related in a precise manner to form a coherent model of the subject.

The whole system and the relationship of any part to the whole remains visible. This means that the environment in which an activity takes place is shown in terms of the effects of other activities and externalities which impinge upon the activity in question.

The graphical language of $IDEF_0$ uses boxes and arrows coupled together in a simple syntax as shown in Figure 1.

Each box on a diagram represents an activity. The arrows that connect to a box represent real objects or information needed or produced by the activity. The

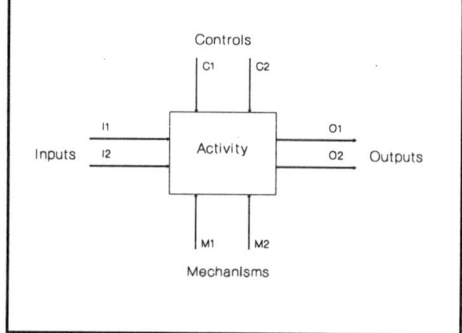

Fig. 1 $IDEF_0$ Syntax

side of the box at which an arrow enters or leaves shows the arrow's role as an input, a control or an output. Incoming arrows (which are shown on the left and top of the box) show the data needed to perform the activity. Outgoing arrows (right of the box) show the data created when the activity is performed. An input is converted by the activity into the output. A control describes the conditions or circumstances that govern the transformation. The bottom of the box is reserved to indicate the mechanisms or means (person, device, computer model etc.) used to carry out the activity.

$IDEF_0$ is a method very well suited to the specification outlined earlier. Its specific strengths lie in that it is a tool designed for modelling processes and in our view it is relatively easy to use. It uses a structured set of guidelines based around hierarchical decomposition, with excellent guidance on abstraction at higher levels. If used well this ensures a good basis for communication and a systems perspective.

6. DESIGN OF NEW PROCESSES

Once a model of a process has been created, a part of the process can be selected for redesign by benchmarking. The process model shows how the part must link in to the whole (in terms of the inputs, outputs and constraints) and provides a boundary within which activities can be redesigned. Thus the model provides the means to identify activities which can be replaced by better activities, the ideas for which may come from benchmarking.

Whilst research is continuing into the identification of standard business processes, it is clear that certain activities must be performed by most companies. This was tested in previous work [Childe 1991] in which an attempt was made to divide production management into its constituent tasks, and then to decompose these tasks into lower level elements, and so on as far as possible. For use in benchmarking, the model provides a structure in which the company can determine that a particular business activity is required, and use benchmarking to help to decide what lower level tasks should be used. This analysis can be applied at any level of abstraction, where the higher level always sets the requirement to be fulfilled by the lower level.

For each of the tasks in the model an attempt was made to determine why the task was required for the particular company. These task determinants aid the use of the model as a template to determine which tasks are required in a company under investigation.

Three types of task were recognised.

Core tasks

Some tasks appear to be necessary in every manufacturing company, in which case the benchmarking exercise could only affect the way in which the task was performed. These tasks were regarded as "core tasks". These included for example "Process orders", "Handle goods inward".

The decomposition of a core task could include optional tasks according to the way in which the core task was performed, particularly the decision whether or not to computerise the task. Thus the critical question for a core task is only how it should be done, which is determined by the selection of lower level tasks of which it is constituted.

Optional tasks

In the cases where the task requirement was seen to depend upon the situation, the task was regarded as "optional", since there would clearly be cases in which the task was not required. Examples of these include "Confirm order to customer", "Inspect goods". Where they could be determined, the particular reasons for optional tasks being necessary were recorded.

Dependent tasks

These were tasks which were found in the decomposition of optional tasks, but which were not themselves optional. These were necessary in any instance in which the parent task was required, thus depending upon the appearance or non-appearance of an optional task. For example, "Report capacity requirements" is an optional task whose decomposition must always include "Aggregate product profiles" and "Identify work for specific time buckets". These tasks are therefore compulsory in the case of the parent task being required.

The design of a new system, or the amendment of an existing system, depends upon being able to take important decisions about the way tasks should be carried out. Inevitably this means identifying the most appropriate set of lower level tasks to fulfil the task, and then for each of the tasks to select the means of carrying out the task, such as by human or mechanical/electronic means.

Work is proceeding on the development of generic process models which can reduce the time taken to produce the initial model for benchmarking. These models will also allow easier comparison between companies.

7. CONCLUSION

Effective benchmarking depends upon being able to identify correctly suitable activities in example companies, through analysis of the activity itself, its performance objectives and measures, and the competitive situation of the business. Good understanding of the role of any activity to the competitiveness of the business can be provided by an approach which sees the activity in the context of a business process. A modelling technique such as $IDEF_0$ can be utilised to provide an understanding of the activities in question and to provide the basis for redesigning the process. Decomposition of the process into its constituent activities, tasks and sub-tasks provides the means to decide at what level the benchmarking activities should be conducted, thus identifying the correct unit of analysis.

REFERENCES

CAMP R, 1989, *Benchmarking - The search for industry best practices that lead to superior performance*, American Society for Quality Control

CHAMBERS, 1988, *Chambers English Dictionary*, W & R Chambers Ltd and Cambridge University Press

CHILDE S J, 1991, The design and implementation of manufacturing infrastructures, PhD Thesis, Polytechnic South West

CIM-OSA Standards Committee, 1989, CIM-OSA Reference Architecture, AMICE ESPRIT

DAVENPORT T H & SHORT J E, 1990, The new industrial engineering: information technology and business process redesign, *Sloan Management Review*, Summer

FINE C H & HAX A C, 1984, Designing a manufacturing strategy, WP # 1593-84, Sloan School of Management, MIT, USA

HAWKINS J M H, 1984, (Compiler), *Oxford Paperback Dictionary*, Oxford University Press

HAYES R H & PISANO G P, 1994, Beyond world-class: the new manufacturing strategy, *Harvard Business Review*, Jan-Feb

HICKMAN L J, 1993, Technology and Business Process Re-engineering: Identifying Opportunities for Competitive Advantage, *British Computer Society CASE Seminar on Business Process Engineering*, London, 29 June

RALSTON D, 1992, Measure for measure, *Proc. 27th Annual Conference of British Production and Inventory Control Society, (BPICS)*, November, pp225-237

SMITH S, WHITTLE S, TRANFIELD D & FOSTER M, 1992, Implementing Total Quality - the downside of best practice, in Hollier R H, Boaden R J and New S J (Eds.) *International Operations: Crossing Borders in Manufacturing and Service*, Elsevier

23

Introducing new technologies in organisations - business model perspective

Jean Bergeron and Jean-Claude Bocquet

Laboratoire Productique Logistique, Ecole Centrale de Paris
Grande Voie des Vignes, Châtenay-Malabry 92295 Cedex, France

ABSTRACT

This paper presents a knowledge model to enhance the process of introducing new technologies in organisations. The central principle is to convert external knowledge, constituting the technology from the supplier (universities, R&D, other organisations), into internal knowledge, constituting the technology integrated in the organisation. Two types of knowledge must be transferred: tacit and explicit knowledge. Four knowledge domains must be represented when introducing a new technology: Technology Core (constituents of the technology), Transformation Processes (transformation processes of the constituents in products and/or services), Products and Services Space (knowledge associated with products and/or services realisable with the new technology) and Organisational Environment (internal and external environment of the organisation). Technological and organisational actors map specific and shared knowledge on these domains. We developed this model as a tool to be used by project management introducing new technologies. It allows to represent knowledge of the process and to coordinate the actor's actions. Introduction of superplastic forming technology illustrates the model.

Keywords: knowledge model, business process modelling, reengineering, superplastic forming, technology introduction.

1. INTRODUCTION

Mastering a new technique, or a new technology has always been decisive for a human organisation. For example, the mastering of iron technology by northern countries participated in the conquest and the final destruction of the Roman Empire. Also, during the beginning of the first industrial revolution, mastering the steam technology to produce mechanical work gave a major advantage to the nations possessing it. Today, at the edge of the 21st century, mastering key technologies, changing at an ever increasing pace, is a matter of survival for industrial organisations, and by causal effects, for the well being of nation's citizens. Consequently, if an industrial organisation is able to build superior capabilities in introducing and in rendering productive the new technology, it will have a decisive advantage over its competitors[1].

In this paper, we propose an organisational approach, based on enterprise knowledge modelling[2], to help introduce a new technology in an organisation. A special attention is being paid to knowledge integration concerning the different actors involved in introducing the technology.

1.1. Knowledge transfer

The key issue in introducing a new technology is to take the reachable external knowledge (knowledge not owned by the organisation) related to the technology and to transform it into internal and usable knowledge. Reachable external knowledge consists of information the supplier of the new technology (R&D or external supplier) can provide about it. There are two types of knowledge to transfer: explicit and tacit knowledge. Nonaka[3] identified four mechanisms used to convert knowledge(learning), from the source to the receiver:

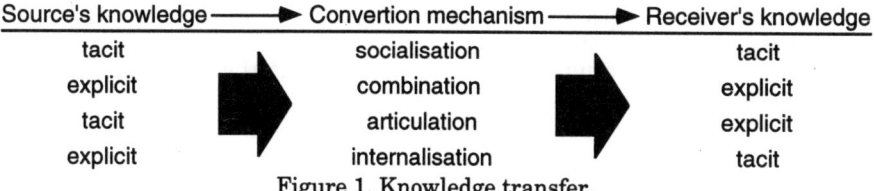

Source's knowledge ⟶	Convertion mechanism ⟶	Receiver's knowledge
tacit	socialisation	tacit
explicit	combination	explicit
tacit	articulation	explicit
explicit	internalisation	tacit

Figure 1. Knowledge transfer.

Usually, transfer and integration of tacit knowledge bring most of the problems in mastering the new technology because this kind of knowledge is not explicitly defined. These problems result principally from the lack of inter-functional skills[4], from the isolation between the different functions (design, manufacturing, marketing, finance,...) or departments in the organisation. Reducing the partition between the different functions of the organisation is of primary concern in the work we realised.

An example illustrates the model that could be applied on introducing superplastic forming technology (SPF) in a typical aircraft manufacturing organisation (Airbus, Boeing, MDD).

> Superplastic Forming: titanium-aluminium and other superplastic alloys, metal forming technology. Aerospace industry uses SPF technology for forming parts which shape requires high deformation (from 100% up to 1000%). This characteristic allows to significantly reduce the number of parts (reached with high deformation potential) and the weight (reached with high resistance and toughness of Ti-Al alloys) of an assembly.

In the following pages an integrated knowledge model of the technology, composed of 4 inter-connected knowledge domains is introduced. The purpose of this model is to bring a <u>coherent perspective</u> of the technological knowledge, tacit and explicit, to the actors involved in introducing the new technology.

2. INTEGRATED KNOWLEDGE MODEL

The aim of building an integrated model is:
- to give a common reference and understanding to the actors;
- to coordinate the actions of the actors in the organisation's environment[5].

The model is based on **four invariants** domains characterising the technology as part of the enterprise's processes. They appear as inter-connected knowledge domains (as shown in figure 2):
- Technology Core (TC), domain of the constituents (physical laws, equipment) associated to the technology;
- Transformation Processes (TP), domain of the transformation processes (operating modes, tasks) required to convert the constituents of the technology into products and/or services (PS);
- Products & Services Space(PSS), domain where knowledge converge on the PS obtainable from the technology;
- Organisational Environment(OE), domain linking the technology to the organisation's environment (social, ecological, organisational, market opportunities, investment's capability).

Each domain is developed in the following pages.

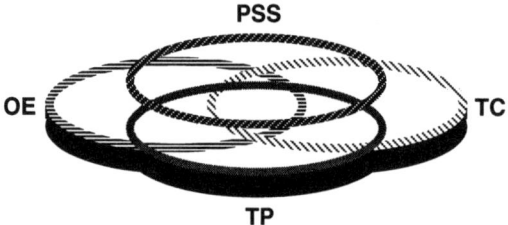

PSS

OE   **TC**

TP

Figure 2. Interconnected knowledge domains.

2.1. Technology Core

Broustail & Fréry[6], classify a technology in one or a combination of these fields: materials, energy (source and energy converter), information, life (biology and living phenomenon control). A consistent knowledge model of a technology must describe any technology belonging to these fields.

The Technology Core knowledge domain is concerned with the elements that the organisation will transform into products or services. Each actor will apply differently this definition to its organisation and to the function he belongs to. It depends on the perspective it has on the technology.

Each actor in the organisation must define which elements (knowledge, tools, information system) will be transformed or used, for the realisation of PS. A typical manufacturing organisation has 9 types of actors directly related with technological processes: design, logistic, maintenance (after-sales services), manufacturing, process planning, procurement, production, quality control and R&D. They are called **technological actors**. The nature and the integration of these functions differ from one organisation to an other. For non-manufacturing organisations, a different decomposition of the actors is necessary, but the fundamental principles remain unchanged.

Knowledge mapping in the domain

When introducing knowledge respective to a new technology in the technology core domain, the actors must acquire specific (respective to their activity) knowledge and shared knowledge about the technology, as shown in figure 3. All actors must map on a common area these two types of knowledge to help determine the cognitive interactions.

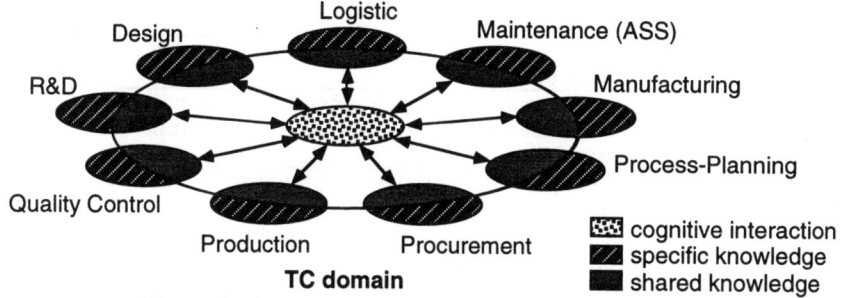

Figure 3. Actors mapping knowledge on TC domain.

Table 1
Technology Core domain of SPF technology

Actors (techn.)	Specific Knowledge	Shared K		
design	CAD systems and other design capabilities available	1	2	3
logistic	transportation resources (trucks, planes,...) available	-	-	3
maintenance	maintenance equipment (ultrasonic probes) available	1	2	3
manufacturing	manufacturing presses available	1	2	3
process planning	automated-process-planning systems available	1	2	3
procurement	procurement systems available	-	-	3
production	production planning systems available	-	-	3
quality control	inspection equipment (non-destructive techniques) available	1	2	3
R&D	testing procedures available to determine range of properties of SPF parts	1	2	-

1- material properties (maximum tensile stress, shear stress, density);
2- communication standards for exchanging geometric model (STEP, SET, VDI);
3- communication standards for production information.

Table 2
Transformation Processes domain of SPF technology

Actors (techn.)	Specific knowledge	Shared K		
design	how to use CAD systems and other design capabilities to produce SPF parts	1	2	3
logistic	how to schedule the transportation resources (trucks, planes,...) conveying the SPF parts	-	-	3
maintenance	how to use maintenance equipment (ultrasonic probes) to determine SPF parts to be changed	-	2	3
manufacturing	how to operate the manufacturing presses to produce SPF parts	1	2	3
process planning	how to operate the automated-process-planning systems of SPF parts	1	2	3
procurement	how to operate the procurement systems with the SPF materials (sheet metal of aluminium-titanium alloy)	-	-	3
production	how to operate the production planning systems with the new charges induced by SPF parts	-	-	3
quality control	how to inspect (with non-destructive techniques) the new SPF parts	1	2	3
R&D	how to apply testing procedures to determine ranges of properties of SPF parts	-	2	-

1- How to transform the sheet metal titanium-aluminium alloy material into a SPF part;
2- How to exchange geometric information concerning the SPF parts with the chosen communication standards;
3- How to exchange production information concerning the SPF parts with the chosen communications standards.

All the information is not necessarily available at the beginning of the constitution of the layer (for example, manufacturing does not know the die characteristics of a press until the part's geometry is defined). The mapping of TC domain is realised concurrently with the other layers (see section 2.5).

The state of the information system (information technology and communication environment between the actors within the firm) determines the quantity and the quality of the knowledge representation achievable. To be effective, management must support team work and create an environment that sustains communication.

Table 1 illustrates a non-exhautive list of TC knowledge (specific and shared) representation of SPF technology introduction in a typical aircraft manufacturing organisation. Three examples of shared knowledge are shown (corresponding to number 1,2 and 3), they are described below the table. The filled spaces mean that the corresponding actor partakes this knowledge. For example: design, maintenance (ASS), manufacturing, process planning, quality control and R&D share the material properties' knowledge.

TC represent which tools, standards and procedures are available for implementing the technology and generating PS. A similar approach can be used for all the technology domains.

2.2. Transformation Processes
The Transformation Processes domain contains knowledge explaining how the building blocks of the technology core layer are transformed in PS. The technological actors map knowledge in a similar form as in the preceding domain.

In this domain, technological actors introduce the know-how: training of the actors on the standards and procedures, the equipment, the different systems to be used. Table 2 presents the TP domain of SPF technology introduction. It contains specific TP knowledge and shared TP knowledge.

2.3. Products & Services Space
The Products & Services Space domain is the "valorising" domain of the technology knowledge model, where PS are developped. At the limit, the technology to be introduced is not important by itself for the organisation, all is important is the result obtainable from it: PS for customers. Usually, when the decision maker selects a technology to introduce, he already supposes there is potential development of PS from it. Therefore PSS layer, in a sane technology development process, is the initiating knowledge layer (focus on customer needs). Initiating knowledge in this layer is partial and the complete PSS layer will result only from the development of the 3 other layers, concurrently with it.

The interest of representing knowledge in the PSS domain, is to give the technological actors a common knowledge platform to maximise the opportunities to develop marketable PS from the new technology. Nowadays, many manufacturing organisations still develop PS punctually, case by case or randomly with a new introduced technology. The designer usually realises a part or a few ones, when the opportunity arrives, with the technology. The organisation does not fully exploit the development space (all the possible PS realisable with the technology) of PS. Systematising and organising the development space of PS, obtainable from the technology, enhances competitive strength of the organisation by increasing the offering of new PS in a shorter delay.

Table 3
Products and Services Space domain of SPF technology

Actors (techn.)	Specific Knowledge	Shared K		
design	determination of design resources (man/hours, material) utilisation to realise the SPF parts	1	2	3
logistic	scheduling the transportation resources (trucks, planes,...) for conveying the SPF parts to be realised	-	-	3
maintenance	verify that maintenance procedures (ultrasonic testing) are usable by the customers (airlines) for the SPF parts to be realised	1	2	3
manufacturing	determination of manufacturing resources utilisation to produce the SPF parts	1	2	3
process planning	determination, with automated process-planning systems, of the process planning to realise the SPF parts	1	2	3
procurement	determination of the supplied goods (aluminium-titanium alloy metal sheet), with the procurement systems, necessary to realise the SPF parts	1	-	3
production	determination of production planning of the new charges induced by SPF parts to be realised	-	-	3
quality control	determination of inspection procedures (with non-destructive techniques) to apply to the SPF parts to be realised	1	2	3
R&D	application of testing procedures to determine ranges of properties of the SPF parts to be realised	1	2	-

1- definition of SPF parts geometry space (realisable geometry);
2- determination of which geometric information (and when) to transfer between the technological actors;
3- determination of which production information (and when), about SPF parts, to transfer between the actors.

2.4. Organisational Environment

In this domains is mapped knowledge linking the technology to the organisation's environment (internal and external). At this level intervene **organisational actors**: top management (Executive Officers), marketing, human resources, finance. The role of these actors, in the model we propose, will be to verify that:
- the organisation can provide financial and human resources to ensure the realisation of PS;
- the PS to be issued from the new technology, respond to market needs, to external constraints (environmental, political, economical) and to strategic objectives of the organisation.

These actors link the PS, by then, the technology, with the reality of the organisation and the market. They must monitor and predict perturbations that can originate from the introduction of the new technology in the organisation.

OE domain is the common place where the technological and the organisational actors interact. Communication between these groups of actors is of primary concern. The establishment of a cognitive model helps to consolidate the two groups by allowing a better vision of the issues resulting from technology introduction.

Table 4 contains examples of specific and shared knowledge related to organisational actors and technological actors (knowledge interface of the two types of actors).

Table 4
Organisational Environment domain of SPF technology

Actors (org.)	Specific Knowledge	Shared K		
top management	strategic objective of improving weight savings by using better materials like SPF	1	2	3
finance	return on investments with the new SPF products enhancements	1	2	3
marketing	market demand for SPF products	1	2	3
human resources	human resources redistribution caused by SPF products development and production (hiring, transfer of personal)	1	2	3
technological actors interface		1		3

1- definition of SPF parts to be realised (integrated frame, integrated door panel, integrated keel);
2- SPF products in accordance with strategic planning of the organisation;
3- determination of which production information (and when), about SPF parts, to transfer between the actors.

Organisational actors must establish a compromise between:
- the amount of allocated resources to be well informed (investigation on politics, economy, ecological consequences, employees motivation);
- and the level of awareness desired, of the consequences (positive and negative) of introducing the new technology in the organisation.

2.5. Development planning of knowledge mapping
The planning of the knowledge mapping, within domains is hard to define because there is no optimal general solution. The solution comes from a case by case analysis and the experience the actors have. Usually, as already discussed, when actors introduce a technology, they anticipate PS development from it. Therefore, it seems logical that the PSS layer initiates the knowledge building process. After the initiation, the actors map in the four domains, including PSS, concurrently and with highly linked coordination between them. Project

management (using the knowledge model) intervenes at this level to coordinate the actions of each actor (technological and organisational) and for optimising the allocated resources (delay, development cost, operating cost, services). The model can be use to represent the knowledge belonging to the technology and the organisation, giving the actors a better understanding of the relations linking them.

2.6. Synthesis of the process

Figure 4 summarises the process of introducing a new technology by using knowledge model representation.

1- At the beginning of the process, the technology is external knowledge (explicit and tacit) to the organisation.

2- During the process, the actors (technological and organisational) map knowledge on the four domains by transforming explicit and tacit knowledge belonging to the technology (external) into specific and shared knowledge (internal). Sufficient resources (human, financial) are prerequisite in order for the project to progress.

3- After the actors have introduced the technology (technology is integrated in the organisation), the organisation can provide PS to customers.

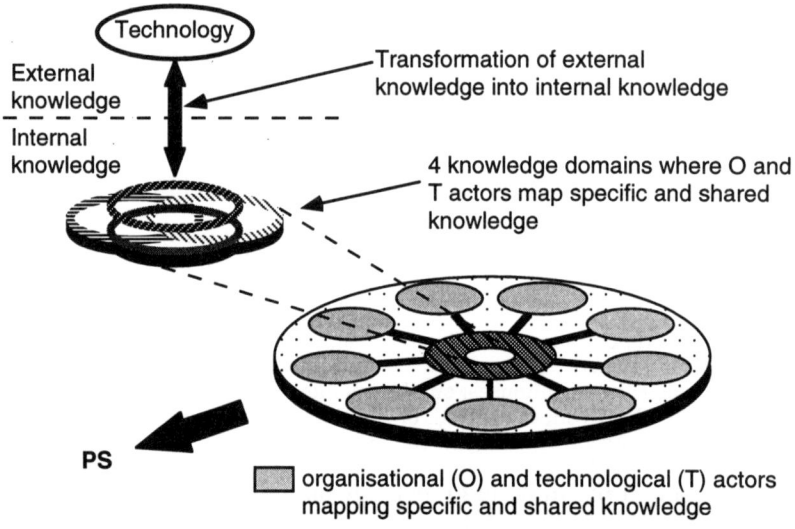

Figure 4. Synthesis.

3. CONCLUSION

Mapping the knowledge belonging to the process of introducing new technology in 4 domains (Technology Core, Transformation Processes, Products & Services Space, Organisational Environment), allows to construct a better view of the actions the actors have to perform. Giving a clearer representation of the knowledge to be developed for introducing a new technology helps the actors to partake it. By formalising and by taking into account the whole process (not only unrelated tasks), the actors will improve coordination and simultaneisation of their activities. The major aspect this integrated model can bring to its users, is the **rise in coherency** of the actions in the process. This coherency results from the establishment of knowledge references (by mapping on the four domains).

Rethinking business processes is the base of reengineering[7]. The knowledge model introduced in this paper stimulates this rethinking by envisioning the introduction of technology as part of the whole business process of the organisation.

REFERENCES

1. Nevens, T.M., G.L. Summe, and B. Uttal, *Commercializing Technology: What the Best Companies Do.* Harvard Business Review, 1990. (May-June).
2. Aranguren, R., P. Eirich, and al. *The Process of Modeling and Model Integration.* in *International Conference on Entreprise Integration Modeling Technology.* 1992. Austin, Texas: MIT Press.
3. Nonaka, I., *The Knowledge-Creating Company.* Harvard Business Review, 1991. (November-December).
4. Bowonder, B. and T. Miyake, *Japanese innovations in advanced technologies: an analysis of functional integration.* Int. J. Technology Management, 1993. (Vol. 8, Nos. 1/2).
5. Duimering, R.P., F. Safayeni, and L. Purdy, *Integrated Manufacturing: Redesign the Organization before Implementing Flexible Technology.* Sloan Management Review, 1993. (Summer).
6. Broustail, J. and F. Fréry, *Le management stratégique de l'innovation.* 1993, Paris: Dalloz.
7. Hammer, M. and J. Champy, *Reengineering the Corporation.* 1993, New York: Harper Collins.
8. Drucker, P.F., *Innovation and Entrepreneurship, Practice and Principles.* 1985, New York: Harper & Row.
9. Foster, R., *Innovation, The Attacker's Advantage.* 1986, New York: Summit Books.
10. De Haan, J. and R. Peters, *The future implementation of advanced manufacturing techniques: experiences from a Dutch Delphi study.* Int. J. Technology Management, 1993. (Vol. 8, Nos. 3/4/5).

24

Modelling Techniques for Benchmarking in Complex Administration Systems: an Approach from a Total Quality Viewpoint

Walter Ukovich   **Franca Zerilli**
DEEI   IT Consultant

University of Trieste
phone: +39 40 676 7135
fax: +39 40 676 3460
e.mail: ukovich@univ.trieste.it

Abstract

This paper deals with enterprise modelling and benchmarking techniques used in a typical non–profit environment, such as a State University in Italy.

Moving from the need to evaluate efficiency and effectiveness of the University as a whole, a frame of reference has firstly settled down to describe the *as is* state of the service supply, to organize and manage the evaluation of efficiency and effectiveness in the central administration and to examine the problems related to the performance inprovement in order to fulfill quality policies or to meet external standards.

The aim of the contribution is to point out the effectiveness and the opportunities provided by some Information Engineering formal representation techniques in setting up and then maintaining models that, first of all, represent business processes needed to provide services and then help in approaching and managing the benchmarking activities.

This work is part of a larger project that, from a total quality point of view, has as a final scope to define strategies and guidelines for quality service supply both in the administrative, teaching and research contexts.

The technique and the models presented can be used both for the quality features definition, in order to fulfill end–user quality expectations, and to set up the maintenance activities — from a supplier and an end–user point of view — also by using benchmarking techniques.

Keywords benchmarking — modelling techniques — public administration — quality of services — university services

Introduction

In November of 1993 the University of Trieste, Italy has started a six–month project devoted to define guidelines and strategic objectives for the production of quality services by the University Central Administration.

The University of Trieste is a medium–size (according to the Italian figures) state University with about 1,250 professors for over 22,000 students in the present Academic Year, 70 Departments and ten Faculties.

This project has been denominated SQUADRA (a multi–meaning Italian word corresponding to "setsquare", "squad" and "team") as an acronym of the motto "Quality Services: a Decision Support for the Head of the University". To date, the first phase of the project has been accomplished, which was devoted to specifying the general approach and methodology for the whole project, and to analize the system of interest. A formal model of the Central Administration of the University of Trieste has been produced and validated.

The scope of this paper is to present the aspects of the SQUADRA project, and in particular the results which bear some relevance for possible benchmarking activities in the specific area of the University Administration, and, more in general, in the wider framework of complex systems providing services.

1   Benchmarking

The benchmarking technique is used to evaluate business processes in an organization through the continuous comparison of its performance parameters with the parameters of other leading competitors, both in similar and in different business areas. The aim in using benchmarking is:

- to spot out weak points and critical issues;

- to scout out opportunities of improvement;

- to plan actions directed to achieve improvements through a process management approach

The benchmarking activities may be structured in phases. Taking a slightly more general approach then usual, they can be described as follows:

1. **analysis of the system**: a formal model of the system (or of part of it) under evaluation is formulated;

2. **design of the measurements**: the relevant parameters for the evaluation are identified and the measurement sessions are programmed; also, other organizations to be used for comparisons are selected; furthermore, whenever possible, appropriate activities directed to acquire other qualitative informations about the the other organizations, such as structure, functions, procedures, and even history;

3. **measurement activities**: the parameters of the previous point are measured; furthermore, whenever possible, the other qualitative informations about the other organizations, as mentioned in the previous step, are also studied;

4. **comparisons**: the measured parameters are compared in order to detect weaknesses, drawbacks, bottlenecks, and to assess target levels for improvement; furthermore, the qualitative information gathered in the previous phase are used to figure out possible ways of improving the performace of the system;

5. **formulation of proposals**: on the base of the comparisons and of the other information acquired in the previous phase, operational action proposals are formulated, aimed to raise the performance levels of the system to the target values.

1.1   Remarks

The role of the phase of analysis must not be underestimated, although it is often not explicitly mentioned; in fact, a suitable model for the system of interest is often available, and the nature and structure of the relevant processes are clear and well known. However, this is not always the case and italian government supported organizations, expecially those providing services, are quite good examples of situations in which more attention is devoted to the procedural aspects of production rather than on how each production step is combined with the others in order to form the business processes.

In such situations building and using an effective model could constitute a key success factor for using benchmarking techniques.

According to our opinion, the model of the system resulting from the analysis phase must be *implementation free*, that is it must describe the process of interest in conceptual terms, focusing on *what* is done rather than on *how* it is done. In this way a wider spectrum of opportunities is guaranteed to find alternative possibilities of implementing the process in order to achieve the desired performance levels.

In the design phase, the operational characteristics of the process are added to the model, which becomes in this way a more faithful representation not only of the specific process of interest, but also of the implementation choices characterizing the considered situation. In this phase the model evolves from a preminently qualitative formulation to a preminently quantitative one, thus providing the basis for identifying the benchmarking parameters.

Concerning the phase of measurement, it must be pointed out that, according to our views, it consists of two different kinds of activities: quantitative evaluation of the relevant parameters, and retrieval of further qualitative information about the comparison organizations.

The latter activity can be interpreted as a modelling activity on the other organizations, which is conceptually analogous to the modelling activity of the analysis phase. Having used an implementation–free model there can be useful also in this aspect of the measurement phase, since, at least in principle, the same approach could be used. Of course, this could be utopistic in most competitive environments, such as in industrial business, but could be very effective in cooperative situations, where several organizations cooperate in an integrated benchmarking project.

2   Contribution of the SQUADRA project to benchmarking modelling techniques

In this section we analyze the contribution of the SQUADRA project within the framework of the benchmarking, as it has been discussed in the previous section.

The aim of the SQUADRA project is to study how to fulfill the perceived (and the imposed) needs of evaluating the efficiency and effectiveness of the processes in the University of Trieste, Italy, and in particular of the central administration of it.

To this end, a quality based approach has been taken in order to identify the services provided and the processes generating them. It is important to point out that defining guidelines and strategic objectives concerning quality constitutes an important part in the organization management policies [5], [6]. According to this approach, the service users and the transversal (with respect to the *vertical* functional lines) flow of activities producing services within the system have been particularly focused on.

The results produced so far in the SQUADRA project pertain to the phase of analysis of the benchmarking technique. They are an example of the application of modelling techniques drawn from Information Technology environments [7], to a complex administrative system. These models are being used with a view to identifying measurable parameters representing the quality features of the system (or part of it), considered as a complex of business processes.

3   Scenario of the SQUADRA project

Universities in Italy, and also in the rest of Europe, must face in the near future situations of increasing complexity with decreasing available resources but tighter competition levels and higher expectations.

This is true for private educational institutions, but also for government supported universities, due to the specific situation of the Italian state economy.

As a consequence, Universities in Italy begin to realize that it is time to define, design and implement tools apt to evaluate efficiency and effectiveness about the use of resources they have. Several committments in this sense have been formulated, both at the national and European levels.

Universities are considered as organizations providing cultural services. The activities within them belong to one of the three areas of teaching, research, and administration. The two former ones are primary activities, in the sense that they directly contribute to the fulfillment of the institutional objectives, while the latter one just supports them.

The part fo the SQUADRA project here presented is devoted to study the administration of the University of Trieste in order to approach efficiency and effectiveness evaluation problems and formulate action proposals to organize and manage them and the performance improvement needed to fulfill internal quality policies or to meet external standards.

4   Motivations and general approach of the project

Roughly speaking, the concepts of efficiency and effectiveness evaluation immediately refer to the idea of measuring features of the relevant product or service, and to elements such as performance evaluators, types of measurements, resource assignment criteria, and so on.

According to our opinion, such aspects, although important, must be embedded within a general framework in order to guarantee that some relevant requirements are met:

- evaluations in different application environments must be performed according to a uniform approach; this requires that they are all conceived and designed within a unique conceptual framework;

- evaluation techniques must be context–free and therefore valid in different application environments;

- it must be possible to iterate evaluations, if necessary, in the same application environment in order to follow its time evolution; moreover, sinergies must result from performing similar evaluations in different application environments; in this respect, result comparability is a necessary feature:

- evaluating must facilitate diagnosing detected inefficiencies and ineffectivenesses in terms of internal coherece of the measured parameters.

In order to show how our approach to evaluating services of the University is able to meet such requirements, we first need to formally specify some concepts.

Definition 1 *A* service *is the effect of the concurrent execution of some functional steps within the organization, connecting an "activator", who triggers the whole precess, to a "client", who gets the products yielded by the process.*

Note that we only consider here the production phase of the whole service lifecycle.

According to the previous definition, we distinguish *external* and *internal* services, depending upon the relative position of activators and clients with respect to the structure providing the service. For instance, services to students are external, while services related to the management of academic status of professors are internal.

The key choice of the SQUADRA project is the quality–based approach to services [1], [2]. This allows to focus on the interaction between the system producing services and its environment, in particular its users, and their expectations on the products provided by the system.

From this quality–based approach we can identify what has to be considered acceptable as a product of the system. As a consequence, we can also fix the guidelines according to which the features of the system that are relevant with respect to efficiency and effectiveness issues can be assessed.

The normative framework of the SQUADRA project is given by the European Standards EN 29000. According to them, quality is defined as follows:

Definition 2 [9] *Quality is the totality of features and characteristics of a product or service that bear on its ability to satisfy stated or implied needs.*

5 Quality of services

According to the stated approach taken for the SQUADRA project, our interest is mainly focused on identifying parameters for benchmarking that can represent quality features. That is [4] [3], parameters that are capable of evaluation against defined standards of acceptability.

Such a capability is the effect of what is called [9] quality policy, that is

Definition 3 *The overall quality intentions and directions of an organization as regards quality, as formally expressed by top management*

In order to determine and implement the quality policy it is necessary:

1. to know what is produced; this requires to analyze the system in order to identify the produced services;

2. to know how production is performed; this requires to build a model of the system, specifying the production steps needed to provide the services;

3. to decide how good the produced services have to be.

In the present study, which refers to the first phase of the SQUADRA project, we are concerned with the above points 1 and 2. As it should be evident, they correspond to the phase of analysis of the benchmarking techique.

6 Results of the first phase of the SQUADRA project

In this section we present some of the results of the first phase of the SQUADRA project, that appear to be of interest with respect to the modelling activities of the benchmarking technique.

6.1 Methodological choices

The cornerstone upon which the project methodology is based is that quality is not an attribute that can be added to the service at the end of its production process; in fact, quality has to be conceived, designed, built, verified and maintained at the same time as the service is conceived, designed, built, verified and maintained.

Such a statement brings the following consequences:

1. **the whole service lifecycle must be taken into account;**
   opposite to the tayloristic, procedure–oriented approaches, which tend to hide business processes, the whole life of the service must be considered

2. **analysis must be conceptually approached, and mostly performed in practice, in a top–down way;**
   the dimension and complexity of the system, and the need to consider its position within its context, require an analysis approach granting:

   • the possibility of modelling a system at various detail levels, without losing the globality of the description

   • the possibility of concentrating the analysis on specific parts of the system, without missing the relations among them

3. **the model of the production cycle of the services must be implementation free;**
   the production cycle of the service (which is only a part of its life cycle) must be represented in terms of functional elements in order to provide the widest choice of opportunities to the qualification of the services in terms of quality; this is of crucial importance not only for the SQUADRA project, but also for using models in the analysis, design and measurement phases of the general benchmarking technique

4. **the transversality of the steps of the service production process, with respect to the functional (i.e.the vertical) structure of the system, must be emphasized**
   in this way the complex of the internal elements concurring to the production of the service, and the relations among them, can be appropriately pointed out, as well as the connections between the system and its context, both in input and output

6.2   Results of the first phase of the SQUADRA project

6.2.1   The models of the service production process

The methodological results of the first phase of the SQUADRA project that bear relevance for the analysis step of the benchmarking technique are:

• use of analysis techniques, in particular the structured analysis of De Marco – Yourdon [8];

• use of rules and techniques of graphic representation, in particular the ones of Yourdon;

• use of automatic tools for preparing project reports which have a widespread diffusion and high level of user friendliness, in order to facilitate exchange within both the local projects and the global benchmarking activity.

The actual activity devoted to the analysis of the system has yielded a number of graphical models, with the features discussed above, that must consitute a solid starting point for the analysis phase of the benchmarking activity.

Owing to the philosophy of the project, an exhaustive and detailed coverage of the central administration of the University was not really important. Rather, assessing the possibilities provided by the adopted techniques was a major point.

Models consist of Data Flow Diagrams (DFD), representing flows involving service production steps and external entities. Service production steps are activities which transform inputs into outputs and, hopefully, add some value to inputs in view of the service to be delivered, with respect its expected quality feature levels. External entities are element interacting with the system providing services, and lie outside the system boundaries. According to the quality–oriented approach of the project, they have been classified into four categories, which are, in order of importance:

- clients: they are the ultimate users of the service;

- activators: they are the external elements that trigger the production of the service; occasionally, they can be the same as the end users of the service, but not necessarily;

- decision units: they are part of the service production process that do not perform any transformation, but can decide on the continuation or interruption of the production process; more specifically, they often have to guarantee the accompliance of the production process to laws and regulations;

- external support units: they provide support services to the internal production process, such as, for instance, banking, mailing, etc.

The first diagram that has been produced is the so called *context diagram*. It represents the central administration of the University of Trieste, i.e. the system under consideration, at the maximum level of aggregation and all the flows with which it interacts with its environment. Such flows are the ones crossing the boundary of the system we are studying.

The second representation of the system is provided at a more detailed level and four main groups of services are identified, concerning services provided by the central administration to the main categories in which the users can be grouped: students, internal personnel and internal structures such as faculties or departments.

These services are the management of financial resources, of personnel, of buying and selling goods and services, and of what relates to students.

The four processes of this second level have been then analized at increasing levels of disaggregation, thus producing further DFDs.

The model produced has been thoroughly discussed with the Administrative Director of the University and the officers in charge of the Department of the University Administration which provide the services considered. A very good level of agreement has been achieved concerning the correspondence between the processes represented in the model and the functional steps of the actual production cycle. Also, it has been verified that, for each service, clients, activators and decision makers have been correctly identified.

In the perspective of benchmarking for the administrative services of Universities, the model formulated in the SQUADRA project can constitute a sound base on which the structure, functionalities and procedures of different organizations can be analyzed and compared. To this end, the possibility of performing verifications and validations based on a formal model is of particular interest, as such activities bear an especially critical relevance for complex organizations.

7   Conclusions

We believe that activities devoted to the assessment of organizations providing services must be performed according to a quality–directed approach. In this way the highest benefit may be achieved from these activities in order to devise how to improve the performance of the system.

As a consequence, also the benchmarking techniques adopted for such activities must be used in a quality–oriented way. This requires for the phases of analysis, design of measurements, and measurement, using analysis and modelling techniques for the organization of interest, and for the other ones, having the capability of representing the key elements and factors that characterize a service. Also, having used formal models and automatic representation tools, the continuous updating and improving of the system representation is made much more convenient. Finally, such techniques are simple and easy to understand, in order to facilitate communication both within and between organizations.

We believe that these results of the SQUADRA project constitute a sound conceptual base to perform benchmarking activities among different organizations, in which each of them compares its own performance with the performance of every other one. For this kind of *polycentric benchmarking* we have also provided a formal methodological approach and practical tools to analize and represent the systems.

References

[1] Philip B. Crosby. *Quality is free.* McGraw–Hill, 1979.

[2] W. Edwards Deming. *Out of the Crisis. Quality, Productivity and Competitive Position.* Cambridge University Press, 1986.

[3] Roger Fournier. *Practical Guide to Structured System Development and Maintenance.* Yourdon Press, 1991.

[4] EN 29004–2. *Quality management and quality system elements — Guidelines for services.* 1991.

[5] Henry Mintzberg. Patterns in strategy formation. *Management Science* 24, May 1978, pp. 934–948.

[6] Martyn A. Ould. *Strategies for software engineering — the management of risk and quality*. Wiley, 1990.

[7] Roger Pressman. *Software engineering: a practitioner's approach*. McGraw–Hill, 1990.

[8] Edward Yourdon. *Modern structured analysis*. Prentice–Hall, 1989.

[9] EN 28402. *Quality — Terminology*. June 1992.

PART FOUR

Tools and Techniques

25

Benchmarking Techniques

Dr.-Ing. Kai Mertins, Dipl.-Ing. S. Kempf, Dipl.-Ing. G. Siebert
Fraunhofer-Institute for Production Systems and Design Technology (IPK)

1. Definition of Benchmarking from Our Point of View

Benchmarking is a continuous process during which services and specially processes and methods of operational functions throughout several companies can be compared /Camp89/. With that differences to other companies as well as reasons for these differences shall be pointed out and possible improvements shall be determined. The comparison regards companies which control the methods and processes to be examined very well.

2. Benchmarking techniques

2.1. General

Soon it was recognized that it was not enough to examine only products as explanation for the different types of costs of the company, but that a fundamental knowledge of the single activities within the added value chain is a necessary assumption. For this purpose methods and tools are needed.

Benchmarking is the device which turns the unstructured process of a permanent improvement into an objective action plan. Benchmarking starts when problems are identified and concentrates on detecting kernel problems to improve the current practice.

For this purpose several techniques are available which partly complement respectivelyand/or base on each other. The selection of the technique is mainly determined by the aim of the increase of added value. For Benchmarking it is important to understand which advantages are to be expected before expensive ressources are invested.

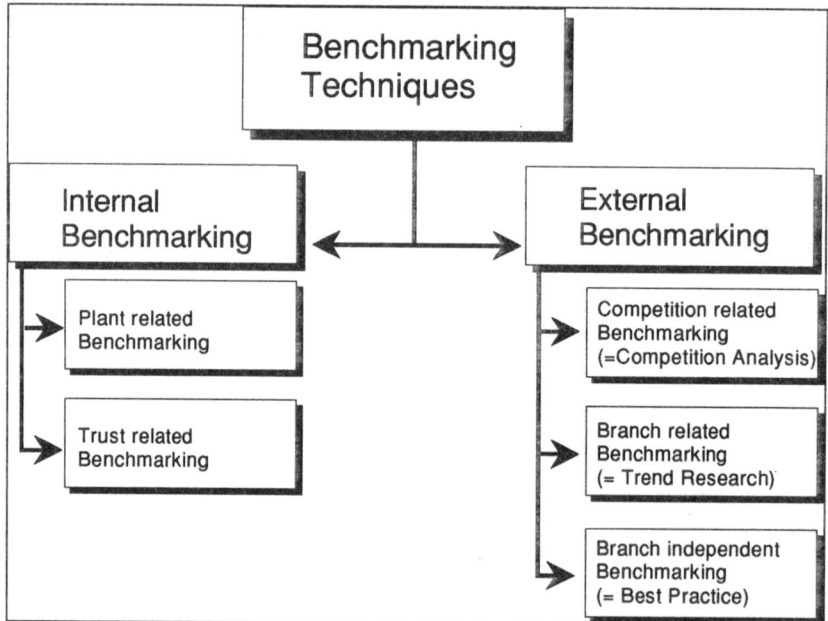

Figure 1: Benchmarking Techniques

2.2.   Internal Benchmarking Techniques

Internal Benchmarking is the simpliest form of Benchmarking because no external limits have to be considered. With internal Benchmarking organizations try to learn from their branches, departments and sistercompanies. During these studies similar processes are examined and compared throughout different areas to obtain detailed information about the provided performance potential. The implementation and data access are easy, but the success regarding a performance increasing change is low because units belonging together have the tendency to submit to cultural and organizational standards. At the internal Benchmarking the look of the management is turned inside before it is turned outside. Current operating sequences and practices shall be registered objectively and shall be understood. This way the necessary details to concentrate the study on the aspects are obtained.

2.2.1. Plant Related Benchmarking

Inside of a company it is possible to find similiar processes which can be compared. Those processes are signed by technological, organizational and personal influences.

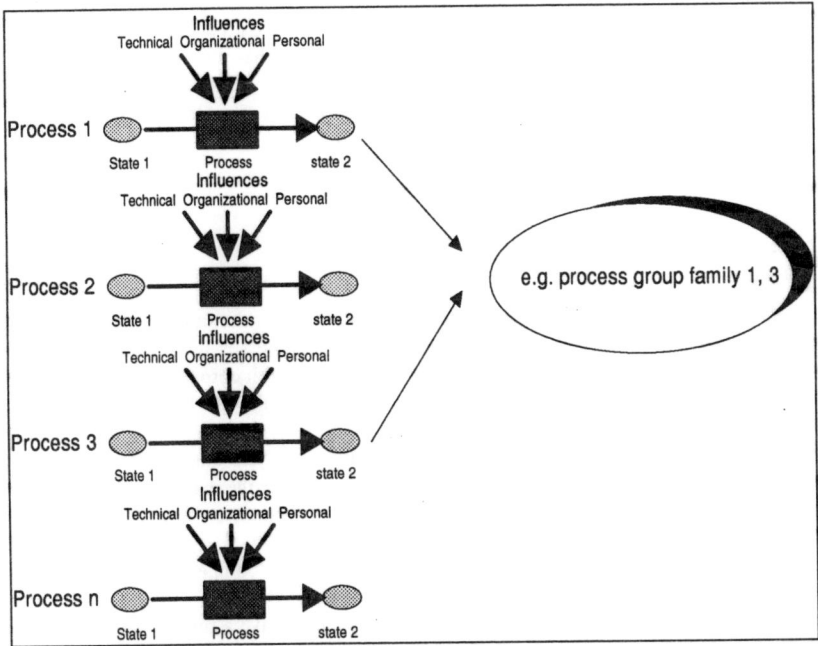

<u>Figure 2</u>: process group family

In this case it is possible to compare those processes by building similar process groups. This is a very specific way of benchmarking.

Definition of a process: A general mark of processes is according to DIN 19222 the mutual influence of operation within a system by which in their total amount matter, energy or information is transferred or stored. Processes are marked by a sequence of activities which again characterized by measurable inputs and outputs and can be part of a higher level processes. Additionally processes have to fulfil following four requirements:

❑ Effectiviness: Processes have to fulfil the tasks and targets given to them.

❑ Efficiency: Processes must fulfil their tasks with a minimum of expenditure.

❑ Ability to test and control: The state of a process should be known at any time and there must be the possibility to make corrections.

❑ Adaptability: Processes must be designed in a way that it can be reacted on changes in their surrounding.

2.2.2. Trust Related Benchmarking

This technique is used for the comparison of several plants within one company. Equal fields are examined, like e.g.: design, manufacturing etc..

This function oriented proceeding is the siemliest and best way to apply own ideas and capacity. Additionally a process is started through this channel of fair competition between the plants that builds the absolute assumption for continous improvement especially in connection with the external Benchmarking.

2.3.   External Benchmarking Techniques

Benchmarking is a technique for looking outside where at the practices of the own company are compared with the external practices. Comparison means that there must be a basis line of similarities. Only similar things can be compared each other. Therefore it is necessary to recognize one's own operations and processes. Only then a valid comparison and the identification of improvement chances are possible. The ideal case is that several companies should take part in such a study,

❑ to minimize the costs,

❑ for easier data collection,

❑ the results can be used on a broad basis.

2.3.1. Competition Analysis

The competitions analysis can be provides information about the current and future market activities of the competitors, their strenghnesses and weaknesses as well as their possible reaction to specific behaviors. It also allows the comparison with companies which control the respective company activities in an excellent way, but it doesn't look beyond surroundings of the competitors.

An important difference between a pure competition analysis and Benchmarking is that functions and services are always measured from the best. While the exclusive comparison with the competitors only makes possible an equal footing with them, the comparison with the best permits to get ahead of the competitors.

Definite functions are often solved in a better way by non-competitors than within equal companies because of the special facilities of the companies.

Branch blindness and the delimit from competitors clearly make a further development more difficult. Benchmarking makes it possible to orient oneself on the best by the to non-competitors and to learn by that how these companies manage to be ahead in definite fields.

2.3.2. Branch Related Benchmarking

Branch related Benchmarking goes beyond the mere comparison of two firms and emphasizes on the search for trends. So it examines the efficiency of a definite function throughout the branch. For that purpose it is necessary to examine a much larget group of companies than it is the case at the competion related Benchmarking.

It is not restricted to the performance of two or three others who sell the same products in about the same market but it is founded on broaded fields. Branch related Benchmarking looks for trends instead of competition positions and serves for the performance analysis of subsystems. Where at the limits to the competition related Benchmarking are fluent because up today there is no answer to the question where a straight study ends and the trend research begins.

2.3.3. Branch Independent Benchmarking (Best Practice)

The key to a long term success in competion is not equality but superiority. One wants to overtake the presently existing best practice an then get ahead of it. Therefore the final form of Benchmarking is aligned with the best. It searches throughout a spectrum of branches for new, innovative practices, independent from their source.

Benchmarking from the best of the class is based on the conviction that the process of added value is based on similar characteristics throughout many different institutions. Therefore the aim is to find the best of the best practices and to use them to alter innovatively the existing practices within the own organization. To define success by external criteria means to develop an idea of best performance and to heighten the company to a new performance level.

Only by that it is possible to leave the competitors further behind and to reach world's class.

3. Benchmarking Problems

Benchmark solutions are always only as good as the respective Benchmarking partner, regardless wether internal or external Benchmark is concerned. From that the range of problems which every Benchmarket is faced with derive.

How do I find the best suited Benchmark partner for my target. This Benchmark partner must meet a number of criteria to be worth the effort for the examination, like e.g.:

❑ Are the operating sequences comparable?
❑ Are the range of problems comparable?
❑ Is the Benchmark partner "best of class" in the field of the problem?

If there is no possible Benchmarking partner known yet who meets the demanded criteria, a new partner must be found. But for external Benchmarking following problems appear at getting the necessary information:

❑ Information about companies often only exist as branch oriented sorting.
❑ Further sorting criteria are often only the size of the company and turnover.
❑ There are no information about company specific strenghnesses or weaknesses available.

Due to this hierarchical information structure the target group soon is limited to the members of the own branch. Information about branch external companies who possibly could be a very well suited Benchmarking partner thus quickly get out of sight and it takes a lot of effort to get them.

For the internal Benchmarking the range of problems are slightly different:

❑ the process sequences are not documented and so
❑ the process sequences are not comparable.

But for a process oriented Benchmark examinations the comparability of process sequences are of special importence because otherwise very similar process sequences from other company functions could be overlooked and thus be cancelled as Benchmark partner.

4. Our Solution

If it is managed to sort company information not any longer in a hierarchical way, but to show all relevant characteristics simultaneously, then it is possible to remove the branch spectacles completely. By that the procedure of finding information can be quickly performed without taking the risk of overlooking important Benchmarking partners.

An approach to solutions of this kind of problems can be the arrangement of company information according to characteristics on a company independent abstract level. As an example a comparison between the electronics industry with the insertion of circuit boards and the food industry with the insertion of chocolate in boxes can be used. These processes are similar in order and control and therefore can belong to one company group. So this arrangement provides branch independent company information that is the ideal basis for the search for Benchmarking partners. Following targets can be achieved by the branch independent arrangement of company information:

❑ save finding of the best suited Benchmarking partner and
❑ reduction of the time expenditure to find this Benchmarking partner.

A similar approach to a solution could be used for the internal Benchmarking. If it is managed to document and classify the process sequences in a way that they can be aggregated in groups of similar process sequences by adequate tools then it is possible to use the potentials offered by Benchmarking also for company internal comparisons, see also figure 2.

In the sense of KVP or Kaizen Benchmarking will be completed as an instrument which managers will use in daily life just like using the telephone. Initiation to changes; instruction of strategy recommendations that must be hardly acquired then; continuous Benchmarking circuits, report of an alteration plan.

The goal of Benchmarking is to reach a competition advantage and to hold this competitive position /Christ/.

References

/Camp89/   Camp, Robert: Benchmarking: The Search for Industry Best Practices that Lead to Superior Performance, Milwaukee, Wisconsin 1989.
/Christ/   Charles Christ, President of the Xerox Reprographics Group, 1980.
/DIN 19222/   Deutsche Elektrotechnische Kommission im DIN und VDE (DKE): DIN 19222, Berlin: Beuth Verlag, März 1985.
/From92/   Fromm, H.: Das Management von Zeit und Variabilität in Geschäftsprozessen, CIM-Management 5/92, S. 7-14.

Curriculum Vitae of Dr.-Ing. Kai Mertins

Dr.-Ing. Kai Mertins was born 1947 in the Federal Republic of Germany. After studying Control theory in Hamburg and Economy together with production technology at the Technical University of Berlin, he became member of the scientific staff of the University Institute for Machine Tool and manufacturing Technology (IWF), Berlin/FRG. Since 1983 he had been head of the department "Production Control and Manufacturing Systems" at the Fraunhofer-Institute for Production Systems and Design Technology (IPK), Berlin/FRG, where he has been Director for Planning Technology (President: Prof. Dr. h.c. mult. Dr.-Ing. G. Spur) since 1988. He has more than 15 years experience in design, planning, simulation and control of flexible manufacturing systems (FMS), manufacturing control systems (MCS), shop floor control systems, and computer integrated manufacturing (CIM). He was General Project Manager in several international industrial projects and gave lectures and seminars at the Technical University Berlin, Polytechnic Nottingham/UK, Czech Republic, Indonesia and China. Special field of interest: Manufacturing strategy development, planning for production systems, shop floor control and simulation.

Dipl.-Ing. Stefan Kempf

Born 1963.
Dipl.-Ing. in Mechanical Engineering at the Technical University of Berlin, Germany.
Mr. Kempf is working for the Fraunhofer-Institute for Production Systems and Design Technology (IPK) at the department Planning Technology since 1990.
His working areas are the planning of production systems, management consulting, application of grouptechnology, management of teams working on large software projects and developing software.

Dipl.-Ing. Gunnar Siebert

was born 1963 and after studying Industrial Engineering at the Technical University of Berlin he is working for the Fraunhofer-Institute for Production Systems and Design Technology (IPK), Berlin/FRG at the department Planning Technology. Since 1992 he is the Manager of the CIM Technology Transfer Center in Berlin/Brandenburg.
His working areas are management consulting, establishing logistics systems, introduction and establishing of CIM-Systems and benchmarking.

26

Customer Value Profiling:
Continuously Benchmarking What Matters Most --
Value Delivered to the Marketplace

Barbara Napier and Steve Crom

INTRODUCTION

In today's rapidly evolving markets, it is no longer sufficient for organizations to rely upon traditional benchmarking data gathered in the traditional manner. In order to provide value to the marketplace and keep a competitive edge, more and more organizations are learning to use feedback from customers as an essential new source of benchmarking data. We propose the use of the Customer Value Profile as a new method of benchmarking market forces from the **customer's** perspective. This new method relies on two key philosophical premises: 1) benchmarking market realities and preferences from the customer's perspective is better than benchmarking the performance of competitors or other industry leaders, and 2) a process of customer value analysis and process improvement that is led by the internal employees is superior to one which is led from an external source. Each of these premises are explored in depth in the following section, followed by a description of a new method of benchmarking customer value, the *Customer Value Profile*.

1. THE ADVANTAGES OF A CUSTOMER PERSPECTIVE IN BENCHMARKING

Most companies have realized by now that *quality* is just the price of admission in the competitive race to become a market leader. As we look past Total Quality to the future, we see that providing better *value* from the customer's perspective is the next step in the evolution toward improved organizational performance. Customer's perceptions of how well an organization is doing as a supplier (of goods or services) is the ultimate performance benchmark -- benchmarking the competition is no longer sufficient. By setting performance targets based on what the competition is doing, an organization risks always being second best. It is only by keeping your eye on the ultimate judge of performance, the customer, that the ultimate performance gains can be had.

There are several advantages of benchmarking customer's perceptions of performance. First, listening to issues raised by customers allows you to determine what they think is important, and what issues are most likely to affect their buying behavior. This information can then be used to calibrate internal benchmarking data to provide ongoing performance feedback to your employees, and to prioritize breakthrough improvement efforts to respond to the customer's needs. This ensures that you are working on improving the most important things, rather than nice-to-have peripheral issues that don't really affect buying behavior or ultimately market share. Second, customers can tell you what needs they have that are not currently being met. From that information you can take this information, possibly in partnership with your customers, and create innovative new services or products they would like to have, that maybe neither you nor your competition currently supply. The company closest to the customer who actively listens best gets the jump on the competition in developing innovations first, rather than hearing about them after someone else brings them to market. Third, customers can give you their perspectives on your relative market placement, so that you know what your strengths and weaknesses are from their perspective. Fourth, the act of listening to your customers will necessarily draw you closer together, and help to cement the relationship and increase customer loyalty. The method of having employees involved in the benchmarking process further cements this relationship; the advantages of employee involvement are outlined below.

2. EMPLOYEE INVOLVEMENT IN CUSTOMER BENCHMARKING

The general phenomenon of listening to the views and opinions of customers is well known, although usually conducted through third party market research. The method used for benchmarking with Customer Value Profiling, however, is different in several important ways. First, Customer Value Profiling is conducted primarily by internal employees of your organization, who identify themselves as such in their interviews with customers. The objectivity of pure data collection which is forfeited by introducing bias (which is not found when "blind" market research is conducted) is outweighed by the many advantages (outlined below) of this new method of data collection. Second, Customer Value Profiling starts with a qualitative component conducted by employees, which explores in the customers own words what their most important issues are. Only after this qualitative experience has been digested and understood by the employees is a quantitative component conducted. Therefore, the benchmarking data has both a quantitative and a qualitative segment, with both enriched by the employees having taken a leadership role in data collection. Third, the Customer Value Profiling process is by definition customer centered; rather than finding out about the marketplace, or about the competition, you are finding out about the customers and their needs, so that you can better align your services to meet and exceed those needs. Fourth, rather than assessing satisfaction with services currently rendered, a key focus of a Customer Value Profile is to uncover innovative new approaches that delight the customers, rather than just satisfy them.

Having employees from a cross section of your organization participate first hand in listening to the customer has a wide variety of benefits over having just one functional department responsible for benchmarking data gathering, or out-sourcing the data gathering to a third party. These benefits far outweigh any bias in pure data collection which is introduced by not having an objective third party interfacing with your customers. First, as employees learn more about the needs of the customers, they also experience first hand the importance of listening to the customer's voice. This helps to ensure that the customers are involved in the decision making and process improvements efforts after the data collection process is finished, since the customer has dozens of new advocates throughout your company. Second, since employees are personally going to customers to find out what is important to them, both the employees and the customers get a strong message that your organization is taking this process very seriously, and that clear and directed changes will result. Third, there is less internal resistance to change since these changes are understood by the employees, and are in fact led by them as well as they participate in task--team process improvement efforts.

Fourth, as mentioned earlier, simply by taking part in the dialogues, the employees play an important role in improving relations with your customers. Fifth, since employees take part first hand in conversations with customers, the resulting benchmarking data comes alive for everyone in a way that isn't possible when they are just handed a data summary. Mean scores and standard deviations are cold and meaningless compared to memories of dialogues and meaningful anecdotes. Sixth, most improvement efforts will involve some level of cross functional cooperation within your organization in order to achieve breakthrough improvement. Having a cross section of employees from as diverse department as sales, operations, marketing, research, and administration involved in the data gathering process begins to model the cross functional cooperation that will make those breakthrough improvements possible.

3. CUSTOMER VALUE PROFILING METHOD

The following sections describe the new method of Customer Value Profiling, as developed by the consulting firm of Rath & Strong Management Consultants, from Lexington, Massachusetts. This benchmarking method has been used by several organizations in the United States and Europe to create a system for benchmarking real-time the performance of essential processes, prioritize breakthrough improvement efforts, and incorporate the voice of the customer in their decision making processes. During this partnership, skills necessary to conduct this benchmarking on an ongoing basis are transferred by Rath & Strong consultants to the clients. This process involves interview skills, quantitative and qualitative data analysis abilities, questionnaire development skills, the creation of benchmark indices, and task-team process improvement abilities. Each of these are learned throughout the Customer Value Profile process.

A Customer Value Profile is a summary of an organization's market placement, measured on dimensions that are described as most important by their customers. Each dimension has, by definition, a "Market Average Value (MAV)." The MAV is, essentially, the industry norm for the quality of goods or services offered to a marketplace, or the marketplace "par." As mentioned earlier, providing excellent quality has become "par" in most venues. In order to succeed today, an organization must strive to be above par, or above the MAV point.

Since the goal is to improve processes, the first step is to identify the most important dimensions to the customers: the Value Descriptors. These Value Descriptors are later operationalized by employees into Value Indices. These Value Indices, when placed on a histogram with an assessment of the MAV, are collectively referred to as a Customer Value Profile. This Customer Value Profile is then used to target breakthrough improvement efforts, determine business strategy alterations, or uncover possible organizational restructuring, training or hiring needs, depending on the outcome. A summary of the Customer Value Profile Process is presented below, followed by some concluding comments.

3.1. Orientation

During the orientation phase, Rath & Strong consultants learn as much as possible about the issues facing our client organization (hereafter referred to as "our organization"): internal strengths and weaknesses, market forces, history, etc. At the same time, we introduce the theory of Customer Value Profiling to employees at various levels of our organization, in a variety of organizational functions. At this point, employees who will be involved in the first stage of interviews are chosen and their role in the process explained.

3.2. Initial Market Contact

Initial market contact is conducted in order to determine, in the customers' own words, what issues are most important to them in choosing a supplier (of goods or services, hereafter referred to simply as supplier). Employees take part in these interviews primarily in an observer role; employees from management, sales, marketing, production, or a variety of other roles may be chosen to take part of these interviews. Interviews are arranged in order to have a sample of participants who as accurately as possible represent the actual range of different types of customers, stratified by size, geographic location, market sector, or any other relevant characteristics. Furthermore, non-customers (potential customers or former customers) are included as well (hereafter, the term "customers" refers to potential as well as actual customers). Typically, 15 - 20 interviews are sufficient, depending on the complexity of the market.

During these initial interviews, open ended questions are asked in order to get a better understanding of what things are most important to customers when they are evaluating current suppliers, or making a decision to choose a new supplier. Questions are also asked to determine if there are current needs that are not being met by current suppliers. A question is included to determine who is the best out of all their suppliers and why. This benchmark question often reveals superior services offered by suppliers in other fields that could possibly be transferred to this market. Other questions are also included as time permits, including questions about general market forces, specific issues facing the marketplace, projection of future changes, etc.

Finally, customers are asked to rate, on a 1 to 5 scale, our organization as well as a competitor on the issues they themselves have identified throughout the interview as being most important to them. This empirical data is used in conjunction with the qualitative data for determining Value Descriptors and getting an early picture of where our organizations' strengths and weaknesses lie.

3.3. Develop and Review Value Descriptors

Value Descriptors are a summary of issues raised by the customers. A thematic analysis of the issues identified throughout the interviews usually yield between 6 and 8 themes; the title given to each of these themes are referred to as a Value Descriptor. Whenever possible, these Value Descriptors should be in the customer's own words. These Value Descriptors are eventually reflected in the Customer Value Profile using internal data for real-time benchmarking of critical processes.

Other analysis of the data collected in the initial interviews includes a summary of relative strengths and weaknesses relative to the competition, mean scores of our organization and the competition given by each customer, and a summary of any salient new information which has emerged by this point (such as information about market trends, new innovations, etc.).

This initial analysis is reviewed with the team of employees who have taken part in the interviews. The thematic grouping of issues is adjusted as necessary to reflect the employees' input. At this point, employees are asked to predict what they think the Customer Value Profile will look like, including strengths and weaknesses as reflected by placement relative to the MAV line.

3.4. Form the Value Profile Team

Once the preliminary data has been collected, the next step is to charter the Value Profile Team. It is important that this team be made up of top management representatives from Sales/Marketing, Operations, Finance, and Human Resources, since any changes that take place as a results of the process will probably involve each of those departments. The formation of this team helps to ensure that there is cross functional support for the entire Customer Value project. The major mission of this team is to guide policy decisions, select team members for future projects, and shepherd the change process.

3.5 Collect Data to Create Preliminary Profile

One of the strengths of the Customer Value Profile method is that it starts with a qualitative segment to determine the relevant issues in that particular market, and then moves on to a more quantitative segment based on that information. The quantitative data gathering procedure involves the use of structured questionnaires administered to a cross section of customers. Interviewers and observers are selected by the Value Profile Team to take part in the data gathering. Participants should have good interpersonal skills, and should come from a broad a cross section of functions and levels of the organization.

A questionnaire is developed with the help of the interview teams that reflects the issues raised in the initial interviews. The Value Descriptors are operationalized into close-ended questions, and various open ended questions are included to reflect buying behavior, demographics, needs currently being unmet, etc.

The interview teams also determine the contacts for interviews that will be made at various customers. Interviews should be stratified as much as possible to reflect the demographic make-up of our organization's customer base as well as strategically picked potential customers. For instance, interviews should span various organizational functions and levels at the customer's organization, various sizes of organizations, etc. Given the types of analyses being conducted with this empirical data and the importance it will play in determining the business strategy, it is important to conduct as many interviews as possible. Although small organizations with small customer bases are not excluded from this type of benchmarking, there should optimally be over 100 responses in this second round of interviews.

Skills training on the do's and don'ts of interviewing is an essential step of the process. Interviewers are trained on the use of probing questions, the importance of body language, and various other aspects of interview techniques. They are then asked to role play the interviewing experience in teams of two. This role playing serves two functions: it pretests the questionnaire, and it provides data about the interviewers' perceptions of the customers' perspectives. The later can be used in subsequent data analysis to determine how accurate the employees were in their preconceived notions about the customers. If possible, the questionnaire should also be pretested on one or two customers and then adjusted as necessary.

3.6. Analyze Results

Results are analyzed in a variety of ways, depending on the data included in the questionnaire. As a minimum, the questions included to reflect the Value Descriptors should be analyzed to see how well they cluster together. Relationships between the Value Descriptors and various buying behavior questions should also be explored to determine relative importance of the Value Descriptors. Thematic analysis of open-ended questions should be conducted, if at all possible in conjunction with the interview teams.

Questionnaires are designed so that interviewees primarily rate the performance of their primary supplier. Since the primary supplier is identified, analysis can then be conducted on the relative strengths and weaknesses of various suppliers, including our organization. This provides important benchmarking data which is then used to form a preliminary Value Profile.

3.7. Review Results with Team and Others

Results of the empirical data analysis are reviewed with the Profile Team as well as the interview teams. The teams envision their ideal Customer Value Profile, and compare this ideal to the preliminary Value Profile. Teams members are asked to select employees who are closest to the processes that are indicated in the preliminary Customer Value Profile who would be most appropriate to help identify which internal metrics are available to measure those processes.

3.8. Identify Required Metrics for Value Indices

At this point, The Value Descriptors begin to be turned into Value Indices. Whereas Value Descriptors are important issues from the customer's perspective, Value Indices are operationalizations of those issues from the perspective of our organization. In other words, if one of the Value Descriptors was "Delivers on Demand," the Value Index might include various internal measures of delivery capability, such as number of rejected deliveries, on-time delivery rates, time elapsed between request and delivery, etc. Information which goes into making up the Value Indices should be readily available, meaningful to employees, and simple. These internal measures are combined into a Value Index.

Once internal metrics have been determined for Value Indices, metrics that must be determined by surveying customers are identified. Depending on the Value Index, short item questionnaires or other feedback procedures may be most appropriate.

3.9. Create Value Indices and Value Profile

Once the metrics for the Value Indices have been determined, the necessary data is gathered together and systems for updating this information are established.   Value Indices, when displayed together with estimations of MAV, form the Customer Value Profile. (MAV estimates are based on data gathered from the questionnaire). The Customer Value Indices and Customer Value Profile should be posted in a visible place and updated periodically -- weekly, monthly or quarterly, as appropriate.

One of the unique features of Customer Value Profiling is the continuous nature of the benchmarking process. This method gives people at the process level information that they can use to take corrective action or make improvements to processes which have been shown to be essential to their organization's success from the customers' perspective. The real-time nature of this process differentiates it from other types of benchmarking, as an organization's performance is continuously being monitored throughout their improvement efforts and beyond.

3.10. Prioritize Processes for Improvement

At this point, the Profile Team has the information they need to prioritize process improvement efforts. From the data analysis, this team has information about which issues most influence buying behavior as well as about our organization's performance relative to our competition. Based on this information and other strategic concerns, the Profile Team prioritizes processes most in need of improvement. Usually two or three Value Descriptors are chosen by the Profile Team for breakthrough process improvement efforts. At this point, recommendations about possible strategic human resource improvements (hiring or training needs, restructuring) are also made.

3.11. Develop Teams for Breakthrough Improvement Efforts

Depending on the issues raised and priorities chosen for process improvement efforts, teams are selected and chartered to take part in those improvement efforts. We recommend focusing on how the team will work together before launching into problem solving efforts. Teams that do not succeed in process improvement efforts usually have great ideas, but cannot get along, or work together to agree on a improvement strategy that everyone will support.

3.12. Communicate the Profile and Key Processes to be Improved

Once the strategic decision of which key processes will be improved has been made, the newly created Customer Value Profile should be communicated to all staff members, along with priorities and rationales for process improvements. The improvement teams should be announced along with their charters for improvement efforts. It is really important at this point to reinforce that every employee has an impact on the services provided to the customer, and ask that everyone take some time to identify ways in which their activities impact one or more of the value descriptors. The Customer Value Profile philosophy and process should be explained to everyone, including systems in place to communicate progress in improvement efforts through regular postings of Value Indices. Each employee should be asked for suggestions on how to reach improvement goals.

3.13. Monitor Value Indices Over Time and Adjust When Necessary

The Value Indices are updated and posted on a regular ongoing basis. In addition, the Customer Value Profile process will need to be repeated on average about once every two years, depending on the changes in the marketplace.

CONCLUSIONS

The ultimate goal of any business is delivering value to the marketplace, thereby increasing market share. By going directly to the customers and determining what value means to them, organizations can improve their abilities to meet and exceed the customers expectations. By focusing your benchmarking efforts on the perceptions of the customers, then internalizing these perceptions into commonly recognized metrics, your entire organization will move closer to the customer. When anticipating their needs then excelling at meeting those needs becomes second nature, increased market share will undoubtedly follow.

Author's Note: The Authors gratefully acknowledge the contributions made by John Guaspari in the co-development of Customer Value Profiling.

27

An Architecture of Automated Consultancy Tools for Benchmarking and
Implementing Production Planning and Control Software

I.P. Tatsiopoulos

National Technical University of Athens, Dept. of Mechanical Engineering,
Section of Industrial Management & O.R., 15 780 Zografos, Athens - Greece

Objective of this study is the development of automated consultancy tools (*ACT*) to
support the evaluation, selection and implementation phases of production planning and
control software (*PPC*). The methodology and the architecture of a *knowledge-based
consultant's workbench* are presented to help solving this time-consuming and costly
problem.   The main aim is the production of tools for identifying the PPC software
configuration and customization requirements combined with the required feasible
organisational changes of the industrial firm. The *ACT workbench* is implemented in the form
of a database-oriented suite of software tools connected to an expert system shell.

1.   INTRODUCTION

The task of implementing manufacturing software is extremely complex and usually
needs the involvement of professional consultants and trainers. As Scheer (1993) reports, the
implementation costs of such systems, including requirements analysis, software package
selection, customization, installation and training amounts to *5 or 6 times* the combined
hardware and application software costs. This problem is crucial particularly for small and
medium sized firms (SMEs), due to their lack of internal expertise and resources.   This
situation leads to the need for time-consuming and expensive consultancy services that tend
to become a serious barrier to the adoption of integrated PPC systems by the majority of
SME's. The only way to overrun this barrier is to provide automated consultancy tools using
current information technology and knowledge engineering.   These tools will help both
consultants and industrial users to cut down the large manpower resources needed, and to
make PPC projects affordable.

The reported efforts in the literature for the determination of PPC systems requirements
and the implementation of manufacturing software have their origin in different science fields
using different approaches. Three main research streams can be identified related to different
problem stages, i.e. the requirements analysis, the software package selection and the system
implementation. The requirements analysis approach is   represented by the CASE
(Computer Aided Software Engineering) tools supporting a variety of well-known structured

systems analysis and design methods   (Yourdon, De Marco, Gane and Sarson, Jackson, SADT or IDEF0, SSDM and Object Oriented Analysis).   More closely to the PPC requirements determination and implementation problem is the CIM-OSA (1991) model, the Scheer (1993) methodology of industry-specific reference models and the GRAI method (Doumeingts, 1989) which puts the emphasis on the decisional rather than the informational structures of a PPC system.

An example of the software package selection approach is the well accepted PPC software evaluation system BAPSY of the TH Aachen (Speith and Brief, 1982),   which uses multi-criteria and value analysis methods to compare different PPC commercial software packages and select the most appropriate for a specific manufacturing firm.   A similar methodology which is less theoretically sound but more practically oriented is the Geitner (1993) PPC market analysis database system.

Representatives of the system implementation stage is the system TWAICE of Nixdorf Computer (Mensel and Michel, 1985) which is mainly a software   configuration system to support the Nixdorf COMET integrated manufacturing   package and the needs of its customers for customization and consultation, as well as the Mertens (1993) tools to regulate parameters of MRP systems and the earlier efforts of this author (Tatsiopoulos, 1990).

Among the ESPRIT research projects, relevance have the projects AMICE and VOICE which promote the CIM-OSA architecture for developing CIM systems and, more closely to this proposal, the project CIMple which produced the FTM (Fast Track Modelling) product to help the definition of CIM requirements.

2.   THE OVERALL ARCHITECTURE

This study adopts the systems engineering concept of dual   architecture of Daenzer (1973) who defines two series of systems engineering phases, i.e. the life-cycle phases that represent the time dimension and the problem-solvings steps which are recursively applied to all the life-cycle steps. However, the definition of phases and steps follows the Checkland's (1983) proposals for the treatment of changes and, moreover, is specifically adjusted to the production planning and control systems environment.

LIFE-CYCLE PHASES
P1.   *Initial Study* (feasibility, root definitions, constraints, strategic decisions)
P2.   *Main Study* (PPC requirements definition, conceptualisation, reference models)
P3.   *Detailed Study* (comparison of software packages)
P4.   *Development of Potential Changes* (software customization and organisational changes)
P5.   *Implementation of Desirable and Feasible Changes*
P6.   *Appraisal* (internal and external benchmarking of PPC function)

PROBLEM-SOLVING STEPS
S1.   *Current situation analysis*
S2.   *Objectives setting*
S3.   *Synthesis/analysis of alternative solutions*
S4.   *Evaluation of solutions*
S5.   *Realisation of chosen solution*

The systematisation of knowledge needed to the consultancy support of each one of the life-cycle phases follows the combination of phases and problem-solving steps. The result is a consultant's workbench consisting of six main knowledge-based systems which correspond to each one of the life-cycle phases, and thirty subsystems corresponding to the combinations of phases and problem-solving steps. There follows a short description of the main systems of the workbench.

P1.   Initial Study
A short-listed 'menu' of alternatives has to be evaluated and presented to higher level management. These alternatives include the *make or buy* decision concerning PPC software packages, the range of budget as compared to the degree of automation (i.e. *low-budget 'hamburger'* solution which meets only the most pressing of the user's objectives, *medium budget 'fried chicken'* solution or a *higher-budget 'chateaubriand steak'* solution) and the Hardware/System Software *platforms* where the PPC application software will run under the constraints of the overall information technology strategy of the company.

P2.   Main Study
It addresses the problem of determining PPC software requirements on the basis of pure technical characteristics expressed in the form of a *criteria catalogue*. A   knowledge-based model is proposed whose objects and relations describe the   type of production system through the representation of a *production systems typology* which leads to a *company typological profile* and the reference model of a generic PPC software package.   The specific criteria out of the knowledge base of criteria catalogues that satisfy the needs of the customer' s industrial firm are sent out to vendors of PPC software packages in the form of a requirements catalogue for bidding on their part.

P3.   Detailed Study
Its main purpose is the comparison of existing PPC software packages in the market on the basis of the answers of their vendors to the requirements catalogue which has been sent as the result of consultancy work at the main study stage. It consists of the *coverage analysis* to determine the degree of covering the company's requirements by the candidate packages' offered features. On the basis of this analysis, an evaluation of packages takes place that leads to the proposed by the consultant solution. We must here take into account that the coverage analysis is done on the basis of the firm' s specific 'ideal' reference model as compared to the generic PPC reference model and the model of features of every software package.

P4.   Development of Potential Changes
Software development is not usually part of the consultant's job. The support that is offered by the ACT workbench at this stage is the detailed determination of inconsistencies between the chosen software package, which now takes the place of the 'ideal' reference model, and the current situation in the firm.

The two extreme decisions are that either the procedures of the selected   software package should be accepted exactly as they are, which should lead   to extensive reorganization within the firm, or the software package should be modified to the extent that it should completely reflect the specific needs of the firm. Both solutions are impracticable in real production management environments. What is really done is to find a solution in between, i.e. to modify both the software package and the organization of the firm to a certain limited extent which represents an accepted total of tangible and intangible costs.

P5.   Implementation of Desired and Feasible Changes
This system tries to transform the potential changes derived during the previous phase (P4) into changes which are agreed to be both desirable and feasible given the resources and the attributes of the organisational environment. As a result the *type* (structural, procedural, attitudinal or environmental) and *amount* of feasible changes are produced both for the customisation and configuration of the software and for the firm's organisation.

Next, we shall describe in more detail, as an example of the methodology, the main study (P2) phase, which also is in a more mature status in the ACT workbench project.

3.   MAIN STUDY

It is suggested here that the software technical evaluation procedure should produce a set of "ideal" PPC requirements at three levels: (a) The functional production type (e.g. job-shop), (b) The industrial sector (e.g. clothing industry), (c) The specific manufacturing firm.

The requirements for all functional production types will be based on a taxonomy according to functional characteristics (e.g. product standardisation, factory layout, etc) and the knowledge domain of production management theory about the needs of a particular production type. The industry sector-specific requirements are imposed by technological processes, special problem areas, objectives, measures of performance and environmental factors, even aliases of production management terms which are characteristic of each separate industrial sector.

For the determination of PPC requirements of a specific manufacturing firm, a *"normative"* approach (Davis and Olson, 1985), using a generic description of an "ideal" generic PPC software package *(reference model)* will be followed at this stage for two main reasons: a) It is quite common that the proposed information system embodied in the production management generic software package may be fundamentally different from existing patterns (in its content, form, complexity, etc.), so that anchoring on an existing information system or existing observations of information needs will not yield a complete and correct set of "ideal" requirements. b) Usually in practice there is not enough time at this stage to proceed to a detailed analysis of the company's current production information processing and organisational procedures.

The systematization of expert knowledge needed at this life-cycle phase (P2) includes the knowledge bases (KBs) which are necessary at each of the problem-solving steps (S1 - S5).

3.1.   Step P2.S1 - Main study's current situation analysis
This includes the description of the firm's production structure and the classification of the firm's production system according to a knowledge base of functional production types and industrial sectors. The following two KBs apply to this situation:

P2.S1-KB1   Taxonomy of Production Systems
P2.S1-KB2   Reference Model of System Variables

P2.S1-KB1   Taxonomy of Production Systems

Many authors have proposed classifications of production systems, e.g.   Schmenner (1981), Schmitt et al (1985) and Kettner et al. (1984). Their   common characteristic is that they adopt a narrow view of the physical   production system dealing only with functional characteristics like shop   layout and process flow classifications, which are not enough to determine production management requirements.   Factors like management objectives and measures of performance, main problem areas, special technological processes and specific environmental constraints which are connected to the various industrial sectors and subsectors are equally important for the determination of PPC requirements.

The proposed taxonomy in this study is of a twofold nature, a functional   taxonomy and a taxonomy based on industrial sectors. For the first one, a   system of classes of parameters and values has its origin in the Schomburg (1980) classification model which was built specifically for engineering firms. However, the taxonomy proposed here is extended and built in a software tool using object-oriented programming concepts. This software tool facilitates the connection of the functional taxonomy to the industrial sectors taxonomy and all relative classes, i.e.   problem areas, objectives, environmental factors and aliases (terminology of production management across industrial sectors) as seen in Figure 1.

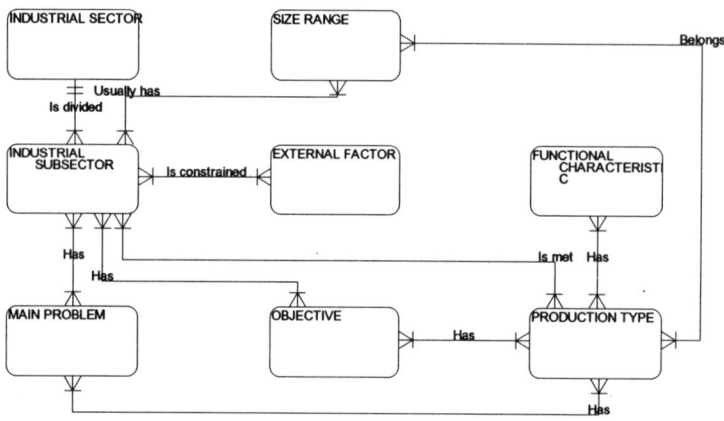

Figure 1. Knowledge Base of Production Systems Taxonomy

As far as industrial sectors and subsectors are concerned, the formal   classification of National Statistical Service has been used.   For the classes of functional production characteristics, the   classification model of Schomburg (1980) enlarged by Tatsiopoulos (1988) is used. This model uses 9 classes, each one of them having several instances.

P2.S1-KB2   Reference Model of System Variables

A structured analysis of the firms's current system would normally include the development of detailed Extended Data Dictionaries including Data Flow Diagrams and Entity Relationship Diagrams. However, for the objectives of a consultancy workbench we consider this procedure extremely complicate and time-consuming. Instead, we base our decription on the concept of the Burbidge's (1984) variable connectance model, which is applied to the following six basic production planning and control functions, likely to be apparent to all manufacturing firms:

PDM = Production Data Management   CRP = Capacity Management
MPS = Master Production Scheduling   SFC = Shop Floor Control
MRP = Materials Management   PUR = Purchasing

For each one of the above functions, a reference model of the following sets of parameters consists the P2.S1 - KB2 knowledge base. The firm's current model can be constructed out of this generic reference model by choosing the relevant parameters.

A *DESIGN PARAMETER* is a variable which can take one or a combination of   a limited number of alternative values, defined in advance. Examples of   design parameters are the 'planning frequency', the 'planning horizon' and   the 'length of planning periods'.

A *REGULATORY PARAMETER* is a variable to which a range of different   values, not defined in advance, can be assigned.

An *OUTPUT VARIABLE* is a variable which can only indirectly be altered by changing parameter values (usually measures of performance).

Examples of Regulatory Parameters and Output variables are: Run   Quantity, Transfer Quantity, Buffer stock (Regulatory), Throughput time,   Storage time (Output). What interests in terms of the production   management requirements is the capability of the PPC system to   provide decision-making for the Regulatory Parameters and to measure the Output Variables.

An *ENVIRONMENTAL (or Input) VARIABLE* is a variable whose value is   imposed by the Environment in which the system exists and over which it has   little or no control. Examples of this kind of parameters are given below:

- Traceability regulations (LOT TRACKING, CONFIGURATION
  CONTROL,SERIAL NUMBERS)
- Stock recording by government contract (YES,NO)
- State tax regulations, e.g. obligation for tracking of
  tax-free materials (YES,NO)
- State import-export regulations (e.g. tax-free raw materials
  if included in export products (YES,NO)
- Foreign currency transactions (YES,NO)
- Number of possible material suppliers (ONE,FEW,MANY)
- Reliability of suppliers (GOOD,FAIR,POOR)
- Number of competitors (NONE,FEW,MANY)
- Strike potential (RARE,POSSIBLE,FREQUENT)
- State regulations regarding hiring, firing and overtime
  (NEGATIVE,RESTRICTIVE,POSITIVE)

3.2.   Step P2.S2 - Main study's objectives setting

The search for objectives is based on the current situation analysis and is crystalised in the form of knowledge bases of hierarchically structured classes of objectives. Important at this stage is that objectives are classified in absolutely *"must"* objectives and *"desired"* but not necessary objectives. A further classification of objectives is undertaken in three dimensions, the *context* (functional, financial and personnel), the *time* (short-term, medium-term, long-term) and the *place* (system internal, external environment).

The objectives setting step has to take objectives of the PPC system, which have been found at the current situation analysis phase in the form of *Output or Performance Variables,* and then classify and structure them in a hierarchical and cosistent way as described above. Only then it is possible to proceed to the following step which is the synthesis/analysis of alternative PPC reference models in the light of the pre-defined objectives. The knowledge base used at this stage is:

P2.S2-KB1   Classes of Objectives

3.3.   Step P2.S3 - Synthesis/analysis of alternative solutions

The potential alternative solutions at this stage include different *reference models of 'ideal' PPC requirements* which would be suitable for the particular manufacturing firm. The chosen reference model should be sent to candidate software vendors in the form of a *requirements catalogue* for bidding.

Alternative reference models are produced out of a generic description of a PPC software package, which is formulated in classes of technical *features,* through the configuration of features to be included in the required PPC reference model for the specific manufacturing firm.

P2.S3.KB1   The Generic Reference Model of a PPC Software Package

Most of the available commercial software modules for production management belong to large integrated packages of manufacturing management software which even though they use slightly different names for their modules, it can be said that the terminology and content of the most functions offered is fairly   standard as, for example, MRP for materials management or MPS for Master Production Scheduling.

The purpose of this study is to build a knowledge base of a generic PPC reference model which contains, in terms of parametric "modules" and "features", all the Knowledge domain of current PPC information technology and production management theory. The typical Data Dictionary for documenting software applications contains the entities Application (Program), Module, Database, File (Table), Data Group (User View), and Data Element. This   study adopts the opinion that such a detailed data dictionary is quite impractical in the case of PPC packages due to their immense size and complexity. Instead, from the level of module downwards, the notion of *"Features"* and *"Parameters"* with their corresponding instances is used for description of the reference model.

For the module parameters, the terms used in this study correspond to the reference model of system variables defined previously in paragraph 3.1.

Module Features (the example of the MRP module)

The concept of *module feature* is characteristic in the world of   commercial software being of a rather verbal nature. Usually it represents   some sort of functionality and how it is performed. Its analytical   description and documentation would require extensive data flow diagrams   and flowcharts which is not practical for our purposes. Therefore, the "feature" is described by a short phrase and a set of values corresponding   to alternative methods of performing its function. In this study an   attempt is undertaken to codify, standardize and store in a knowledge base   all known PPC software features and their alternative values. In terms of   the requirements of the production management system, first we are   interested in the very existence of a specific feature and then in its   value, i.e. the method used. Next is given an example of codified   features for the MRP module.

```
01. MRP operation (BATCH, INTERACTIVE, BOTH)
02. Form of MRP explosion (REGENERATIVE, NET-CHANGE, BOTH)
03. Requirements presentation (BY PERIOD, BY DATE, BY JOB)
04. Calculation of lead times (BY ITEM, BY PARENT-SON RELATION, BY
    BATCH PROCESS TIME)
05. Form of orders issued (WORK ORDER, FLOW ORDER, BOTH)
06. Pegging of Requirements (SINGLE LEVEL, FULL LEVEL EXPLOSIVE,
    FULL LEVEL IMPLOSIVE, MRP BY CONTRACT)
07. Selection of depth of requirements explosion (YES,NO)
08. Simulation net-change capabilities (YES,NO)
09. Lot sizing techniques (USER INPUT, PROGRAMMED RULES OF THUMB,
    PROGRAMMED OPTIMIZING)
10. Phantom items (YES,NO)
11. Blanket material acquisitions for large orders (YES,NO)
12. Allocations (BY PRODUCTION ORDER, BY CONTRACT, TIME-PHASED,
    AUTOMATIC MULTI LEVEL, ALL, NO)
13. Allocation of safety stocks (USER INPUT, PROGR.RULES OF THUMB,
    MATHEMATICAL)
14. Backflushing (YES,NO)
15. Government contract support, i.e. purchase and production
    orders by contract (YES,NO)
16. Lot tracking (YES,NO)
17. Variable Yield (YES,NO)
18. Configuration control (YES,NO)
19. Support of co-product and by-product creation (YES,NO)
```

In order to decide on the selection problem of module features and   parameters, the production systems taxonomy and the generic PPC software   knowledge bases have to be combined and inferenced under the logical schema of Figure 2.

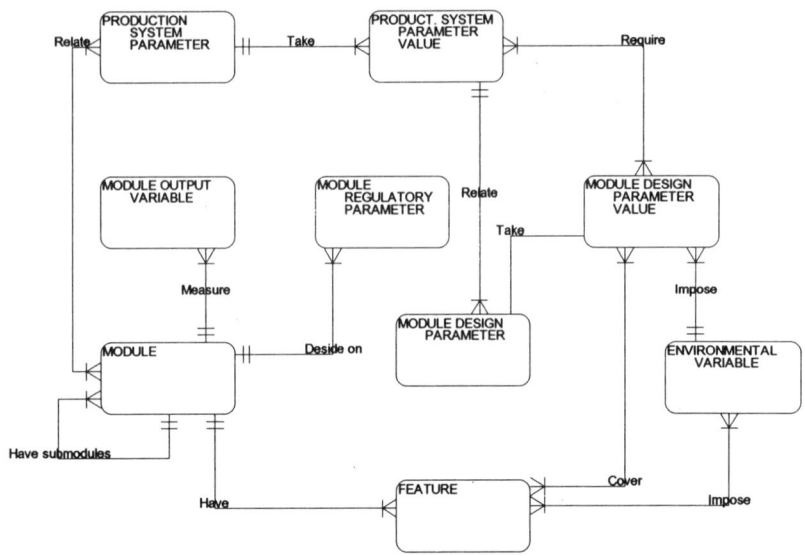

Figure 2. Generic Reference Model of a PPC software package

A PPC software structure derived from a certain production system type can be covered by the existence of a particular software module (MOD) and/or the value of a module feature (FTR). For example the module MRP has as a submodule the "KNOW ABOUT THE CONSEQUENCES OF A LATE PURCHASE ORDER ON   THE PROGRESS OF A CUSTOMER ORDER" which is required by the PSP parameter value "Demand Dependent on the Customer Order" and is covered by the   software module feature MRP-06 (FULL LEVEL IMPLOSIVE).

3.4.   Step P2.S4 - Evaluation of solutions

There is a broad spectrum of "Suitability" of a particular software   module feature from very good to very bad. For this reason there is often a   temptation to use quantitative methods like value analysis which specify   the decision on a numerical scale, say from 1 to 100. However, the factors   talked about in this study are general guidelines, often not very well   specified, and in a real-world situation they may not be known with great   accuracy. This suggests that a 0 to 100 decision may be a meaningless   breakdown given the fuzziness of the factors that go into making the   decision. A better choice would be to limit ourselves to a three-way   decision: *CRITICAL, DESIRED, NOT NECESSARY*.

3.5.   Step P2.S5 - Realisation of chosen solution

This step of phase P2 (main study) means the consolidation of a criteria and requirements catalogue in the form of a questionnaire and its sending to all candidate PPC software vendors according to a mailing list database of their names and addresses.

> *P2.S5.KB1   Mailing list of PPC software vendors*

REFERENCES

CIM-OSA, Open System Architecture for CIM, Esprit Consortium AMICE, Belgium, 1991.
CIMple, Esprit Project No. 5424, 1992.
BURBIDGE,J.L., *A Production System Variable Connectance Model*, Cranfield Institue of Technology, Cranfield, 1984.
CHECKLAND, P., *Systems Thinking, Systems Practice*, Wiley, UK, 1983.
DAENZER,W, *Systems Engineering*, io Verlag, 1973.
DAVIS, G.B. and OLSON, M.H., *Management Information Systems*, McGraw Hill,1985.
DOUMEINGTS,G. et al, Knowledge-Based System for the Design of Production Management Systems, in: J. Browne (Ed), Knowledge- Based Production Management Systems, Elsevier (North-Holland), IFIP, 1989.
GEITNER, U., PPS Marktuebersicht, *FB/IE, REFA*, December, 1993.
KETTNER,H., J.Schmidt and H.R.Greim, *Leitfaden der systematischen Fabrikplanung*, Carl Hanser Verlag, Muenchen, 1984.
MENSEL,G.und J.MICHEL, Moeglichkeiten des Einsatzes wissenbasierter Systeme in der Fertigung, *ZwF* 80 11, pp. 495-500, 1985.
SCHEER, A.-W., Knowledge-Based Optimization of CIM-Systems Using Industry-Specific Reference Models, in: Pappas,I. and Tatsiopoulos,I. (Eds), *Advances in Production Management Systems, IFIP Transactions*, North-Holland, 1993.
SCHMENNER,R.G., *Production/Operations Management*, SRA, Chicago, 1981.
SCHMITT,T.G., T.KLASTORIN and A.SHTUB, Production classification system: concepts, models and strategies, *Int.J.Prod.Res.*, 23 (1985) 3, pp. 563-578.
SCHOMBURG,E., *Entwicklung eines betriebstypologischen Instrumentariums zur systematischen Ermittlung der Anforderungen an EDV-gestuetzte   Produktionsplanungs- und-steuerungssysteme im Maschinenbau*, Dissertation,   Aachen, 1980.
SPEITH, G. and BRIEF, U., BAPSY - ein Instrumentarium zur Beurteilung und Auswahl von Produktionsplanung - und steuerungs - Systemen, *FIR - Mitteilungen* (1982) , TH Aachen, Nr. 43 (Juni).
TATSIOPOULOS, I.P., A Systematization of Knowledge for the Selection and Implementation of Materials Management Software, J.Browne (Ed.), Elsevier (North-Holland), IFIP, 1989.

28

Productivity Measurement

Methods and tools developed in TOPP

(A productivity program for manufacturing industries)

Research manager Bjørn Moseng

SINTEF Production Engineering

SUMMARY

The paper describes four methods (self audit, extended audit, self assessment, benchmarking) for measuring of productivity and competitiveness. The methods are developed in the Norwegian Productivity Program - TOPP, and is used in approximately 50 companies participating in the program. The self audit and external audit are developed mainly for mechanical industry (including shipyards, offshore and electromechanical industry) while self assessment and benchmarking are more general methods. The productivity is described by indicators given points on a scale from 1 to 7 (1 is far behind international competitors, 7 is "best practice").

1. PRODUCTIVITY MEASUREMENT

Measurement of productivity and competitiveness in industry companies is difficult and raise several questions. What is productivity? Does productivity means the same for different people and organisation? Different viewpoints and positions give different answers.

The classic definition of productivity is produced goods pr unit of production factors. This definition indicates a focus on the value adding and physical production process. However to produce a lot of goods with max. profit if there is no market has no sense. Customer satisfaction is an extremely important factor in obtaining competitiveness.

Measuring of performance could be looked at in 3 dimensions (fig. 1).

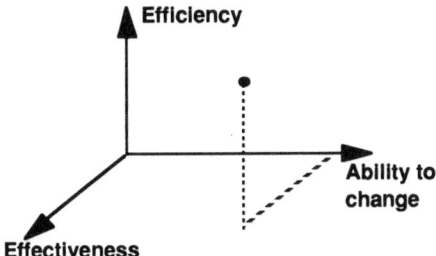

Figure 1 Performance measurement

In business terms the 3 views could be described as follows:

Effectiveness   to which extent are customer needs satisfied

Efficiency   to which extent are the total resources in the company used in an effective and economic way

Ability to change to which extent are the company prepared to handle changes in surrounding conditions (strategic awareness)

When establishing a measuring methodology it is important to have in mind all 3 dimensions. A lot of surrounding factors will influence the competitiveness of a company. Some important factors to consider are shown in figure 2.

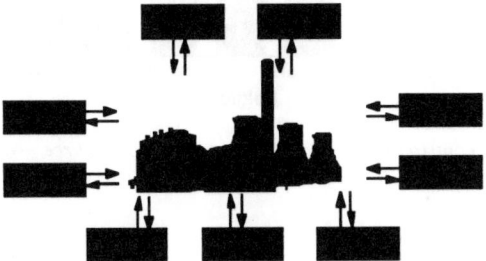

Figure 2. Company surroundings

2. THE TOPP PROGRAM

Introduction Norwegian economy is very much based on export of raw materials and semi finished products (fig. 3). Companies working on the international market are often sub suppliers to multinational customers. The contribution from industry to the gross national product (BNP) is also very low, and much lower than countries we like to compare with (fig. 4).

Those factors were important arguments to look at productivity and international competitiveness in Norwegian industry. To meet this challenges the productivity program TOPP has been launched in Norway.

Exports	%	Imports	%
Mineral oil	23,0	Ship	9,5
Gas	10,1	Clothes and accessories	4,9
Aluminium	8,6	Computing equipment	3,9
Fish	5,7	Instruments	2,6
ship	4,3	Paper	2,5
Paper	3,7	Telecommunication	2,4
Raw materials for plastics	2,5	Cars	2,3
Pig iron	2,5	Furniture	2,2
Nickel	2,4	Raw material for plastics	2,1
Total	62,8	Total	32,4

Figure 3 Norway - export/import (Source: Federation of Norwegian industries)

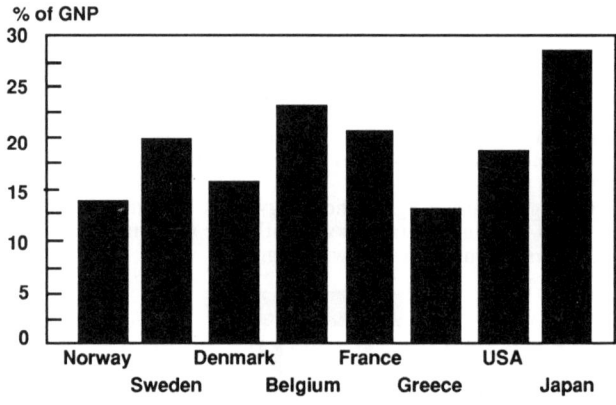

Figure 4 Industry contribution to gross national product (source: Norwegian Bureau of statistics 1993)

Goal The overall goal of TOPP is

to focus on total productivity for the whole enterprise and stimulate an industrial climate that improves international competitiveness.

Important objectives and key issues in TOPP are
- time to market
- quality
- flexibility
- total cost

The program involves co-operation between
- The Federation of Norwegian Engineering Industries
- Norwegian Institute of Technology (NTH)
- Industrial companies (~ 40)

and is sponsored by

- The Research Council of Norway

The TOPP program is planned for the period 1992-95. Following subprograms are scheduled:

1. Analysing company productivity and competitiveness (self audit, extended audit, benchmarking etc.)
2. Implementing actions for industrial productivity improvements (industrial projects, seminars, courses, industrial networks, etc.)
3. Generating new knowledge (research projects analysing productivity data, etc.)
4. Long term competence program (education , courses, dr.ing. programs, master degree, etc.)

3. METHODS DEVELOPED IN TOPP

Taking into consideration different views and dimensions in productivity and performance measurement, following approaches (methods) have been developed

a) Self audit   (questionnaire)
b) Extended audit   (experts)
c) Self assessment   (continuous improvement)
d) Benchmarking   (breakthrough)

The different methods will be briefly described in the following.

3.1 Self audit

Methodology The self audit is based on a questionnaire answered by the companies. The questionnaire consist of 3 parts asking for different types of information.

Part 1   Fact data
Part 2   General evaluation of functions and system variables (by individuals)
Part 3   Detailed evaluation of functions and system variables (by group)

A lot of fact data or estimates have to be given in part 1. This consist of general information about the company, product data, cost distribution, maintenance cost, quality cost, strategies, customers, sub suppliers, systems, financial and economic data etc.

In part 2 individuals (~20) in the company are asked to give their overall evaluation of different functions (primary and support) and system variables (facilities, equipment, personnel, organisation etc.). This evaluation is given confidentially and is later compared with information given in part 3 (see fig. 6).

Part 3 is the largest and most detailed part of the questionnaire. The company is asked to build groups to evaluate a lot of detailed questions concerning all functions and system variables. Example on questions to evaluate the design function is shown in fig. 5.

Functions and system variables are evaluated on a scale 1-7 where 1 is far behind, 4 is on same level as most important competitors, 7 is "best practice".

By design we mean the function responsible for the production support either for order production or in connection to product development. Typical activities will be analysis, calculation, work out drawings etc.

6201   Evaluate the following areas/aspects as to current situation, and give a realistic possibility for improvement by estimating the status after 2 years. Judge how important the factors are for the company's competitiveness   (N - No importance, M - Medium importance, G - Great importance).

Very bad			Medium			Very good
1	2	3	4	5	6	7
			Status today	Realistic status in 2 years	Importance for the company	

		Status today	Realistic status in 2 years	Importance for the company
a)	Tools and technical **facilities**	1--2--3--4--5--6--7	1--2--3--4--5--6--7	N--M--G
b)	Design competence in own company	1--2--3--4--5--6--7	1--2--3--4--5--6--7	N--M--G
c)	Design investigation and work through	1--2--3--4--5--6--7	1--2--3--4--5--6--7	N--M--G
d)	Laying down of product parameters (function, form, material, dimension etc.)	1--2--3--4--5--6--7	1--2--3--4--5--6--7	N--M--G
e)	Presentation of production support (drawings, piece lists etc.)	1--2--3--4--5--6--7	1--2--3--4--5--6--7	N--M--G
f)	Quality safeguarding system for design work	1--2--3--4--5--6--7	1--2--3--4--5--6--7	N--M--G
g)	Procedures for treatment of **change orders**	1--2--3--4--5--6--7	1--2--3--4--5--6--7	N--M--G
h)	Systematic registration and reuse of empirical dates	1--2--3--4--5--6--7	1--2--3--4--5--6--7	N--M--G
i)	Co-operation with production/production planning	1--2--3--4--5--6--7	1--2--3--4--5--6--7	N--M--G
j)	Market and customer contact	1--2--3--4--5--6--7	1--2--3--4--5--6--7	N--M--G
k)	Co-operation with contractors	1--2--3--4--5--6--7	1--2--3--4--5--6--7	N--M--G
l)	Co-operation with purchase	1--2--3--4--5--6--7	1--2--3--4--5--6--7	N--M--G

Figure 5: Question to evaluate the Design function (example)

Report - analysing of data: The data is collected and stored in a database. Different types of reports can be generated automatically. Individual reports based on data for one company or reports based on average data from a given number of companies can be generated. The report layout is exactly the same which means that companies easily can compare own data with data from groups of other companies. The report consist of 3 parts

   a) Main report with key indicators (10 - 15 pages)

   b) Enclosure for comparison (~ 50 pages)

   c) Enterprise specific enclosure (detailed economic and financial data) (~15 pages)

An example of evaluation of primary functions is shown in fig. 6.

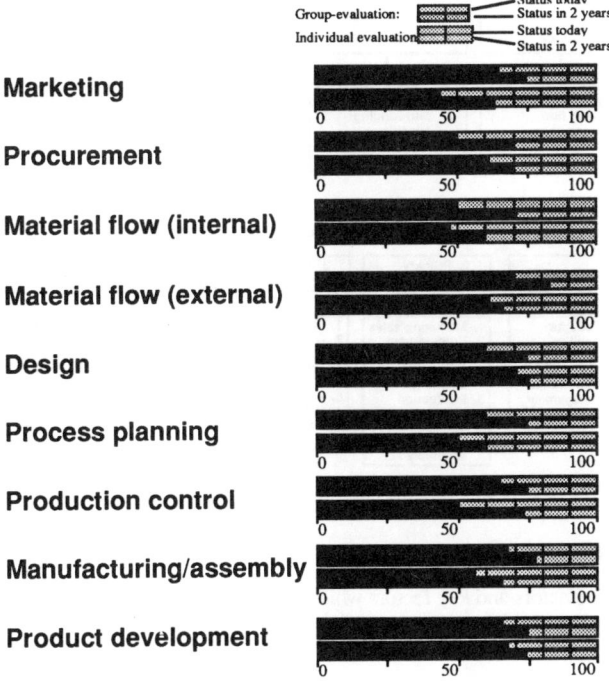

Figure 6. Evaluation of primary functions

3.2 Extended audit

Methodology: The extended audit is performed by experts analysing the company. Measuring of productivity and competitiveness is done on 2 levels.

a) Company level - using indicators and key factors with focus and viewpoint on the whole enterprise

b) Company divided in parts - using indicators and key factors with focus on parts and separated areas in the company

An overview of the different areas (37) in the model is shown in figure 7.

Model study at Company level

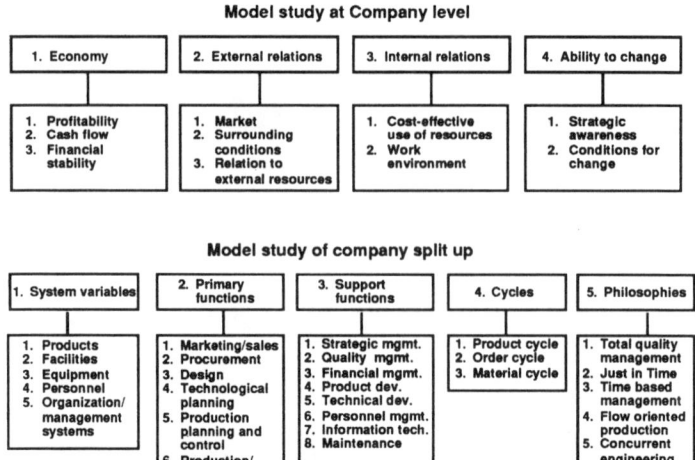

1. Economy	2. External relations	3. Internal relations	4. Ability to change
1. Profitability 2. Cash flow 3. Financial stability	1. Market 2. Surrounding conditions 3. Relation to external resources	1. Cost-effective use of resources 2. Work environment	1. Strategic awareness 2. Conditions for change

Model study of company split up

1. System variables	2. Primary functions	3. Support functions	4. Cycles	5. Philosophies
1. Products 2. Facilities 3. Equipment 4. Personnel 5. Organization/ management systems	1. Marketing/sales 2. Procurement 3. Design 4. Technological planning 5. Production planning and control 6. Production/ assembly	1. Strategic mgmt. 2. Quality mgmt. 3. Financial mgmt. 4. Product dev. 5. Technical dev. 6. Personnel mgmt. 7. Information tech. 8. Maintenance	1. Product cycle 2. Order cycle 3. Material cycle	1. Total quality management 2. Just in Time management 3. Time based management 4. Flow oriented production 5. Concurrent engineering

Figure 7. Analysing areas - extended audit

Characteristics for the analysing areas are:

Economy - indicators and key factors which describes the economic conditions and potential. An important issue of this analysis is to evaluate the economic capacity for future investments and change.

External relations - indicators and key factors to measure the ability of the company to exploit and take advantage of surrounding conditions. This includes customers, competitors, use of external resources, strategic alliances, etc.

Internal relations - the ability to use existing internal resources (products, machines, personnel etc.) in a time and cost effective way, and to take care of the internal milieu.

Ability to change - the ability to foresee and be prepared to meet new trends and quick changes in the environments.

Systems variables - primary resources and conditions necessary to produce products

Functions - Activities in the product life cycle. The functions are divided in primary and support functions.

Cycles - to follow and analyse flows between functions. Flows could be of different types (material , information etc.). The objective is to measure co-operation and infrastructure between functions.

Philosophies - to analyse overall production and management philosophies used in the company

All 37 analysing areas are broken down in a hierarchy of sub indicators which are given points and weighted. Check list for all levels exist. Indicators for evaluation of the TQM process are given in fig. 8.

No.	Description sub indicators	Points	Weight
1	Goals and strategy	3	10
2	Know-how and competence	5	10
3	Organisation and management	2	5
4	Information and analysis systems	2	5
5	Strategic quality management	2	15
6	Use of human resources	2	5
7	Quality control and qualtiy improvements	3	5
8	Results of quality work	4	20
9	Customer satisfaction	3	20
10	Cooperation and information	4	5
	Calculated	3,15	100%
	Total score	3	

Figure 8. Evaluation of TQM (example)

Report: The methodology offer check lists of all 37 analysing areas. All areas are analysed, but focus is given to areas of main importance in the company. This means that in most cases only one philosophy (see fig. 7) is evaluated.

The results are presented in a report organised in 3 parts.

1) Analysis of main productivity key factors and indicators. Comments are given by the experts (~15 pages)

2) Proposed actions for the company to obtain better results. Evaluation given by the experts (5-10 pages)

3) Detailed description of sub indicators for all analysing areas (~50 pages)

Figure 9 shows an example of a profile chart of indicators for system variables and functions.

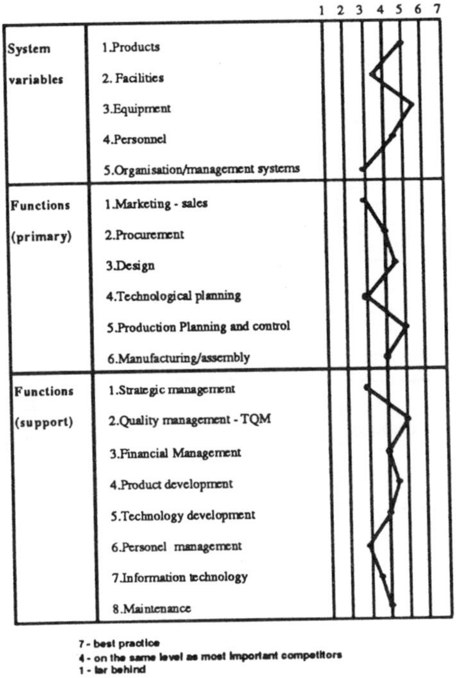

		1 2 3 4 5 6 7
System variables	1.Products	
	2. Facilities	
	3.Equipment	
	4.Personnel	
	5.Organisation/management systems	
Functions (primary)	1.Marketing - sales	
	2.Procurement	
	3.Design	
	4.Technological planning	
	5.Production Planning and control	
	6.Manufacturing/assembly	
Functions (support)	1.Strategic management	
	2.Quality management - TQM	
	3.Financial Management	
	4.Product development	
	5.Technology development	
	6.Personel management	
	7.Information technology	
	8.Maintenance	

7 - best practice
4 - on the same level as most important competitors
1 - far behind

Figure 9. Profile chart - indicators

   TOPP Self assessment is a modular system to help the companies to build their own tool for measuring of own performance. The self assessment focus on measuring of trends and is a tool to check if a business process is in control and continuos improvement (fig. 10) is taking place.

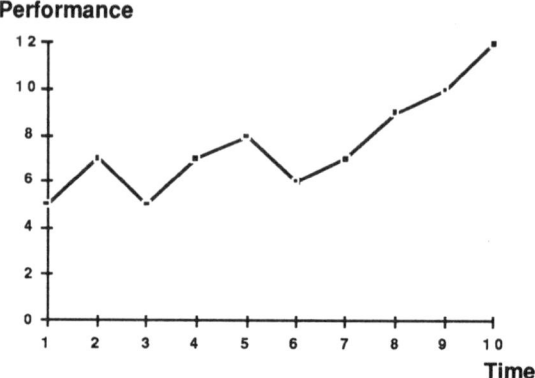

Figure 10. Measuring of trends - self assessment

The methodology consist of following steps:

a) Identification of critical and important business processes in the company to be evaluated

b) Selection of analysing areas and indicators to measure the business process

c) How to organise the self assessment

d) Collection of data

e) Presentation of results

f) Evaluation of results, actions

   An overview of the methodology is shown in fig. 11. 22 different business processes with analysing areas, indicators and measuring techniques are described.

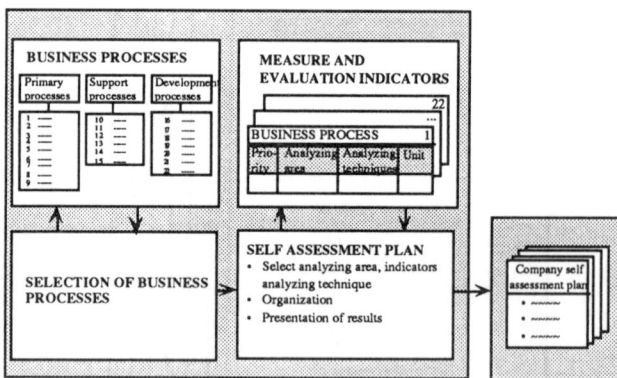

Figure 11. Self assessment

3.4   Benchmarking

Benchmarking is the process of continuously comparing and measuring an organisation against business leaders anywhere in the world to gain information that will keep the organisation take action to improve its performance (APQC[1] 93).

To use benchmarking to compare own processes with best practice and to learn and take advantage of others experiences, will often result in real breakthroughs (fig. 12). Best practice is often found in other branches or in different types of companies.

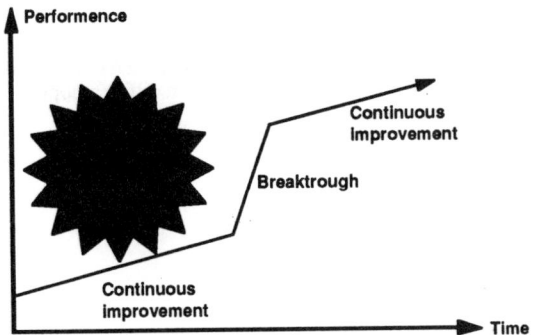

Figure 12. Benchmarking - breakthrough

Some reasons and benefits from using benchmarking are

- Establish effective strategic goals and objectives
- Understand and meeting changing customer requirements
- Striving for excellence through industry best practices
- Create a better understanding of own processes
- Establishing reference points for measuring of own performance

The different steps in the benchmarking process is shown in fig. 13

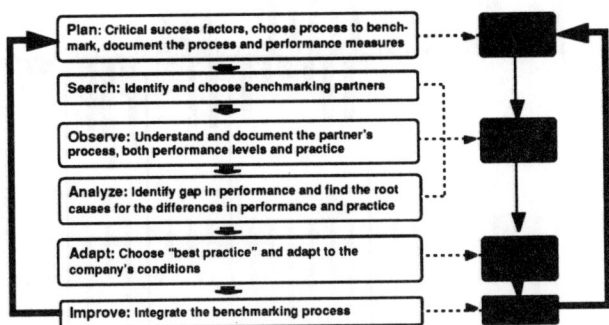

Figure 13. Benchmarking process

---

[1] APQC - American Productivity and Quality Center, Houston

4   ANALYSING RESULTS

All methods have been used in the TOPP program. The number of companies involved are:

a)   Self audit   ~ 60 companies

b)   Extended audit   ~ 40 companies

c)   Self assessment   individual use

d)   Benchmarking   2 companies (in progress)

A lot of data about industry productivity and competitiveness are collected in the audits. A special research project has been performed to analyse those data, trying to find some general results and conclusions concerning the status of productivity and competitiveness of Norwegian industry. 8 different reports have been written focusing on different topics. The reports are:

1.   General characteristics and possibilities

2.   Analysing methodology

3.   Analysing of procurement

4   Analysing of quality

5   Analysing of production planning and control

6   Analysing of product development

7   Analysing of marketing and sales

8   Analysing of strategy development

Figure 14 show a gap analysis from evaluation of different functions in the companies (self audit questionnaire).

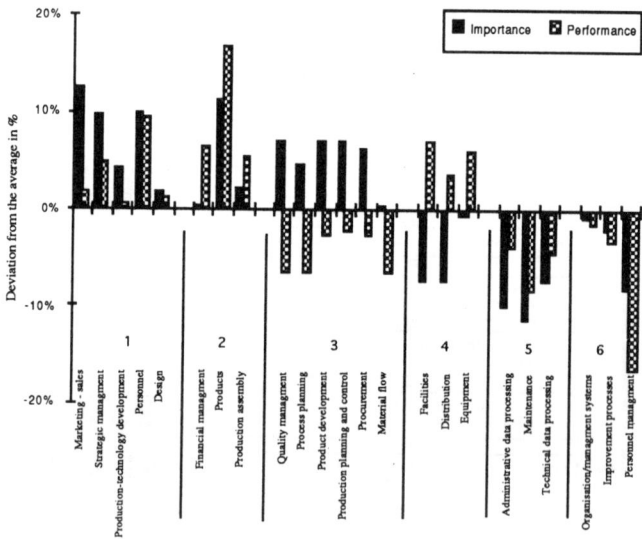

Figure 14. Gap analysis concerning performance and importance of different analysing areas (37 companies)

The analysing areas are sorted in 6 groups. For each area the deviation from the average of all areas is shown. Group 1 shows areas where both performance and importance are better than the average. However the importance is greater than the performance, which indicates that the area is good, but there is still a lot to achieve.

This gap analysis is based on average values from 37 companies. However the companies are quite different (size, products, production methods etc.). Another grouping of the companies in more homogeneous groups would perhaps have given other results.

Figure 15 show some results from analysis of the design function.

Design	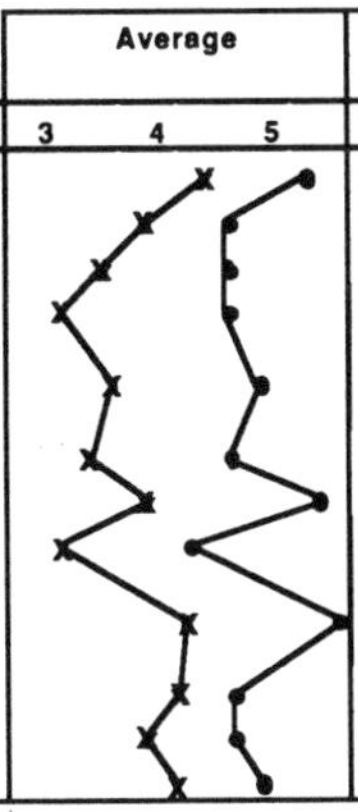 Average 3	4	5	Bad 1	2	Medium 3	4	5	Good 6	7
a) Tools and technical equipment					3	2	11	9	8	
b) Design competence				1	1	5	17	5	4	
c) Design review					2	16	8	5	1	
d) Define product parameters (function, form, material, dimensions, etc.)				1	3	15	7	2		
e) Development of technical document (drawings, bill of materials, etc.)					5	5	14	4	3	
f) Quality assurance of design				2	3	11	9	4	2	
g) Change order procedures					5	7	10	4	6	1
h) Systematic registration and reuse of experience databases				1	8	10	9	2	2	
i) Co-operation with manufacturing/production					1	4	11	13	2	
J)Market and customer contact					4	5	7	11	3	1
k) Co-operation with suppliers						4	13	9	3	
l) Co-operation with procurement						8	13	6	5	

x—x— Status today   ●—●— Realistic status in 2 years

Figure 15. Detailed analysis of the design function.

Areas to be improved seems to be

- design review
- definition product parameters
- quality assurance of design
- systematic registration and reuse of experience data

The 8 reports mentioned earlier and special branch reports for groups of companies (self audit) presents a lot of other analysis.

5   CONCLUSIONS

Measuring productivity and competitiveness for a whole enterprise is difficult. To find indicators, key factors and analysis areas which are relevant to different types of companies, and to harmonise this into a common methodology is even more difficult.

The methodology described is based on different approaches. However in any case there is much individual judgement. It is difficult (or impossible) to measure total productivity by summarising local measurements of sub areas. This is the reason for analysing the enterprises from different viewpoints. History has shown that economic success one year does not guarantee a competitive edge. Many perspectives that measure different factors have to be used to get a better understanding of competitiveness.

REFERENCES

[1]   TOPP Questionnaire for measuring of competitiveness (54 pages)

[2]   TOPP Self audit company report (~75 pages)

[3]   TOPP Extended audit - method handbook

[4]   TOPP Extended audit - operation handbook

[5]   TOPP Extended audit - company report (~70 pages)

[6]   TOPP Self assessment handbook

[7]   TOPP Benchmarking handbook

[8]   Moseng, Bredrup: A methodology for industrial studies of productivity performance, production planning and control, 1993 vol 4 no. 3

The references 2, 3, 4, 5, 6 and 7 are written in Norwegian.

29

Compare Your Performance and Practice with the World Excellence:
A Prototype of a Benchmarking System

Jens O Riis[1], Frank Gertsen[1] and Hongyi Sun[2]

1 Dept of Production, University of Aalborg, Fibigerstraede 16, 9220 Aalborg, Denmark
2 Dept of Business Administration, Stavanger University Centre, PoB 2557 Ulland-haug, 4004 Stavanger, Norway

To find a partner company is a critical element of benchmarking, partly because it requires the establishment of a win-win situation, partly because of the magnitude of resources and effort needed. In this paper we propose to use the results of an international survey as a basis for reference points and to use the questionnaire as an instrument for assessing the company's goals, performance and practice. A benchmarking system is proposed which indicate relevant factors to benchmark. The way in which reference points are established is demonstrated by means of examples.

1. INTRODUCTION

A recent survey has revealed that 50% of the 600 sample companies from 20 countries claimed to have used benchmarking *). This signals the relevance of benchmarking and a significant interest in working with this method. However, as benchmarking as a management method emerges, it is essential to clarify its role and to address practical issues of benefit versus the effort used.

We see the significance of benchmarking as providing a reference for an enterprise as a vehicle for generating momentum within the company to improve its performance.

In carrying out benchmarking the selection of a partner company has proven to be critical. Although the benefits of comparing both performance and practice with a well renowned enterprise is without doubt, the mere acceptance of the company partner and willingness to serve as a partner is often difficult, e.g. because of fear of disclosing sensitive data. Often, the only chance is to create a "win-win" situation

---

*) International Manufacturing Strategy Survey (IMSS). We would like to thank Dr Per lindberg at Chalmers University and Professor Chris Voss at London Business School for the initiation of the IMSS project. We also give gratitude to Danish Technical Research Council (STVF) for its partial financial support of this research through Integrated Production Systems (IPS) project.

which requires that the initiating company itself may offer a significant contribution in the benchmarking process. In addition, the effort needed is considerable.

In shaping the content of the theory and practice of benchmarking it appears relevant to seek alternative ways than bi-lateral benchmarking. As an outcome of participating in an international survey on manufacturing, we have developed a method for displaying the individual respondent in comparison to the best in the population, the median company, and the weakest. In this paper we shall first introduce the survey and use it for proposing a benchmarking system of factors to incorporate in benchmarking. Then examples will be presented to demonstrate the way in which an individual enterprise may compare one factor. At the end of the paper we shall discuss the merit and limitations of the approach.

2. A PROTOTYPE OF A BENCHMARKING SYSTEM

In 1993, an international survey was conducted covering 600 industrial enterprises from 20 countries (Argentina, Australia, Austria, Belgium, Brazil, Canada, Chile, Denmark, Finland, Germany, UK, Italy, Japan, Mexico, Netherlands, Norway, Portugal, Spain, Sweden and USA [1]). Most of the companies are producers of electrical and metal products. The size distribution of the IMSS survey is shown in table 1.

Table 1   Size distribution of participating enterprises in the IMSS survey

Size	N	%
< 50	18	3%
51 - 500	332	55%
501 - 1000	130	22%
1001 - 5000	91	15
> 5000	14	2%

The IMSS survey studies three categories of factors: goals, performance, and practice. Goals include management statement of goal priority (e.g. quality, cost, delivery, etc.) and the extent to which manufacturing strategy is linked to corporate strategy. Performance factors are concerned with business performance, such as return on investment, inventory turnover, etc., and manufacturing performance, such as cost, quality and delivery. Practice is divided into three groups of factors: technology, organization and programs.

In general, it is a complex matter to define relevant factors to benchmark. Having worked with the questionnaire during interviews of top executive, we have observed that the factors dealt with are considered relevant. Furthermore, it is possible with an effort of two to three hours to fill in the questionnaire. For this reason, we shall propose a benchmarking system with a list of relevant factors covering goals, performance and practice be adopted from that of the IMSS survey. It is shown in table 2.

Table 2   Factors in the proposed benchmarking system

GOALS &	**Company goals**	Priority in quality, cost delivery service etc.
STRATEGY	**Market feature**	Market coverage, focus and development
	Mfg. Strategy	Linkage of manufacturing strategy and goals
PERFORMANCE	**Business**	Return on investment
	performance	Sales on Investment
		Return on sales
		Profit per person
		Sales per person
		Market share
	Manufacturing	Quality
	performance	Manufacturing cost
		Unit cost improvements
		Cost structure
		On-time delivery
		Procurement lead-time
		Design lead-time
		Manufacturing lead-time
		Distribution lead-time
PRACTICE	**Organisation**	Human aspects
		Education, suggestions, teams, etc.
		Integration aspects
		Co-ordination between department
		Structure
		Level, span
	Technology	Production technology
		NC, FMS, Robots, CIM etc.
		Engineering technology
		CAD, CAD/CAM, CIM
		Management technology
		MIS, network, CIM
	Programmes	Quality
		Total quality control, zero defects, ISO
		Production plan
		Just in time, MRP
		Design
		Concurrent engineering, design for
		manufacturing/assembly etc.

3. EXAMPLES OF BENCHMARKS

As Danish participants of the IMSS survey, we have used the benchmarking system to analyse Danish data and developed a set of benchmarks at the Danish national level (Gertsen, Sun & Riis, 1993). The next step is to extend our work to the international level and implement international benchmarks in two or three case companies.

For the sake of comparability, 180 metal-product (International Standard Industry Classification, ISIC 381) companies were selected of the IMSS database. We have chosen to show three different examples of benchmarks, which may be used for benchmarking by all companies in the metal product industry. They illustrate the potential outcome of a cross-sectional comparison. The only effort needed is the time to fill in the IMSS questionnaire. The first benchmark relates to performance, the second to practice, whereas the third to goal priority.

3.1 Detailed benchmark of return on sales

Statistically, benchmarking is called "position statistics". The position of an individual company may be expressed in relation to the 25% fractile and the 75% fractile, forming three groups of companies: the top 25%, the middle 50%, and the bottom 25%. The scale varies due to purpose and variable. Benchmarks can take on several forms such as column chart, table, line etc.

Figure 1 shows a detailed benchmarking for return of sales (ROS, n = 69). This reveals how much profit a company is able to get from each unit of sales. According to the survey, on the average, a company achieves 5% from sales. The position of a company in this benchmark will tell how good the company is in terms of ROS compared to other companies of the industry.

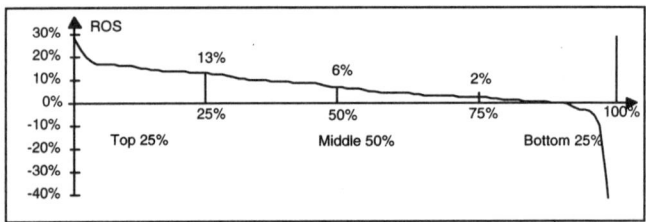

Figure 1   Benchmark of Return on Sales (ROS) in the metal product industry

Another way of displaying the data is shown in figure 2 where the score of the Danish sample companies is displayed on the left side and the percentiles of the international data is shown to the right.

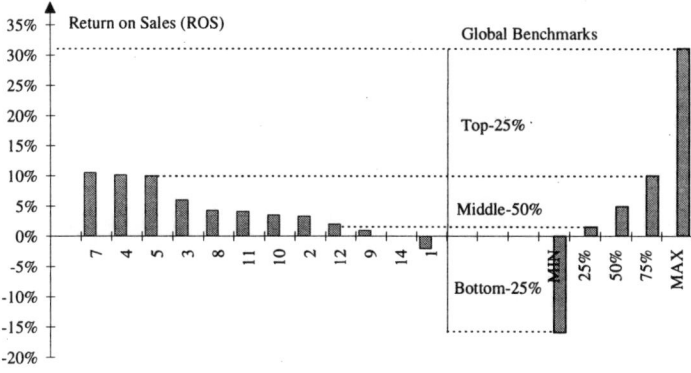

Figure 2 The Danish sample benchmarked on Return on Sales (ROS)

The average ROS of the Top-25% is able to achieve US$ 31 from US$ 100 of sales. Most of the Danish sample companies are in the middle 50% group.

3.2 Simple benchmark for team utilisation

A benchmark can also be a simple table with a five point scale. For example, the tables below show two types of benchmarks for team utilisation. Table 3-1 is a percentile benchmark measured by percentage of employees working in teams, while table 3-2 is a 1-5 scale benchmark measured by percentage of companies. We propose that more teams leads to better performance. A company can easily find the figure of its own, and its position in the table, as an indication of its utilisation of teams compared with others.

Table 3-1 Benchmarks by percentile (n=152)

Benchmark (% of sample)	Min	Top 75%	Top 50%	Top 25%	Max
% of employees	=0%	≥35%	≥20%	≥50%	100%

Table 3-2 Benchmarks by 1-5 scale (n=133)

Benchmark (1-5 Scale)	None	Little	Some	Much	A lot
% of companies	12%	16%	27%	29%	16%

3.3 Benchmark for company goals priority

One type of benchmarks aims to show tendency. For example, figure 3 is a benchmark of the priority of different goals of the metal product industry in 1992-1993 (On the scale a "1" indicates Not Very Important and a "5" Very Important). Similar benchmarks may be made for future goals and market tendency etc. If a company makes the same chart of its own and compares it with the benchmark, it may obtains a basis for deciding its business goals.

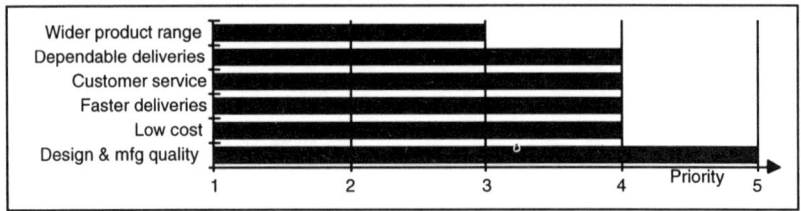

Figure 3   Benchmark for goal priority in the metal product industry 1992-93

4. DISCUSSION

The above mentioned examples of benchmarks for goals, performance and practice of industrial enterprises enable a company to position itself in relation to companies in the same country or internationally. In addition, the survey has also been subjected to nation-wise analyses.

One of the advantages of this approach to benchmarking is that an industrial enterprise within the time and effort it takes to fill in a questionnaire may get access to comparisons with other companies on a broad variety of factors covering goals, performance and practice. The benefit of benchmarking, in our view, should be measured in terms of the surprise (shock) and constructive energy it creates within the company. One of the limitations lies in its lack of details and the difficulty of obtaining a more holistic picture of one or more companies attractive for comparison.

The approach to benchmarking may be used as a first screening of factors relevant for further analysis. Experience shows that it is not easy to define relevant indicators or to develop appropriate measurement tools.

Benchmarking as a management method still has to find its place in relationship to Continuous Improvement, Total Quality Management, Total Productive Maintenance, etc. which may call for a distinct, more narrow role of benchmarking as that of providing a reference which may inspire management and employees in a company and generate time, energy and will-power for improvement. On the other hand, we need to develop and test tools for benchmarking that take into account the practical constraints imposed on enterprises. It is our hope that the approach to benchmarking related to an international survey may stimulate this discussion.

REFERENCE

Gertsen, F., H. Sun and J. O. Riis (1993) *Compare your company with the best: a preliminary report of IMSS*, University of Aalborg, ISBN: 87-89867-34-3.

30

Determination of what to benchmark : a customer-oriented methodology

S. Kleinhans [a], C. Merle [a] and G. Doumeingts [a]

[a] L.A.P. / GRAI - University Bordeaux I - 351, Cours de la Libération , 33405 Talence - FRANCE - Tel : 33-56-84-65-30 - Fax : 33-56-84-66-44

Abstract :

Managers , in their quest for efficient management tools, attach more and more importance to benchmarking techniques as action-oriented démarches to set functional goals. Yet, benchmarking having essentially been developed by Industry, different approaches coexist and still need to be federated and made coherent.

This paper stresses the lack of tools in one of the stages in the benchmarking process, which is considered by experts as a key one : the determination of "what to benchmark".

A methodology aiming to identify the process - or the activity - to benchmark is presented.

Key-words :

Benchmarking - Methodology - Order Winning Criteria - GRAI method - modelling techniques - objectives - key drivers - performance indicators

1. INTRODUCTION

For the last ten years, the Industrial World has paid more and more attention to benchmarking techniques. E.E. Sprow for instance, [15] provides an interesting figure from a 1992 survey undergone among American firms : 79 % of them feel they must benchmark to survive.

J.A. Schmidt also speaks about recent studies by Towers Perrin, the Massachussets Institute of Technology and the American Productivity and Quality Center, indicating that, "by 1995, most major US companies will have some type of benchmarking program" [13].

The growing interest brought to these techniques essentially lies in the pragmatism and applicability of such approaches, especially in a time when Management suffers from a crisis in its theory and concepts.

Historically, benchmarking has first been implemented by Industry before being conceptualised and theorised *a posteriori* by academic researchers (the opposite process is much more commonly encountered).

A positive aspect of such a process is that it stresses the real need and interest of Industry for Benchmarking.

A negative aspect is the lack of methodology it induces : today, one cannot find the "benchmarking toolkit". Still taken from the above quoted 1992 American survey [15], another key figure indicates that "95 % of companies admit they do not know how to benchmark".

In [2] , two definitions are provided, one of them giving prominence to the central theme of the current paper.

"Benchmarking is the continuous process of measuring products, services, and practices against the toughest competitors or those companies recognised as industry leaders to achieve superior performance" (Rank Xerox)

"Benchmarking is the ongoing task, at all levels of our business, of finding and implementing world best practice **in the key things we do that deliver customer satisfaction**" *(PA Consulting Group)*

This last definition focuses on what we believe to be the key point : in order to be efficient, the benchmarking process should be applied only on those activities in the company that have impact on the products, services and practices characteristics contributing to the customers' satisfaction. The identification of the benchmarking object is considered by Camp [4] as one of the most difficult phases in the whole process.

This paper is structured as follows :

In a first part, we recall the main principles in the overall benchmarking démarche, stress the importance of the "identification of what to benchmark" sub-phase and provide a State of the Art of the existing methodological elements in this domain.

Then, in a second part, we propose a methodology aiming to support managers in determining what should be benchmarked in their company.

Finally, we conclude on the necessity of undergoing such benchmarking improvement actions continuously and consequently, of updating regularly the gathered data.

2. IDENTIFICATION OF WHAT TO BENCHMARK

2.1. The overall benchmarking process

Despite of the different approaches, most authors now agree upon depicting four main stages in a benchmarking study, plus a fifth one called "maturity".

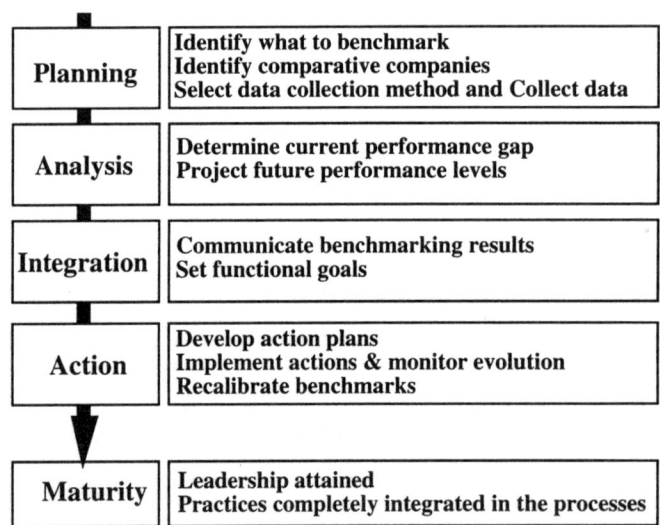

Fig. 1 : Steps in the benchmarking process [4]

Concurrent approaches exist (see for instance AT&T's 12-step and Alcoa's 6-step processes in [1]). Yet, all of them include the "very heart of benchmarking" as depicted hereafter.

- an internal analysis of the company, ending in the identification of those processes, activities or functions that constitute "improvement urgencies";
- an environmental survey for determining the leading enterprises (i.e. having the "Best Practices");
- a calculation of the "performance gap", by comparing performances obtained in the company and in the previously defined leading enterprises;
- a planning and implementation of improvement actions, chosen and agreed by all actors in the company.

2.2. Importance of determining the right processes to benchmark

Indeed, benchmarking can help companies in their continuous improvement process. But benchmarking's efficiency is proportional to the attention paid to the determination of the activity to benchmark. Unfortunately, as J.A. Schmidt points out :

"Much of the benchmarking done today is irrelevant to making money or besting competitors. Rather than focus on what's truly important, companies benchmark what they can" [13].

First of all, costs of a benchmarking study should not be neglected.

According to AT&T's benchmarking group [8], a "medium size" benchmarking study (4 to 6 months) costs between $60.000 and $80.000 (including manpower).

The International Benchmarking Clearinghouse surveyed its membership and got these numbers [15] :

> Benchmarking training : $ 1.000/employee/year,
> Total costs for one study (in average) : $ 53.145,
> Hosting site visits (7 per year) : $ 62.720.

Yet, direct cost of benchmarking is not the main reason why benchmarking should be grounded on a precise definition of "what are the key processes".

A parallel can be made with OPT (Optimised Production Technology) : when Goldratt was proposing improvement actions in the physical system of industrial firms, he insisted a lot on the fact that focusing on non-bottlenecks machines (i.e. machines that do not limit the production flow) was a complete waste of time and money.

Undergoing a benchmarking process on some activities in the company that do not *a priori* require improvements is a similar waste. Moreover, it might end in a strategic failure, for all efforts in the company are thrown towards a wrong direction.

So, the benchmarking process should be undergone only in the areas where changes are required, in those activities that "contribute in providing competitive advantage" [14].

Nevertheless, this identification phase remains difficult, as is shown by the following litterature survey of the main techniques, tools and methods used.

2.3. State of the Art

Through our litterature survey, we have identified four sub-tasks that should appear in the identification phase :
- identify global objectives,
- split them up into sub-objectives according to the internal structure of the firm,
- locate the improvement areas,
- prepare indicators on which to ground the comparison with other companies.

2.3.1. An overall point of view

Most authors insist on the necessity for the benchmarked process to contribute to the improvement of the company's overall performance, by :
- reaching a global objective at the company level (corporate strategy objective, competitive advantage, etc.);
- providing customer's satisfaction;

- solving a global inconsistency (delays, inventories, costs, quality, etc.).

For instance, J.A. Schmidt identifies three primary forms of benchmarking: strategic, customer and cost [13]. Each of them includes this preliminary phase :
- Strategic benchmarking, by essence, takes into account the company's strategy;
- Cost benchmarking aims to solve cost problems at a company level by focusing on productivity and direct cost structures;
- Customer benchmarking begins with the determination of the attributes that influence customer value perceptions.

For customer benchmarking, J.A. Schmidt proposes a method based on customers' interviews or written surveys for determining and classifying (each year, for instance) the value attributes according to their relative importance for the customers. An importance/satisfaction matrix is used (see figure 2), on which the position and evolution of the value attributes (for the company and for its competitors) can be viewed. The same matrix can be found in [10].

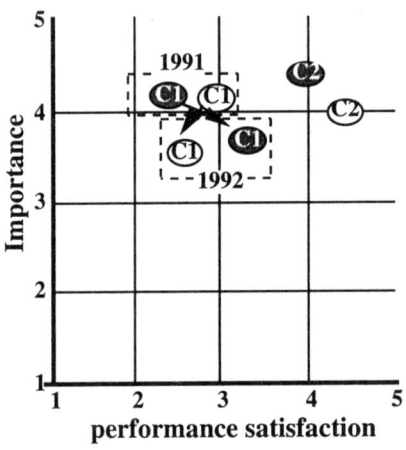

Fig. 2 : **Importance/satisfaction matrix** [13]

R.C. Camp , in [4] as well as in [5] also advises to initiate the benchmarking process at a global level. His experience is that the product or output of the business function should be identified either by "developing a clear mission statement detailing the reason for the organization's existence, including typical output expected by customers" or "pose a set of questions that might reveal current issues facing the function" [5].

IBM Rochester, 1991 Baldrige quality award winner, has undergone its benchmarking approach in order to reach such goals as : "be the undisputed leader in customer satisfaction, be first with the best products,[...]" and "be the best-of-breed (BOB) in C.I.M." [7]. The consistence between the former set of goals and the latter results from a strategic analysis preceding the implementation of the benchmarking process.

K. Bemowski, in [1], quotes Alcoa's way of deciding what to benchmark by answering questions related to "customer's satisfaction, consistency with Alcoa's mission, business needs, significance in terms of costs or key nonfinancial indicators".

Another question asks if "the topic is an area where additional information could influence plans and actions". This question is interesting because it situates benchmarking as a weapon to build the Strategic Information System of the company.

2.3.2. A decomposition phase

The competitive advantage or the overall inconsistency has now been identified globally, at the company level. But benchmarking solutions cannot be performed on such a large scale : they involve processes, not global functionings. The objective (resp. the problem) must then be split up into sub-objectives (resp. sub-problems) according to the internal structure of the firm.

Y.K. Shetty [14] proposes to describe the organization of the company by adapting Porter's value chain analysis [12].

This approach clearly stands for a dis-aggregation from a macro-level description (performances or problems in the whole enterprise) to a micro-level description (contribution of each function to the overall outputs).

Moreover, Shetty proposes to use Porter's value chain for the identification of the "key activities".

Another decomposition technique is given in [16] :
- Determining the company's mission is the first milestone of the method
  (e.g. : to be the industry leader in innovation);
- This mission is translated into more finite goals for each functional area;
  (e.g. :   engineering department mission :   to develop new concepts effectively,
              marketing department mission :   to improve responsiveness to a new
                                                  concept development request)
- For each functional area, Key Quality Characteristics (KQC) are determined
  (KQC are "those process outputs most important to the customers of that process");
- Critical Success Factors (CSF) are associated to those KQC. CSF are defined as "the contributing process elements that affect KQC";
- Finally, indicators are built to measure the level of the KQC.

Vaziri insists on the necessity of giving a clear operational definition to these indicators in order to allow further comparisons with similar functionings in other companies.

An equivalent approach is developed in [7] : a method called AHP (Analytical Hierarchy Process) supports the decomposition of the overall goal (e.g. be the BOB CIM site) into sub-goals (Quality, responsiveness, flexibility and cost). Each sub-goal is weighted according to its relative importance for the attainment of the global goal. Success factors are then identified and associated to the four sub-goals. Finally, because benchmarking needs more "granularity", each success factor is decomposed in "requirements". Success Factors as well as requirements are weighted.

2.3.3. Locating the improvement area

The previous step did not aim to determine directly the area to benchmark. Its purpose was only to provide a picture of the company's functioning with enough detail ("granularity") for launching a benchmarking project.

Yet some authors propose methods and tools which simultaneously identify key activities.

For instance :

• Porter's value chain analysis [14] provides in the same time a tool for the description of all activities and a way of identifying key issues.

• Analytical Hierarchy Process (AHP) [7], by weighting sub-goals, success factors and requirements, also locates the improvement areas (the more weighted are candidates to benchmarking actions). Figure 3 gives an example of application.

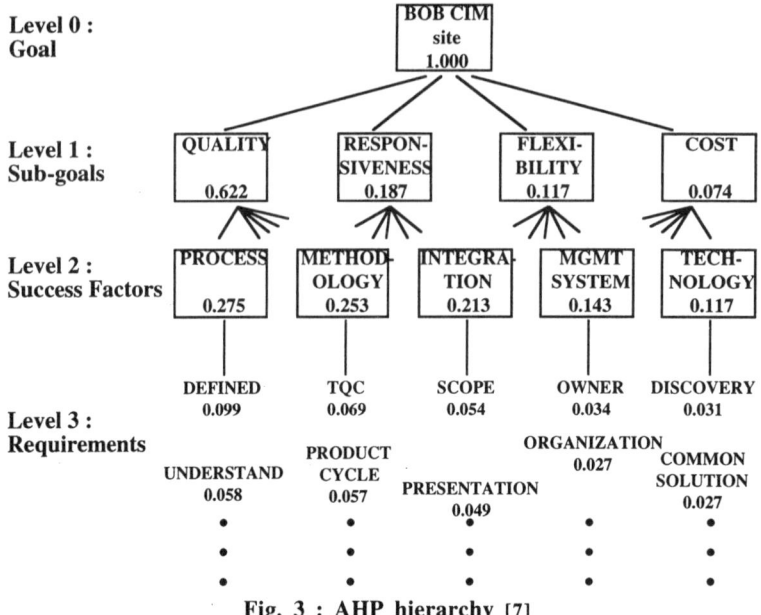

Level 0 :
Goal

Level 1 :
Sub-goals

Level 2 :
Success Factors

Level 3 :
Requirements

Fig. 3 : AHP hierarchy [7]

Nevertheless, generally speaking, the description of the activities of the firm and the identification of key processes are two distinct steps.

R.C. Camp [5] advises to use Ishikawa cause to effect diagrams. The causals (components) are candidates for benchmarking.

Vaziri [16] applies a Pareto analysis to select among Critical Success Factors and Key Quality Characteristics "the 20% that account for 80% of the effect".

2.3.4. Preparing the comparison

Even if the first phase of benchmarking lies in an internal process-oriented analysis, it should not be forgotten that the aim is to compare one of the company's practices against external "Best Practices". The comparison can - and should be - carefully prepared in order to ensure validity and significance of the results.

Vaziri proposes to determine an indicator for each identified KQC.

The "Best Practice" will be the one having the highest (resp. the lowest) value for this indicator [16].

IBM Rochester has developed a sophisticated "Maturity Index" (MI) for each requirement previously determined : a five-point scale gives a "baseline" for comparing IBM Rochester to other companies. The results of the comparison can easily be visualized by drawing two curves in a two-dimensional space (requirements/value on the scale) [7].

3. Proposed methodology

The litterature survey has shown that some tools exist for identifying the "key activities" or "key processes" candidates for benchmarking. As far as we could see, the use of all these tools and techniques can be given more structuration. Moreover, in some areas - the internal modelling of the firm and the determination of adapted indicators for instance - improvements can be brought.

We hereafter present a tentative methodology that integrates most issues discussed above.

3.1. First step : determining the Order Winning Criteria (OWC)

As Kenichi Ohmae says : "Of course, it is important to take the competition into account, but in making strategy that should not come first. It cannot come first. First comes painstaking attention to the needs of the customers."([11] quoted in [10]).

As a matter of fact, corporate strategy as well as benchmarking should be grounded on the Order Winning Criteria (OWC), that is to say on those elements that bring customer's satisfaction and allow to win orders and gain market share.

The objective of this step is to determine what are the OWC for a specific business.

We propose to pose a set of questions to the customers about *the outputs from our company (products and services characteristics)*. The aim is to get an evaluation of how customers rate these outputs in term of importance as well as in term of degree of satisfaction.

Positioning all the outputs on the matrix shown in figure 4 will allow to determine the OWC, that is to say those outputs considered as important by the customers, but for which the customer's degree of satisfaction is low.

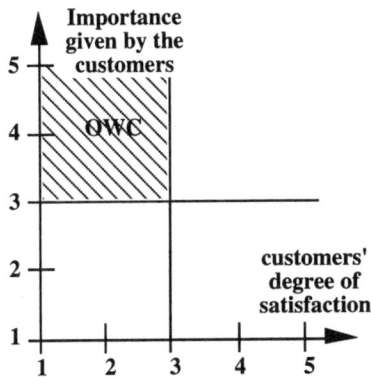

Fig. 4 : Determining the Order Winning Criteria

It is also possible to pose internally the same set of questions to the company's managers, in order to check if there is any gap between the customers' needs and the perception of these needs in the company. This additionnal action is by no means compulsary but many lessons can often been drawn from it.

3.2. Second step : modelling the internal production process

Undergoing the benchmarking project towards the attainment of OWC ensures coherence with the company's overall objectives. Yet, identifying the OWC is not sufficient, for benchmarking cannot be initiated at such a macro-level. So, modelling the processes in the company is the following necessary step.

One could argue that modelling only the part of the company that is relevant for the attainment of the OWC is enough. This is not true for a very simple reason : the company is a system, all elements interact permanently, and it is impossible to identify a priori "the part of the company that is relevant for the attainment of the OWC" : the visible symptom is not the cause.

For the last decade, the GRAI/LAP laboratory has developed several methods for modelling production systems. The GRAI method [6] aims to model the production system along its decisional dimension. The method comprises a structured démarche, a reference model and two modelling tools : the GRAI grid and the GRAI nets. Figure 5 shows a GRAI grid.

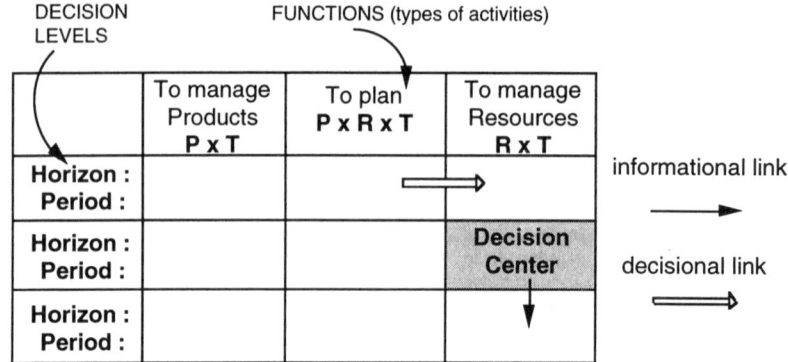

Fig. 5 : the GRAI grid

The GRAI method has been integrated into a more global methodology, GIM (GRAI Integrated Methodology) which allows to depict completely a production system by describing three sub-systems in it :
- the decisional system, thanks to the GRAI tools;
- the physical system, thanks to the IDEF 0 formalism;
- the informational system, thanks to the entity-relationship formalism.

Coherence between the three types of models is ensured through validation phases.

We propose to use GIM to model the whole company in order to visualize how the global OWC are decomposed and take place in elementary activities.

Despite of its relative "heaviness", this modelling step provides some company models :
- that can be helpful in the "integration" phase of the benchmarking process (fig.1), thanks to their explanatory capabilities, to communicate internally the benchmarking results ;
- that can be reused as such for further benchmarking applications in other activities.

Moreover, the building process of these models induces by itself implication and participation of all actors within the company. Initiating the benchmarking démarche on such bases can ensure a better understanding of the whole improvement project.

3.3. Third step : extracting and structuring objectives and key drivers per activity from the models

From the GIM models, we can extract directly the list of objectives and key drivers for all activities. From the first step, we also have at our disposal the list of Order Winning Criteria.

Consequently, the objective of this third step consists in linking sub-objectives (with the associated key drivers) to the global OWC by cause-effect relations.

The most adapted tool here seems to be Ishikawa cause-effect diagram :

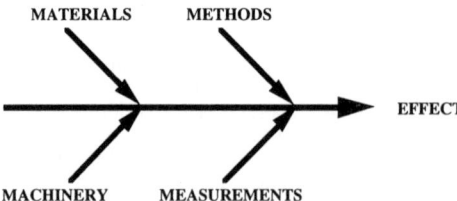

Fig. 6 : Ishikawa cause-effect diagram [9]

Applied to our field of interest, Ishikawa diagrams can allow to identify the main areas - and the main drivers - contributing to the attainment of the OWC :

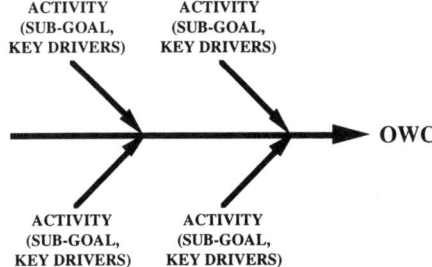

Fig. 7 : Use of Ishikawa diagrams to identify key activities

3.4. Fourth step : identification of the indicators

As stressed by Vaziri in [16], a set of indicators should be built up in order to ensure validity and significance of the comparisons with "Best Practices", by providing common measurements. Camp also insists a lot on the importance of "only comparing comparable things" [4].

The ECOGRAI method, developed by the GRAI/LAP laboratory [3] allows to build a coherent system of performance indicators for industrial systems, based on the decisional structure identified in the GRAI grid. The basic steps are :

S1- to build the performance indicators (P.I.) system by decomposing overall objectives into sub-objectives;

S2- to define, for each sub-objective, the associated key drivers (elements, resources having impact on the attainment of the sub-objective);

S3- to build the performance indicators, by considering them as a part of a triplet compound of (sub-objective, key driver(s), performance indicator(s));

S4- to integrate the performance indicators system into the informational system of the company.

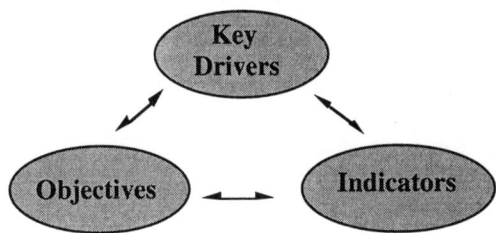

Fig. 8 : ECOGRAI triplets

The ECOGRAI method :
- ensures that Performance Indicators effectively measure the contribution of each function to the company's objectives;
- ensures coherence in the (sub-objective, key driver(s), performance indicator(s)) triplets. Most methods define directly indicators from the objective, without taking into account the drivers. This can lead to build non-operational indicators, because when measurements are performed and provide bad performance values, managers do not know what drivers should be modified for correction.

We propose to adapt the ECOGRAI method for the benchmarking process.

Steps S1 and S2 of the method are already performed : the OWC have been split up into sub-objectives, and the key drivers have been listed. The only missing element in the triplet is the Performance Indicator. Only by applying the S3 step, we can build a set of indicators on which further comparisons with "Best Practices" will be grounded.

Moreover, the designed PI system can also be used for the internal control of the company.

4. Conclusion

According to a study quoted in [8], the number of benchmarking applications will increase significantly in the ten next years. Undoubtlessly, new dedicted tools and techniques will be developed, and existing tools will be improved. For the time being, the methodology we propose for the identification of "what to benchmark" ensures that :

- the benchmarking process is undergone with respect to the customers' expectations and consequently, with the strategic objectives of the company;
- the internal activities on which the benchmarking process is undergone are the key ones in terms of impact upon the OWC obtainment.
- the designed performance indicators are coherent with both objectives and key drivers, and they allow a valid comparison with "Best Practices".

The "critical success factor", for this methodology as well as for the whole benchmarking process is *to create a dynamic of change and improvement* : the end of a benchmarking action does not mean the end of the change process. Continuity in the improvement actions is required. For instance, the GIM models and the performance indicators system designed should be permanently updated as soon as strategic objectives are modified or structural evolutions occur in the production system. By this way, managers can get an exact picture of their company whenever they need it.

Mentalities are evolving from the idea of day-to-day management to the idea of management of change, and benchmarking is a powerful technique to support such a challenge.

REFERENCES

[1]   K. Bemowski, "The Benchmarking Bandwagon" - Quality Progress - pp 19-24 - January 91.
[2]   J. Bergström and I. Kivimäki, "Benchmarking" - Seminar in Industrial Management - Helsinki University of Technology - 1993.
[3]   M. Bitton, "ECOGRAI, méthode de conception et d'implantation de systèmes de mesure de performances pour organisations industrielles" - Ph.D. thesis - GRAI Laboratory - University Bordeaux I - 1990.
[4]   Robert C. Camp, "Benchmarking - the search for Industry Best Practices that lead to superior performance" - ASQC Quality Press 1989, Milwaukee, Wisconsin.
[5]   Robert C. Camp, "Learning from the Best leads to Superior Performance" - Journal of Business Strategy, May/June 1992.
[6]   G. Doumeingts, "Méthode GRAI, méthode de conception des systèmes en productique". Ph.D thesis - GRAI Laboratory - University Bordeaux I - 1984.
[7]   H.G. Eyrich, "Benchmarking to Become Best of Breed" - Manufacturing Systems, pp 40-47, April 1991.
[8]   D. Garvin, "Building a Learning Organization" - Harvard business Review, July/August 1993.
[9]   K. Ishikawa, "Guide to quality control" (Authorized French translation) - Asian Productivity Organisation, 1976.
[10]   K. Jennings and F. Westfall, "Benchmarking for strategic action" - Journal of Business Strategy, pp 22-25, May/June 1992.
[11]   K. Ohmae, "getting back to strategy" - strategy : seeking and securing competitive advantage, Harvard Business Review Book Series, p 61, 1991.
[12]   Michael E. Porter, "Competitive Advantage"- The Free Press - New York - 1985.
[13]   Jeffrey A. Schmidt, "The link between benchmarking and shareholder value" - Journal of Business Strategy, pp 7-13, May-June 1992.
[14]   Y. K. Shetty, "Aiming high : competitive benchmarking for superior performance" - Long Range Planning, Vol. 26 Nr 1, pp 39-44, Pergamon Press Ltd, 1993.
[15]   E.E. Sprow, "Benchmarking : it's time to stop tinkering with manufacturing and start clocking yourself against the best" - Manufacturing Engineering, pp 56-69, September 1993.
[16]   H. Kevin Vaziri, "Using competitive benchmarking to set goals" - Quality Progress, pp 81-85, October 1992.

31

The Lean Analysis of the Business - Lean-Factors and Lean-Characteristics

Dipl.-Ing. Richard Kugel

General Manager, macils. management-center gmbh, Endelbangstr. 18, 70569 Stuttgart, Germany

1. THE BASIS OF THE LEAN-ANALYSIS

Lean Analysis serves to point out the possibilities and measures of a business so that the required time-, cost- and quality targets in the preparation and manufacturing of products in line with real market conditions can be put into practice most favorably.

The basis of this analysis are the surroundings and the market with its future developments, the business with its current structures and strategies as well as an analysis system consisting of efficiency-determining lean-factors and lean-characteristics. In the course of this analysis a target system specific to a certain business is developed out of these 3 elements and the present state is described. As this target system strongly depends on the surroundings, the business itself and the correlation between these two, it has to be adjusted during the analysis in loops of recurrence parallel to the ascertainment of the present state. After the final description of this state, the necessary potentials and measures can be identified and determined concerning their meaning.

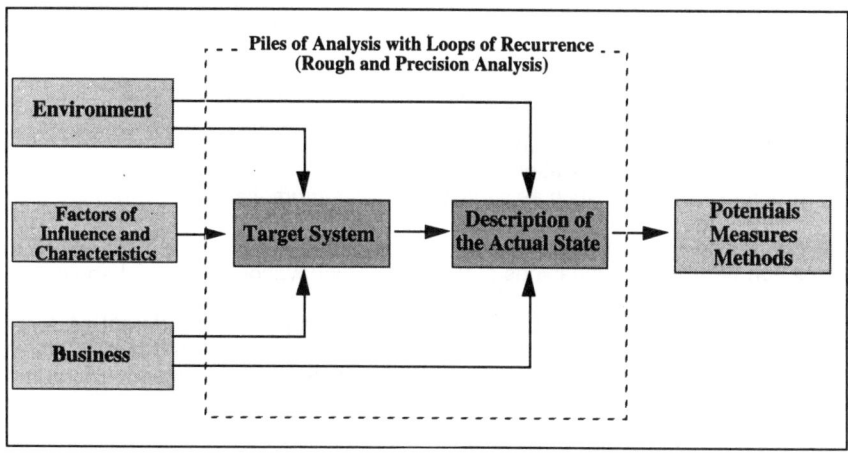

Process of Analysis

When desribing the target system and the present state of the business, it has to be distinguished between factors of influence and characteristics. The factors of influence can be changed directly by using suitable measures and methods whereas the characteristics serve as a measuring system for the description of the actual processes in a business.

Thus e.g. the quality-initiating time characterizes the problems occurring at the interface between product development and production process as well as the penetration of a preventive quality securing.

Both groups of value of the " lean-analysis" are only significant as absolute references compared with the extensive data material of the business line or of similar businesses, which have almost achieved the optimum structure. For this reason, potentials can be deduced more easily if the classification of values is oriented at the target variables of the business in question.
It must be aimed at measuring against the requirements of the future market, not against the average business of the respective line. The target variables of the factors of influence depend decisively on the outward influences of the market and the structures of a business.

The influencing factors have to be designed according to the factors determining the success of special sectors and interfaces in the whole process. The complete target system is compiled together with the data gathering for the representation of the actual state. This takes place in loops of recurrence and represents the actual part of the analysis.

2. STRUCTURE OF A TARGET SYSTEM SPECIFIC FOR A BUSINESS

In order to get an explicit idea of the target system significant for the business, the outward influences of the market, the technologies, the product and the production process itself as well as of the general surroundings have to be stated and with regard to the development of the future influence be assessed. It is essential that not a "flashlight" photo of the actual situation serves as measure, but the dynamic of the factors which are considered.

There are 3 ways for a business to gain advantages in the market:

- by means of produkt innovation
- by means of process innovation
- by structure innovation such as service and sales promotion

The importance of the different sectors and interfaces in a business depends on the correlation between the meanings of the types of innovation mentioned above.

The description of success responsibility and importance of the different sectors of the business depend on the optimized design of all parameters and factors of influence. For most parameters this expression is oriented at the demands of the market.

There is for example a dependency of

- the R&D-expenditure on the innovation pressure of the market

- the importance of the interface between development and sales and the technical form concerning the share of customer projects and products made to customer's specifications.
- the necessary market and procedure flexibility of the business on the dynamic of the market,
- the optimum manufacturing and development depth on the number, the dynamic and the half-life period of the corresponding technologies.
- the method of integrating the suppliers in the phases of product development up to the "just-in-time" concepts on the product and process structure.
- the optimum innovation speed concerning domestic products on the competition structure
- the importance of an explicit modularization of products on the market dynamic and structure of clientele
- the necessary closeness to the market on the competitive situation
- the importance of the interface between development and production and assembly on the number of parts produced and on the lot sizes
- the organizational structure and the importance of the integration on the sectors and the expression of project management

These contexts can be reproduced in portfolios which show how a business can react to the requirements of the market.

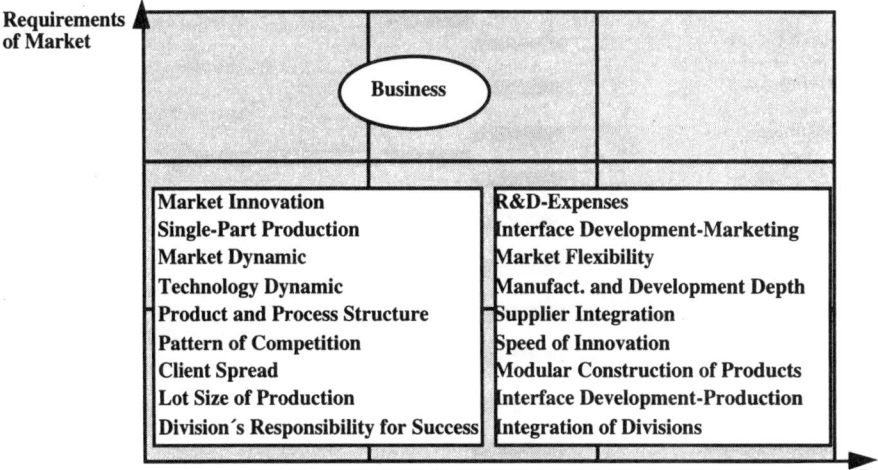

Connection between requirements and factors of influence

3. ASCERTAINMENT OF THE ACTUAL SITUATION OF THE BUSINESS

In the following, the particular lean-factors in their form of appearance must be collected and evaluated with regard to the identified and fixed target structure. Furthermore, the lean-characteristics will be considered and presented. These can, for the most part, only be described quantitatively. An expressive illustration, however, requires also an assessment of the quantitative values concerning their importance for the business.

This data gathering can be supported by questionnaires which are used for the ascertainment of larger basic units. After having terminated the data gathering, the lean-factors and lean-characteristics for the businesses can be described.

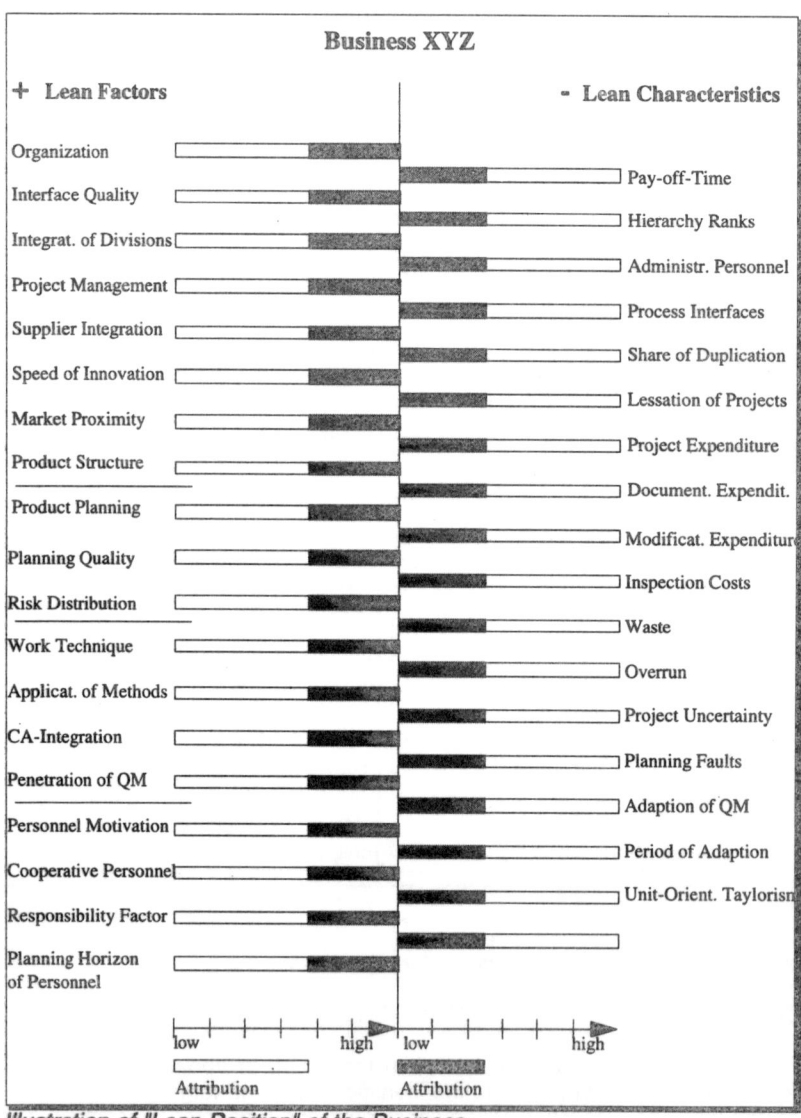

Illustration of "Lean Position" of the Business

32

An Integrated Approach to Software Systems Planning and Selection based on CIMOSA-Models

Georg Naeger

University of Karlsruhe, Institute for Real-Time Computer Systems and Robotics,
Department of Computer Science, University of Karlsruhe, Germany

Abstract

Strong international competition forces enterprises to reduce organizational overhead and cost. Today's industrial practices allow, in general, only the selection of whole planning and control systems for CIM (e.g., CAD systems, PP&C systems, ...) These systems can be adapted only in a limited range to the user requirements. With the introduction of executable enterprise models to the industrial environment, it is expected that it will be possible to design tailored CIM systems built of standardized software modules which will lead to higher complexity of the planning process for software systems. Thus, today's planning and selection approaches will face more and more problems. In order to master the increasing complexity of the planning of software systems for enterprises, new planning methods and tools are required. An integrated approach to software system planning and selection based on CIMOSA enterprise models was developed. The underlying idea of the approach is the conversion of the planning problem into an equivalent constraint satisfaction problem. The conversion process can be accomplished fully automatically.

1 Introduction

Because of the globalization of the competition, the industrial nations have to compete with countries with low wages. This enforces additional reduction of costs and overhead. Today's industrial practice can be characterized by highly automated islands which are only loosely coupled, usually via file transfer. Computer Integrated Manufacturing is broadly recognized to be an approach to overcome these problems through the integration of automated islands [1].

Today's industrial practice allows, in general, only the selection of whole planning and control systems for CIM (e.g. CAD systems, PP&C systems, ...). These systems can be adopted in a limited range to the user requirements. Usually, the different CIM systems of a company are loosely coupled via file transfer. The software imposes a number of constraints on the organization of a company. Therefore, a company must adapt to the requirements of the software which more or less meets its needs. Tailored software is very expensive and difficult to maintain.

A solution to this dilemma is seen in the introduction of executable enterprise models into the industrial environment. The basic paradigma of this approach is the separation of functionality and behaviour. In this approach, the functionality of a CIM system has to be provided by the software and the behaviour of the CIM system has to be provided by the enterprise model. During operation, the enterprise model decides which function has to be performed next. The 'operation system' of this CIM system calls the software packages providing the functionality. With this approach, it is expected that it will be possible to design tailored CIM-systems built of standardized software modules.

We also have to take into account that the automation islands are provided by a number of vendors and that they are used by a number of users. Integration would be much easier if there is a generally agreed upon solution for the required architectural structure [2]. CIMOSA is a new open system architecture developed in the framework of the ESPRIT[1] project AMICE aiming at the provision of such a generally agreed upon solution based on executable enterprise models.

Today's planning activities principally use a two step approach for the selection of software systems: First, an experience driven selection of appropriate methods to support a specific enterprise function is made. The methods selected are not independent of each other because they use data that have to be generated by previously applied methods. There is software support for this specification process available, but usually no systematic consistency checking is performed. Furthermore, the specification process itself is not independent of the

---

1   ESPRIT: "European Strategic Program for Reseach and Devolopment in Information Technology"

software resources available. During the specification process, software resources available have to be taken into account. In the second step of the ordinary approach, a suitable software system is identified through an economic efficiency analysis. The result of the economic efficiency analysis is a software system that is the best compromise but it is not clear which requirements are met by the selected system and which are not.

With this approach, the person who makes the specification has to have a good overview of the enterprise, available methods and the software system market. This approach is used today to select extensive software systems (like CAD or PPC systems), the data exchange of which is easily understandable. Even in this case, the specification process would be facilitated if the user is supported in consistency checking and systematically informed about available software systems in a systematic manner. If the software systems become more open and integrated or executable enterprise models are introduced, the ordinary approach will face serious problems. In this case, software supported method specification and consistency checking is required. Furthermore, it is desirable to develop an approach for software system planning and selection that takes the functionality of available software systems in account.

After an overview of CIMOSA, we will introduce an integrated method which addresses these problems.

2 CIMOSA - an Overview

In this paragraph the major features of the CIMOSA enterprise model are introduced. This description concentrates on those aspects of the model which are required for the understanding of this article.

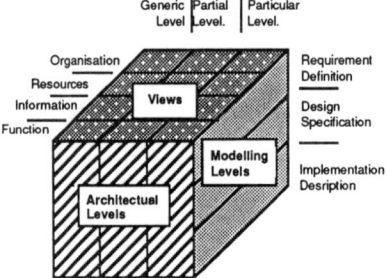

Fig. 1:   The dimensions of CIMOSA

The basic structure of the modelling framework developed by the AMICE project can be visualized with the help of a cube (Fig. 1). The model is divided into three dimensions:

The Architectual Level dimension describes the generality of the model. The Generic Level provides the constructs out of which the models are made. The Partial Level provides a number of Reference Models, which are basically a library of enterprise models which allows to take advantage of the simularities between enterprises of the same branch and size during the modelling process. The particular model is that of an individual company which describes exactly the behaviour and the functionality of this enterprise.

The Modelling Level dimension covers the stages of development from the requirements definition to an executable model. For our considerations, it is sufficient to have a closer look at the Requirements Definition Modelling Level.

The View dimension of the aims at the reduction of complexity of the model by dividing it into different Views. Each of these Views covers a special aspect of the model and there are links defined between the four Views. In further explanations, we focus on the Function View and explain the other dimensions as far as it is required for comprehension.

The separation of functionality, behaviour and structure is the basic paradigm of the Function View. This view decomposes the functionality and behaviour of an enterprise in a hierarchical manner (Fig. 2). A Domain contains one or more Domain Processes. The relationship between Domains are described with the help of Domain Relationships. The Domain Processes are the root of the decomposition tree. It employs Business Processes which are in the middle of the decompositon tree. The leaves are named Enterprise Activities and are employed by Business Processes or, if there are no Business Processes, by Domain Processes. We also use the term Enterprise Function if we talk about a Domain Process, a Business Process or an Enterprise Activity. The input and output of data is described with the Object View construct of the Information View.

Domain Processes and Business Processes contain a so-called Procedural Rule Set. There, it is described in which sequence the employed Enterprise Functions have to be executed. This sequence can depend on the result of a previously executed Enterprise Function.

In CIMOSA, four types of these Procedural Rules are defined. In principle, a Procedural Rule Set consists of a collection of rules similar to the following example:

WHEN (Ending Status of Enterprise Activity X = Value 1) DO (Execute Enterprise Activity Y)

Depending on the type of the rule, CIMOSA allows to Enterprise Functions to be executed mutually exclusivly or in parallel.

As opposed to Domain Processes and Business Processes, an Enterprise Activity does not have a Procedural Rule Set, since there are no more detailed Processes to be controlled. But on the level of Enterprise Activities, the requirements for resources have to be defined.

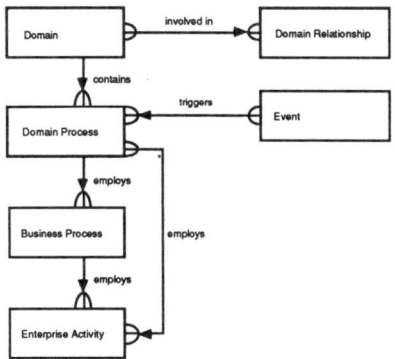

Fig. 2:   Structure of a CIMOSA domain

Therefore, the Enterprise Activities have a part called "Capability Section". In this paragraph, we explain only this parts of the Capability Section, which we use during the software system planning process. It consists of a textual description and a Description of Features. The Description of Features is defined by a non-emtpy set of socalled Feature Elements. They consist of an locally unique Identifier and a name of the Feature Element. We use the Feature Elements to model the methods, with the help of which Enterprise Activities can be supported by software systems. In addition to the CIMOSA attributes, we introduced some additional attributes to the Feature Element. First, we introduced method restrictions. This means that a specific method requires that another method has to be applied by a previously executed Enterprise Activity. In the Reference Model, it is represented by a list of Features/Feature Elements. Second, we added so-called input restrictions to the Reference Model. An input restriction makes sure that a method is previously applied which produces a specific output. This specific output is represented with the help of a list of Object Views. Third, we introduced an attribute to the Feature Elements which represents the Function Output, which is produced by the method.

Resources which provide the functionalities are modelled in the Resource View. The Resource construct is the basic construct used to describe an enterprise object which provides Capabilities. In the Resource construct there is an identical Capability Section defined in order to make a match between the requirements defined in the Function View and the Capabilities provided by possible Resources . The Resource construct has some other attributes, like identifiers, textual description etc. In addition, there are also attributes to describe the relationships between

several Resources. But for our considerations it is not required to introduce these parts of the Function View. The interested reader may refer to [3].

3   An Integrated Approach to Method Specification and System Selection

At the beginning of a software selection process, an as-is-analysis of the enterprise is performed. The result of the as-is-analysis is a kind of enterprise model. Based on this enterprise model, the specification process is performed. In our project we used the CIMOSA enterprise model. As introduced above, CIMOSA also provides constructs to model resources and we added an opportunity to model consistency criteria of methods. Therefore, all the knowledge required for the software planning and selection process is represented in our CIMOSA model.

This representation is suitable to model the planning problem but not to solve it. On the other hand, Artificial Intelligence provides a number of standard procedures to solve planning problems. In order to solve our planning problem, we will convert it into an equivalent constraint satisfaction problem.

Therefore, the planning system consists of two "worlds". The CIMOSA models are used to define the planning problem in a way which is easily unterstandable for the user. The constraint network is used to solve the planning problem. In the next paragraphs, we will introduce the architecture of our system which is depicted in Fig. 3.

The kernel of the planning tool for software systems consists of the following components:

- The *Reference Model* plays an important role in the planning tool. It contains at least the Function View, the Information View and the Resource View. In the Function View, each Enterprise Activity contains a Capability Section. This Capability Section is instantiated while the Resources are modelled. The modeller of the Resources has to specify which Enterprise Activity can be supported by which Feature of a Resource. Therefore, the Function View of the Reference Model contains a collection of methods for each Enterprise Activity. These methods can support the execution of the Enterprise Activity. In CIMOSA terminology, this means that the methods provided by Resources are modelled as Features and Feature Elements in the Capabiltiy Section of the Enterprise Activity.

- The *individual enterprise model* represents the enterprise for which the software system has to be planned. It is derived from the Reference Model. Therefore, we can identify the corresponding component of an Enterprise Function in the Reference Model. The constraint variables are the set

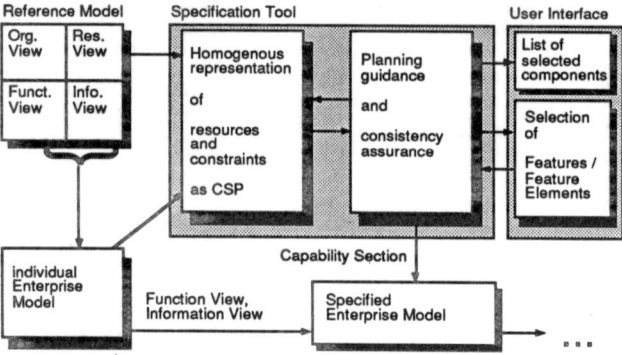

Fig. 3:   Architecture of the software system planning tool

of Features which are necessary to describe the requirements for Resources of the Particular Enterprise Model and a variable representing a combination of resources supporting the Enterprise Activities of the Particular Model.

- The *specification tool* contains the constraint network which represents the planning problem in an equivalent constraint network. Through the user interface, the system offers the available and still employable methods. The user enters his/her selection. If required, the Capability Section can be added to the individual enterprise model for further processing. Afterwards, it is named a Specified Enterprise Model.

Several conversion steps are required to transform the CIMOSA representation into an equivalent constraint network. This conversion process will be explained in more detail.

The process starts with the instantiation of the Particular Model (Fig. 4). It is derived from the Reference Model in such a way that the origin of the Enterprise Activity can be identified in the Reference Model. As already mentioned above, the link between Enterprise Functions and Resources is provided in the Reference Model.

In the next step, the dependencies between the methods applicable in the Particular Model are converted into a constraint network. The representation of dependencies between the software methods works as follows: We assume that a prerequisite can be a specific Function Input which must be provided by a previously executed Enterprise Activity, or it can be a specific method which must be applied by a previously executed Enterprise Activity.

Third, the allowable method combinations which are provided by Resources are also converted into a constraint network. The constraint network is now complete and a

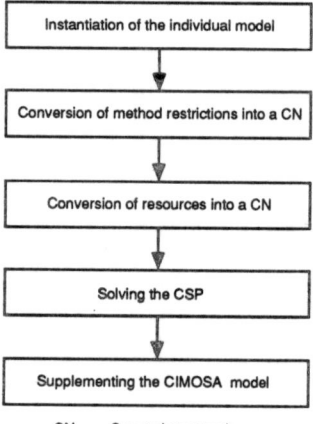

| CN | Constraint network |
| CSP | Constraint satisfaction problem |

Fig. 4:   The course of software system planning and selection

problem-solving algorithm controls the instantiation of the constraint variables. If the user selects a method which cannot be provided by a Resource, a constraint will be violated and the instantiation of the method has to be taken back.

Finally, the solution found can be added to the individual Enterprise Model.

The most difficult task is to generate the equivalent constraint network. How this can be accomplished will be decribed in the next paragraphs. Afterwards, we will explain the various conversion steps required to generate the constraint network. Before summing up, we will present the planning guidance.

3.1 The Conversion of the Particular CIMOSA Model into a Contraint Network

The conversion task requires two steps: First, we have to find out in which sequence Enterprise Activities are executed. This step requires an analysis of the Procedural Rule Sets and the Events of the individual enterprise model. Second, we have to consult the Reference Model in order to find out which Feature/Feature Elements impose which constraints on other Enterprise Activities of the Particular Model.

As we mentioned above, in the first step we have to find out in which sequence the Enterprise Activities have to be executed. The Procedural Rule Set defines the sequence in which Enterprise Activities of Business Processes have to be executed. The sequence can depend on termination states. For our purposes, it is sufficient to extract the static precedence relationships which is independent of the actual ending state. Furthermore, we have to distinguish between the parallel execution and the mutually exclusive execution of Enterprise Functions because this has a strong influence on the constraint generation. We will explain that using an example (Fig. 5):

A method of Enterprise Function 4 imposes a constraint on a previously executed Enterprise Function (here: 2 or 3). If the Enterprise Functions 2 and 3 are executed mutually exclusively, the required method must be provided by both, since only one Enterprise Function is executed at any one time. If the Enterprise Functions are executed in parallel then it is sufficient if only one Enterprise Function provides the method required by Enterprise Function 4, since both Enterprise Functions are executed in any case.

Following this train of thought, we developed an algorithm which analyzes the Procedural Rule Set and converts the static structure defined by it into a precedence graph. The nodes of the graph represent Enterprise Functions. The procedure starts at the level of the Domain Processes and refines the precedence graph until its nodes represent only Enterprise Activities which belong to the lowest decomposition level on the

Requirements Definition Modelling Level. With the help of the precedence graph, we can easily determine the predecessors of an Enterprise Activity.

In the second step of the conversion process, the constraints must be generated which represent the dependencies between the methods available. During this conversion process we have to consult the Reference Model and the precedence graph. The algorithm which does this works principally as follows:

Mutually exclusive executed Enterprise Functions:

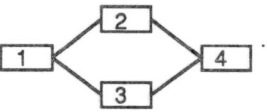

Enterprise Functions executed in parallel:

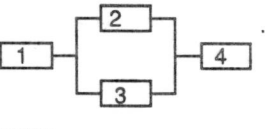

▢ Enterprise Function

G020

Fig. 5: Two Examples for precedence relationships of Enterprise Functions

For each Enterprise Activity of the Particular Model, the algorithm determines by which methods this Enterprise Activity can be supported in the Reference Model. For each of the methods, it is checked whether it imposes a restriction on another Enterprise Activity. The precedence graph and the Reference Model are consulted in order to find solutions which do not violate these restrictions. A constraint is generated which represents the allowable solutions. In this way, all restrictions of the Particular Model are worked off and all the constraints which represent the existing dependencies between the methods are generated.

3.2   The Conversion of the CIMOSA Resource Model into a Constraint Network

In the conversion steps we have described up to now, we formalized the dependencies between methods which can support an Enterprise Activity.

As mentioned above, the variables of the constraint network are the set of Features which are used in the Particular Model to describe the Required Capabilities. The available Resources provide a specific combination of methods which restrict our method selection during the specification process. Therefore, we have to represent the admissible combinations of methods. But several Resources are required to provide all the methods required. Therefore, one constraint represents an admissible combination of methods provided by several Resources. Because of these reasons, we need one constraint for each possible combination of Resources.

The algorithm for one Resource works principally as follows: for each Feature of the Particular Model we determine which Resource provides a method to support that Feature. For each Resource we generate a constraint and add the Feature of the Particular Model as a constraint variable. The methods provided by a Resource are the allowable values for the constraint variable and, thus, we add these values to the constraint. The last constraint variable we add represents the name of the Resource. In the specification process, this variable will be instantiated as the last variable. The remaining allowable values of this variable are the names of the Resources which can provide the methods required.

When we have generated the resource constraints, our constraint network is complete. In the next paragraph, we explain how the constraint satisfaction problem can be solved.

3.3   Planning Guidance

There are several methods available for solving constraint satisfaction problems [4]. The most important class of problem-solving methods are the backtracking algorithms. The differences between the various algorithms of this class consist of the different powerful constraint propagation methods. In the simplest case, there are none. In more complex cases the constraint propagation methods remove all values from a variable domain which can cause conflicts. The more powerful the constraint propagation algorithm is, the more time-consuming its execution will be. In our case, we decided to use a backtracking algorithm with forward checking. This means that the constraint propagation method filters the domains of variables in such a way that the remaining values do not violate constraints (with the current variable instantiations).

The sequence instantiation of variables is another important aspect, and we have to consider whether the instantiation sequence is comprehensible to the user. This restricts our freedom in choosing the instatiation sequence as follows:

1. Features belonging to one Enterprise Activity should be instantiated as a group.

2. The instantiation sequence should follow the flow of information in the enterprise model. In order to make the method selection process more goal-oriented, Grabowski (Grabowski, 1992) recommends starting with the instatiation process at the end of a chain of Enterprise Activities and to follow the flow of information in a reverse sequence. This sequence has the important advantage that the user defines the "last" method he wants to apply first and during the specification process he knows exactly what is required by the successors. This makes the specification process more goal-oriented and it will be applied in our system.

We have now described how the constraint network will be generated and how the planning guidance works in principle. In the next paragraph, we will summarize the most important features of the integrated approach.

4   Summary

In Section 1 we mentioned two major planning problems which arise with the introduction of higher integrated software systems into the industrial environment. First, we mentioned the requirement to develop a planning process which supports the user to select methods in a consistent manner. Second, the specifications must be made in a way that there are Resources or a combination of Resources available which can provide the functionality required.

With our approach, it is possible to model restrictions caused by methods as input requirements or as methods requirements. These requirements are converted into an equivalent constraint network automatically. In a similar way, method combinations provided by Resources are automatically converted into a constraint network as well. The problem solving procedure ensures that only methods can be selected which are provided by Resources and it ensures that there are Resources which meet all the requirements defined during the specification process.

5. Acknowledgement

This research is carried out at the Institute for Real-time Computer Systems and Robotics of the University of Karlsruhe. The authors would like to thank the other partners involved in the project. It was supported by the State Government of Baden-Wuerttemberg, Germany.

References

[1] Normungsausschuß Maschinenbau (NAM) (1991) *European Prenorm: Computer Integrated Manufacturing - Systems Architecture - Framework for Enterprise Modelling*, Beuth-Verlag, Berlin

[2] AMICE (1989) *Open System Architecture for CIM*; Springer-Verlag Berlin, Heidelberg, New York, London Paris Tokyo, Hong Kong

[3] Formal Reference Base (FRM) (1993), *CIMOSA - Open System Architecture for CIM - Technical Base Line*, Version 2.0, ESPRIT-Consortium AMICE, Bruxelles

[4] Zahn (1993) *Entwurf und Implementierung eines Expertensystemkerns zur Konfiguration - unter Verwendung eines zu entwickelnden Constraint-Erfüllungs-Systems*, Master Theses Institute for Real Time Computer Systems and Robotics, University of Karlsruhe

[5] Grabowski H., Schäfer H. and Krepinski A. (1992) *Instrumentarium zur methodisch unterstützten Planung und Integration betriebsspezifischer CAD/CAM-Verfahrensketten*, CIM-Management 2/92

33

THE "MATERIAL CONVERSION CLASSIFICATION"

John L Burbidge

Much of the present research literature on Production is of little help to industry because it fails to relate its findings to the types of industry where it is applicable. For the same reason, many attempts at benchmarking fail because they neglect the need to limit the targets they set to particular industrial situations.

There is a need for a classification of industries which brings together types of industry requiring similar production systems which can also accept the same targets for benchmarking. The classification recommended here is the "material conversion classification" (MCC) based on the ratio of material to product varieties. It finds the four main classes of Process, Implosive, Square and Explosive industries. These classes are further sub-classified by "flow type" and "organisation".

These classification systems are not new [1] and [2], but they need restating and elaboration. Together they provide a good foundation for the selection of production systems and for the specification of targets for benchmarking.

1.   THE MCC CLASSIFICATION

The "materials conversion classification" (MCC) is illustrated in Figure 1. It finds four main classes, based on the ratio of the number of material varieties

	PROCESS	IMPLOSIVE	SQUARE	EXPLOSIVE
	m ⬭ p	m ▽ p	m ▢ p	m △ p
Key:- m = No. mtl. varieties p = No. prod. varieties				
Mtl. - Input:- Product - Output:- Mtl. flow type:-	Bulk mtl. Bulk mtl. Line flow	Bulk or Gen mtl. Gen mtl. or comp Batch or line flow	Components Components Batch flow	Gen. or sp. mtl. Assemblies Batch flow
Examples:-	B Cement E Ore treat (mine) F Milk F Sugar F Distilleries C Gases O_2; N B Bricks B Timber F Breweries E Tanneries E Paper	E Foundries D Potteries D Glass C Dec. laminates T Spinning, fibres E Brake linings D Printing T Knitting F Bakeries E Rolling Mills E Wire drawing	E Jobbing m/c (some) D Dry cleaning E M/c overhaul T Dyeing textiles T Finishing textiles E Heat treatment E X Ray E Painting E Electro-plating E Metal spraying E Polishing	E Automobile E Electronics D Consumer durables E M/c tools E IC engines E Electrical T Weaving D Clothing & shoes C Chem. dye stuffs D Furniture E Welding

Key:- F = Food   C = Chemicals   T = Textiles   E = Engineering   B = Building   D = Domestic

FIGURE 1   TYPES OF INDUSTRY

used to the number of products produced.

(a)   **PROCESS industries** are those in which a small number of material
        varieties is converted into an equally small number of product varieties.
        Examples are found in factories making cement, spirits and many
        chemical and food products. Both the raw material and the product
        generally take the form of bulk material, normally measured in standard
        units (SUs) of length, weight, area or volume, although packaging may
        give added product variety in some cases.

        This is a unique and homogeneous class because such factories are
        easily recognised; they can generally be laid out for continuous line flow
        (CLF); they are normally suitable for automation based on the automatic
        transfer line; because there are few varieties of material and product,
        Production Control is simple and statistical forecasting of future sales is
        more reliable than it is in other industries which make a greater variety
        of products and use a greater variety of materials.

(b)   **IMPLOSIVE industries** are those in which a very small number of
        material varieties is converted into a large number of product varieties.
        Examples are found in foundries (pig iron), potteries (clay), glass works
        (sand), bakeries (flour) and others. The raw materials are generally bulk
        materials (e.g. pig iron, clay, sand, flour etc.) and the products are
        components (e.g. castings, pottery items, glass works products and
        cakes) or partly processed raw materials (e.g. rolled bar or wire).

        This is again an easily recognised unique and homogeneous class. A
        tendency to use the same processes in the same sequence gives simple
        material flow systems, leading to simple production control in which
        material supply can again be based, with some accuracy, on statistical
        forecasting [3].

(c)   **SQUARE industries** are those in which a large number of material items
        are converted into an equal number of products. They are normally
        service industries where the customer supplies the materials which are
        returned to him after processing. Examples are laundries; dry cleaners;
        textile dyers and "finishers"; and painting, electro-plating, X-ray
        inspection and heat treatment sub-contractors.

        This class is again homogeneous and easily recognised. Materials and
        products are always the same items; the material flow systems inside
        the factories are generally simple, and Production Control is again
        simple because there is no material supply problem. Production Control
        is concerned mainly with scheduling and loading.

(d)   **EXPLOSIVE industries** finally, are those in which a large number of material varieties are converted into a relatively small number of product types. Most Explosive industries are assembly industries which often make some, or all the parts they use. They generally manufacture in two or more stages (e.g. component processing followed by assembly).

This is the most complex type of industry. The material flow system is intrinsically complex but can be greatly simplified by using Group Technology and be automated to form Flexible Manufacturing Systems (FMS). Production Control can, in this case, also be extremely complex. Traditional methods based on multi-cycle ordering (e.g. stock control - based on re-order levels - or MRP) are unreliable due to their complexity; to the dynamic distortion of the "industrial dynamics" [4] and "surge effects" [5]; and with MRP, due to explosion based on long term sales forecasts, which are always unreliable.

The explosive industries' class is homogeneous because it brings together the assembly industries, which tend to suffer similar problems (e.g. long lead times, heavy stocks, high obsolescence and low flexibility), for which similar solutions tend to be needed (e.g. Group Technology, JIT, single-cycle ordering etc.).

(e)   **GENERAL** In general, this four class MCC classification is reliable in practice and simple to use. Most factories fall mainly into one of the four classes. A few factories may contain more than one (e.g. an Engineering Assembly factory (explosive) with its own foundry), (implosive) and others may be at a transitional point between process and implosive (e.g. a pork pie factory which is planning to introduce sausages, scotch eggs and paté as additional products, using mainly the same materials). In general, however, the MCC is a reliable classification for which the worst that can be said is that it doesn't go far enough for some purposes.

2.   MATERIAL FLOW TYPES

The MCC classification is a classification of Material Flow Systems. It brings together industries with similar "maps", showing the routes followed by materials between the places where work is done on them to convert them into products. It does not consider the way in which the materials flow through the material flow system. Figure 2 shows a sub-classification into four main flow types of jobbing, batch, continuous flow and OKP production which contributes to this need. Not all combinations of classes in these two classifications are feasible, but adding this flow classification increases the number of types from 4 to 10.

MCC Class:-	PROCESS - P	IMPLOSIVE - I	SQUARE - S	EXPLOSIVE - E
Flow type	m ▭ p	m ▽ p	m ▢ p	m △ p
Jobbing J	-	IJ	SJ	EJ
Batch B	PB	IB	SB	EB
Continuous C	PC	-	-	EC
OKP O	-	-	-	EO

FIGURE 2   MCC & FLOW TYPES

(a) **Jobbing** factories make jobbing products; which are special to each order; are normally ordered in small quantities; and are unlikely to be re-ordered again to exactly the same specification. Such products cannot be ordered "for stock" and must be made "to order" in the quantities specified by the customers. Jobbing production is unlikely to be found in process industries, but it can be a sub-class of Implosive, Square or Explosive industries.

(b) **BATCH production** is a type of production in which standard parts and products are made intermittently in batches. The products produced must be made to standard designs, either "for stock" or "to order". Batches of different parts share capacity on the same machines and other facilities. "Batch production" can be a sub-class of any of the "MCC" classes. It might be said that batch production is a type of continuous flow production in which manufacture is intermittent because it is faster than consumption.

(c) **OKP - one of a kind production** is a variant of jobbing production, making large one-off products such as ships, special machine tools, aeroplanes, buildings or small batches of such products [6]. These products usually have a long lead time and their overall regulation can be seen as a case of "project management". OKP companies will normally have departments making components such as machine shops; sheet metal shops; plate shops, cold forge and welding departments; and others, which fit easily into the explosive batch industry category.

(d) **CONTINUOUS production** is a type in which a single product, with perhaps some variants, is made continuously. Production in this case takes place at the same rate as consumption. Continuous production is widely found in the process industries and can also be found to a limited and reducing extent in the Implosive, Square and Explosive industries, in the form of component processing or assembly lines.

(e) **GENERAL.** Two or more of these flow type classes may sometimes qualify the same MCC class but most factories can generally be covered by one single two digit class.

3.   TYPE OF ORGANISATION

One other classification which has a profound effect on the production system and on its performance is the method of organisation adopted.

Organisation is concerned with the way in which the people in an enterprise are formed into organisational units or teams for supervision and control. In production there are two main types of organisation:-

(1) **"Process organisation"** is a type in which organisational units specialise in particular processes. e.g. All lathes and lathe operators together; under one foreman; all milling machines together, and so on.

(2) **"Product organisation"** is a type in which organisational units complete products, assemblies, components or parts and are provided with all the machines and other facilities they need to do so. e.g. One organisational unit or Group makes a list or "family" of gears. Others complete other sets of parts.

It should be noted that "Process organisation" and the "Process Industry" type are names for two entirely different concepts, even though they both use the word "process" in their names.

There are again two main types of Product organisation:-

(1) **"Continuous line flow" (CLF)** in which production operations are carried out in line, always in the same sequence. Examples can be found in most "Process industries" and to a limited extent, with mass production products in "Explosive industries".

(2) **"Group Technology" (GT)** [7] in which different sets of machines and other processing facilities each complete all the components in their own particular "family" or list of components.

The effect of adding this classification of organisational types to the "MCC plus flow type" classification of ten types (illustrated in Figure 2), is illustrated in Figure 3. It is postulated that although Process organisation was generally used in the past and can still be used today, it is always possible to change to Product organisation if one so wishes. [8] Because Product organisation is much more efficient, Process organisation is obsolete.

As long as Process organisation is still used in industry, the ten types in Figure 3 with the "-PO" suffix, will still be needed and the full classification will include 20 types.

4.   THE MCC CLASSIFICATION AND THE CHOICE OF PRODUCTION CONTROL

If one knows the type of industry in a company, one has a clear idea of the type of system which should be used to regulate material flow. Consider first

INDUSTRY TYPE MCC + flow	PROCESS ORG (PO)	or PRODUCT ORG.	
		CLF	GT
PB	PB/PO (X)	PB/clf	-
PC	PC/PO (X)	PC/clf	-
IJ	IJ/PO (X)	-	IJ/GT
IB	IB/PO (X)	-	IB/GT
SJ	SJ/PO (X)	-	SJ/GT
SB	SB/PO (X)	-	SB/GT
EJ	EJ/PO (X)	-	EJ/GT
EB	EB/PO (X)	-	EB/GT
EC	EC/PO (X)	EC/clf	-
EO	EO/PO (X)	-	EO/GT

Key: - = Impossible, or very unlikely.
(X) = Possible but obsolete.

FIGURE 3 INDUSTRY TYPE AND ORGANISATION

the Process industries. In both the "continuous flow" and "batch flow" modes, these provide the simplest case for the regulation of material flow. Simple statistical forecasting, based on Moving Annual Totals (MAT), can be used to forecast future sales and material requirements. A supplementary control of stocks makes it possible to adjust output to match changes in demand.

The Implosive industries next, in both the Batch and Jobbing modes, are also simple to manage and require similar types of Production Control system. Supply of the small number of material varieties can be based again on statistical forecasting. Production programmes for manufacturing can be based on the regular periodic accumulation of orders received, with smoothing to eliminate excessive variations in demand rate between periods.

The Square industries, in both the Jobbing and Batch modes, require only simple Production Control methods. Programming for Manufacture can again be based on the regular periodic accumulation of orders received, with smoothing to eliminate excessive variations in demand.

The Explosive industries require the most complex Production Control systems. They need sales programmes for finished products; which are converted with "smoothing" into production programmes to regulate assembly; which are exploded to find the requirement for parts and materials. Similar systems can be used to regulate production in most Explosive industries.

It will be seen that if one knows the MCC class of industry of a company, one has made a major step forward towards finding the best type of Production Control system.

5.   THE MCC CLASSIFICATION AND BENCHMARKING

The measures appropriate for benchmarking tend to be similar for different companies with the same MCC classification. The values assigned to benchmarking targets for these measures will tend to fall into narrow bands for each MCC class.

There is no published research which has attempted to establish benchmarks for different measures in different types of industry. It is submitted here that such research would be valuable to industry.

As an example of what is needed, Figure 4 shows a speculative range of values for each MCC class for "Rate of Stock Turnover". In this case, the type of ordering system (multi or single cycle) also has an important effect on stock turnover and the "MCC + flow type + org type" classes have been further divided into multi and single cycle sub-classes.

	Code	PROCESS ORG (PO)		PRODUCT ORG - (clf)			PRODUCT ORG - (GT)		
		Multi-cycle	Single-cycle	Code	Multi-cycle	Single-cycle	Code	Multi-cycle	Single-cycle
Process	PB/PO	8 - 15	25 - 40	PB/clf	20 - 25	50 - 55	-	-	-
	PC/PO	50 - 60	80 - 90	PC/clf	60 - 65	100 - 120	-	-	-
Implosive	IJ/PO	12 - 15	20 - 23	-	-	-	IJ/GT	20 - 23	25 - 27
	IB/PO	14 - 16	25 - 27	-	-	-	IB/GT	26 - 28	32 - 35
Square	SJ/PO	14 - 18	18 - 21	-	-	-	SJ/GT	20 - 23	31 - 33
	SB/PO	17 - 19	20 - 23	-	-	-	SB/GT	17 - 20	26 - 28
Explosive	EJ/PO	3 - 5	6 - 9	-	-	-	EJ/GT	6 - 10	18 - 22
	EB/PO	4 - 7	10 - 13	-	-	-	EB/GT	10 - 14	28 - 32
	EC/PO	30 - 35	45 - 50	EC/clf	30 - 35	45 - 50	-	-	-
	EO/PO	14 - 18	16 - 20	-	-	-	EO/GT	10 - 14	28 - 32

KEY: Code = scc code; Multi-cycle = multi-cycle ordering system; Single-cycle = single-cycle ordering system; clf = continuous line flow; GT = Group Technology.

FIGURE 4 "SPECULATIVE" BENCHMARKS FOR "RATE OF STOCK TURNOVER (turns p.a.) BY MCC CLASSIFICATION CLASS

Note: These rate of stock turnover figures are speculative. They are not based on research and are intended only to indicate the kind of distribution expected by the author.

6.   SUMMARY

The "Material Conversion Classification" (MCC) was originally designed to find classes of industry which need the same or similar Production Control systems. Its extension to include sub-classes for flow type and organisation type makes it suit this purpose consistently.

It is submitted here that this classification, with perhaps some further sub-classification in particular cases, should bring together classes of factory for which common benchmarking targets can be specified within narrow bands.

REFERENCES

1.   1962. Burbidge, J.L. The Principles of Production Control (4th Edition 1978) Macdonald and Evans, Plymouth, UK.

2.   1968 Idem. Group Technology. Turin International Centre, Turin, Italy.

3.   1994 Idem. The use of PBC in the Implosive Industries. Production Planning and Control. Vol. 5, No. 1. Taylor and Francis, London, UK.

4.   1958. Forrester, Jay W. Industrial Dynamics. Harvard Business Review, Harvard, USA.

5.   1991. Burbidge, J.L. Period Batch Control with GT - the way forward from MRP. BPICS 26th Annual Conference. 14-16 Nov. BPICS, UK.

6.   1993. Burbidge, Falster and Riis. Reducing delivery times for OKP products. Production Planning and Control. Vol. 4, No. 1. Taylor and Francis, London, UK.

7.   1988. Group Technology the state of the art. Proc. IIS Seminar. Elektivni Proizvodni Sistemi, Dubrovnic, Novi Sad University, Jugoslavia.

8.   1992. Idem. Change to GT. Process organisation is obsolete. IJPR. Vol. 30, No. 5. Taylor and Francis, London, UK.

34

Fuzzy Logic in Team Design

A. Stickley*, B. Grabot**

*TQM Research Engineering Design Centre, City University
Northampton Square, London, EC1V O HB, England

**ENIT, Laboratoire Génie de Production BP 1629 - 65016 Tarbes, France

Present team selection do not assure effectiveness. Using existing teams, team profiles can be constructed from interpersonal combinations of member's main intrinsic motivators, called Drivers. With their team effectiveness ratings, these profiles provide a model to aid team design. The uncertainty and imprecision of subjective measures will be reduced by using fuzzy logic and the theory of possibility as a modelling tool.

1. INTRODUCTION

Teamwork is an area of product design and development where there is much interest by industry and academics and a lucrative one for management development consultants. However it is an area which is not well researched. Teams are a variable concept in the reality of industry with little common ground on what constitutes a team, or more precisely what is the composition of an effective team. Team members are chosen primarily on the basis of their availability and skills, and whether someone would make a "good team member" on vague subjective criteria In the new quality era, functional areas too have to be represented though sometimes restricted to "where appropriate", and there is a growing awareness of the value of including customers, users and suppliers. What passes for good team design is team building or team development using models such as team roles [1], psychological traits [2] or Process Communication [3], and the evidence of any team success is merely a subjective claim or is anecdotal. There has also been no descriptions found so far in the literature of quantitative evaluation methods for the effectiveness of teams, except for that of one of the authors [4], who has developed a Team Effectiveness Rating from research work described in [5].

This paper looks at Team Design from a sound theoretical basis of Transactional Analysis Drivers, a five category system of a person's habitual learned responses and expectations of him- or herself and others [3]. These Drivers form interpersonal compounds, links of underlying understanding or expectations, with the Drivers of all other team members which can be can be summed and combined to form a Team Profile. This team profile is made unique by assigning it a performance indicator got from the Team Effectiveness Rating questionnaire above which were completed for each team by the team leader, project manager and one other involved manager, preferably a stakeholder in the team's work.

The subjective measures of both an individual's Drivers and the managers' team rating are full of imprecision, ambiguity and uncertainty. This applies also to team composition, which in turn affects the clarity of the Unique Team Profile and its usefulness in prediction of a team's success. Statistics alone does not resolve these problems which lie in the realm of possibility rather than probability. Fuzzy logic can be used as a tool here. The ultimate aim is to develop a model of teams with a prediction of a better chance of success, which can help project managers make decisions on team design from the mix of people available in addition to the other criteria they already use.

2. THE UNIQUE TEAM PROFILE MODEL

A team profile can be constructed as a histogram of interpersonal compounds. An interpersonal compound is a one-to-one link between a "Driver" in one person to a "Driver" in another. A Driver is a complex belief system which can be considered as an habitual unconscious expectation we have of ourselves and others. When a person is being influenced by one of his Drivers, others can observe the behaviour belonging to that Driver, and agree on the category of the Driver. Checklist questionnaires can also be used to determine a person's Drivers [6].

2.1 Drivers

Drivers, are a development of the Transactional Analysis theory of life scripts, "a life plan made in childhood, reinforced by the parents, justified by subsequent events, and culminating in a chosen alternative" [7]. The Drivers are the initial behaviours observed when entering our mini-script, the short-term reinforcement patterns of our life-script. It does not follow that we must always enter completely into the pattern described above. This is important in team building and team development, as well as in the attendant individual member's personal growth, because awareness and recognition of the Drivers in ourselves and others gives us options to remain in a healthy interpersonal cooperative state.

There are five Driver categories here described briefly with their descriptive names.
1. Please Me, (or Be Pleasing), can have warm relations with others, but may at times seem self-centred or manipulative.
2. Be Perfect, (or Be Right), do well in areas of careful reasoning and detail, but can also get bogged down in details instead of working on the important aspects. They can also be highly critical.
3. Be Strong can show responsibility and be cautious, but can also at other times seem slow or withdrawn.
4. Hurry Up can be decisive and dynamic, but can also be impatient and demanding or slow themselves down by making mistakes or having accidents.
5. Try Hard can undertake challenging and important work, but may also make mountains out of molehills, or put more energy into going things our own way rather than in achieving the goal. He can also be rebellious.

The Driver Checklist used in the team investigations was originally developed and validated in [8]. There are two parts to the checklist, the first part (cf. figure 1) required that the individual team member ticked the list along with someone else who knew him well, and was willing to be frank. He added his ticks as he perceived the individual whether there was agreement or not. This was to help overcome the problem in self-reporting since someone else can often be more aware of our habits than ourselves. None or all of the nine elements in each driver section may be ticked. The second part was a one choice tick out of five concerns that the individual had himself and which he filled out on his own.

DRIVER CHECKLIST

1. BE PLEASING
1. smile or laugh a lot when I am talking to someone
2. leave when things go wrong instead of facing them now
3. nod my head when I talk or say "You Know?"
4. dress skilfully to show my own special style
5. laugh to smooth things over when I am a little nervous
6. say nice things before asking for something
7. act cheerful to cover my bad feelings
8. automatically give first priority to others
9. am usually restless when I am by myself.

2. BE RIGHT (PERFECT)
1. point out the mistakes of others, or challenge them to justify
2. am often quite early
3. tidy up and put away, or don't use a clean waste bin
4. move with a very erect posture
5. criticise what people think or do
6. collect, or display, interesting things or information
7. get things just exactly right
8. show, or hide, a lot over what is not very important
9. do not trust others to do things well enough

3. BE STRONG
1. am outwardly calm even when upset
2. consider long before deciding
3. think how to do without things
4. carry around more than I need
5. make the best of a bad situation for much too long

6. do things for others that they should do for themselves
7. am extra cautious
8. use my face to hide my feelings
9. can be physically uncomfortable a long time and not notice

4. HURRY UP
1. hurry when it does not matter
2. do not get around to buying clothes
3. say "Are you ready? Let's go", and leave quickly when I start
4. tap my fingers, wiggle my feet, or jiggle my knees up and down
5. do too much too fast
6. bump into people or things
7. interrupt to hurry people along, or start leaving before they finish
8. pace back and forth while waiting
9. walk fast, eat fast, or talk fast

5. TRY HARD
1. have trouble finishing
2. realise that I have done it the hard way
3. tell myself that this time I will do it right, and then don't
4. have difficulty with things that go smoothly for others
5. don't get around to important things for too long
6. ease off and delay when I get close to finishing
7. have many things disorganised, or let dirty dishes accumulate
8. am sometimes quite late, or do not get there at all
9. delay too long before starting

Figure 1. Driver Checklist

Results have shown that this second part may not be reliable in all cases in our present organisational culture. Hazell's data were from volunteers. The Team Design research data came from teams within UK engineering industry, with projects managers administering the questionnaires. Individuals were not necessarily volunteers, nor committed. This methodology was an unavoidable reality, and may have let to confidentiality issues. Some discrepancies between the two parts were noticed and investigated which showed there were mismatches. For that reason the second part results are not being used at present.

Scoring was two points for numbers 1, 3 and 5, and one for the others.

2.2 Team Profiles

Although we all show aspects of all five drivers, usually only two are considered important to us. This intrapersonal pair forms links, interpersonal compounds, with all of the other team members, each team pair forming four links of underlying understanding, or perhaps misunderstanding. Two people with similar main Drivers are considered to have a better chance of understanding each other and work together because they have the same expectations. Dissimilar drivers may or may not have the same effect. The absence of a main interpersonal compound of similar Drivers, for example "Be Right - Be Right", would make an unbalanced team, with less chance of being effective. A team of five is the minimum number in this model needed to avoid the certainty of a missing main interpersonal compound, although it remains possible with larger teams.

For a team of five, there are ten member pairs making a total of forty compounds, and a team of nine thirty-six pairs and one hundred and forty-four compounds. It is the distribution of these interpersonal compounds into the possible fifteen that makes up the team profile histogram. An example is given in figure 2 of an existing team.

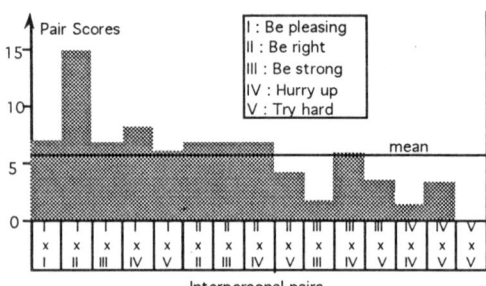

Figure 2. Team Profile

Some individual's results showed close scores in the second and third top drivers. In order to take this into account, and to address the imprecision inherent in a subjective questionnaire, the one line team profile was changed into a band. This was achieved by adding (-1) to the two top drivers and (+1) to the others for the whole team and recompiling the team profile. This profile was the overlaid onto the original.

However, only considering the two top drivers means that much information is ignored.

A. End Results Rating (very poor, poor, fair, good, very good)

1 technical success	2 on time performance	3 on-budget performance
4 commitment and results orientation	5 innovation and creativity	6 concern for quality
7 willingness to change	8 ability to predict trends	9 involvement and energy
10 capacity to resolve conflict	11 communications effectiveness	12 team spirit/cooperation
13 mutual trust	14 membership self-development	15 interface effectiveness
16 high achievement needs		

B. B. Team aspects during Project Elements (very poor, poor, fair, good, very good)

1 clear objectives	2 sufficient resources	3 no recurring power struggles & conflict
4 involved management	5 job security	6 clear goals and priorities
7 interesting work	8 recognition	9 experienced manager
10 good direction and leadership	11 qualified team personnel	12 professional growth

Figure 3. Team Performance Rating Questionnaire

2.3 Performance Evaluation

In order to make a unique team profile, a Team Performance Rating checklist was used (cf. figure 3). This checklist was constructed from [7]. The first sixteen elements are result- and people-oriented measures of project performance which are correlated with the twelve enhancers and barriers to team effectiveness within the team itself. Enhancers (B. nos 7-12) are used here to avoid confusion with their term drivers and Transactional Analysis Drivers. In constructing the checklist Thamhain's original barriers have been changed to their opposite to maintain a consistent rating scale.

An overall figure was obtained by addition of all the elements from the three managers who rated the team, taking their mean and expressing that value as a percentage of the possible total. This value was then plotted on the right hand side scale of the Unique Team Profile, and the previous profile aligned to it by the mean of the interpersonal pairs. This allows different teams to be compared graphically for patterns.

3. IMPROVEMENTS OF THE MODELS THROUGH FUZZY LOGIC AND THEORY OF POSSIBILITY

As pointed out in the previous section, there is a need for modelling tools in order:
i) to take into account more than two drivers if their evaluations are close, and
ii) to express the possible inconsistencies between the various sources of the performance evaluation.

Fuzzy sets and the theory of Possibility allow us to model both the imprecision and uncertainty of information, and can be used either to avoid threshold effects, or to express the consistency between data. The principles of these tools are shortly described in the next part, and their use in the case of team profiles and team performance evaluation is then described.

3.1 Fuzzy sets and the theory of Possibility

The membership of an element u to a fuzzy set is not expressed though a binary value (0 or 1) but through a value $\mu(u) \in [0,1]$ [9]. Such a set is then completely described by its membership function μ that associates to each element of the set its membership degree. A proposition that contains imprecise aspects can be modelled by a fuzzy set by mean of the choice of an attribute, and of an evaluation scale of this attribute. In that way, a proposition such as "John is young" can be translated by "John belongs to the set of the young persons".

Let us assume that John is 28 years old and that its membership degree to the set of the young people is 0.5. The postulate of the theory of Possibility is that the membership degree of John to the fuzzy set of the young people can be interpreted as the possibility that John is 28 years old, knowing that "John is young" [10]. The membership function μ is in that case considered as a distribution of possibilities.

With a unique formalism, it is then possible to express imprecision (through membership functions) and uncertainty (through distributions of possibilities).

3.2 Application of the theory of possibility to the team profile

Consider a team of four members (denoted A, B, C, D) as an example (teams usually have 7±2 members) and the evaluation of their respective Drivers given by the table of figure 4.

The choice of two main drivers for each person raises some problems for A, because drivers III and IV are very close, and for D because II and IV are equals. Let us consider the table of figure 6, where each line has been divided by the value of the maximum driver of the line. Each number can be considered as the possibility that the corresponding driver is the main one for a person. The possibility that a link exists between two main drivers can then be expressed by the minimum of the possibility degrees of the two drivers considered. In fact, it is not necessary to take into account all the drivers, and a threshold can be chosen in order to eliminate the lowest ones.

		\multicolumn{5}{c}{DRIVERS}				
		I	II	III	IV	V
	A	9	3	6	5	2
TEAM	B	6	6	4	3	3
MEMBERS	C	1	1	7	5	1
	D	3	6	7	6	3

	I	II	III	IV	V
A	1	0.33	0.66	0.55	0.22
B	1	1	0.66	0.5	0.5
C	0.14	0.14	1	0.71	0.14
D	0.42	0.85	1	0.85	0.42

Figure 4. Evaluation of the Drivers members of a team

Figure 5. Possibility that a driver is the main one

Let us consider that it is not worth taking into account the influence of a driver below a possibility degree of 0.5. The possibility of links between main drivers is shown on figure 6 with the original models number of links on the left hand side (same driver part links, e.g. IIxII are multiplied by 2 to adjust for combination counting):

Number of links *(original model)*	Interpersonal Pairs	Possibility links
2	IxI	1
3	IxII	1,0.85,0.85
5	IxIII	1,1,1,0.66,0.66
2	IxIV	0.85,0.85,0.71,0.71,0.55
0	IxV	0
2	IIxII	0.85
5	IIxIII	1,1,0.85,0.66,0.66,0.66
2	IIxIV	0.85,0.71,0.71,0.55
0	IIxV	0
6	IIIxIII	1,0.66,0.66,0.66,0.66,0.66
2	IIIxIV	0.85,0.71,0.66,0.66,0.66,0.66,0.55,0.55,0.55
0	IIIxV	0
0	IVxIV	0.71,0.55,0.55
0	IVxV	0
0	VxV	0

Figure 6. Possibility of links between main Drivers

These degrees of possibility define a discrete distribution of possibility on the number of links between drivers, e.g. the distribution of possibility of the number of links between driver II and driver III which is illustrated in figure 7 where the discrete distribution of possibility is represented as continuous in order to be easier to understand. The interpretation is then the following: it is entirely possible that there are two links IIxIII (possibility=1), another link is possible with a degree of 0.85, and three other are possible with a degree of 0.66.

The possibility distributions of the links can be expressed in the team profile using the thickness of the bars for the degree of possibility of the number of links, as shown in figure 8, and the team profile becomes in that way the summary of the possibility distributions of links between drivers. This profile is much more informative than the standard one, which is indicated in the same figure in grey when the drivers II and III are chosen for D.

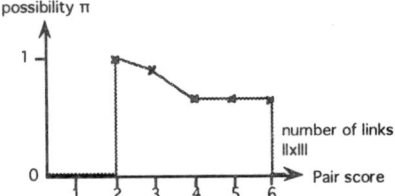

Figure 7. Distribution of possibility of the number of links between two drivers.

We can see, for example, on figure 8 that the link IIIxIV is possibly much more important in this team than the link IIIxIII, on the point of view of the number of possible links, because the distribution of possibilities take into account the third or the fourth drivers of the members of the team if their score is large enough. The link IVxIV, which does not appear in the bargraph, is now present here with a maximum degree of possibility of 0.71.

Since the effect of threshold in the choice of the main drivers is attenuated, the possibility distributions described here take into account much more information than previously, and in a more synthetic way.

3.3 Application to the performance evaluation

The evaluation of the performance of the team raises the problem of the combination of information concerning the same datum, coming from different sources. The association of a note to a linguistic appreciation, and the calculation of the mean of the three evaluations is clearly not a satisfying answer. This problem can be shown by these two examples:

- the technical success of a project is considered as "fair" by the three sources that we consider here. The average result of the evaluation corresponds then to the note associated to "fair".

- the technical success is considered as "poor" by the first source, "fair" by the second, "good " by the third. As in the first case, the global evaluation is the note corresponding to the "fair" appreciation.

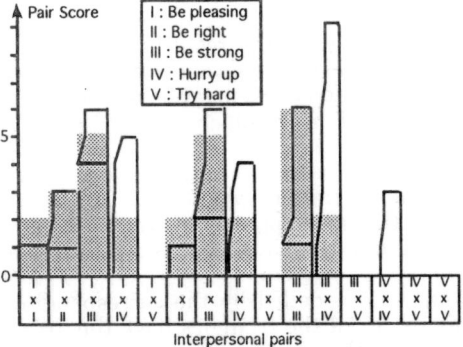

Figure 8. Team profile using possibility distributions

It is possible, thanks to the theory of possibility, that two different appreciations can be partially consistent if they are close. Figure 9 shows a classical way to model linguistic appreciations on a numerical scale. The degree of overlapping of the triangles models the partial consistence of two consecutive appreciations. If several appreciations concerning the same information are available from different sources, two cases may arise [11]:

- the appreciation of the sources are partially or totally consistent. They can then be aggregated through an operation such as:

$$\pi_{12} = \frac{\pi_1 * \pi_2}{\sup(\pi_1 * \pi_2)}$$ where π_1 and π_2 are the possibility distributions of the appreciations of respectively source 1 and source 2, and $*$ an operation of intersection of two fuzzy sets like the minimum. The division by $\sup(\pi_1 * \pi_2)$ is used here in order to normalise the resulting distribution of possibility. This normalisation is justified by the consistency of the sources.

- If the appreciations are not consistent, it is possible to combine the two distributions of possibility through a union operator, but the association of a notation to this combined distribution (for example through the calculation of the centre of mass of the shapes) has the same inconvenience as the calculation of the mean when a precise note is associated to the linguistic appreciation.

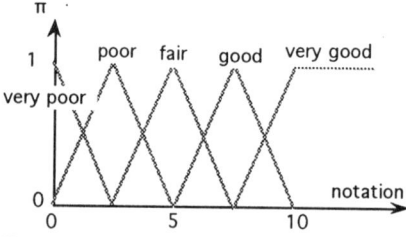

Figure 11. Fuzzy models
of linguistic appreciations

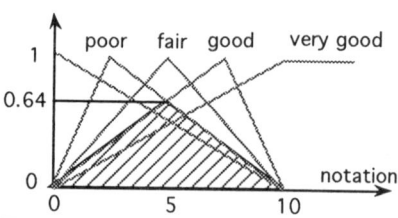

Figure 12. Fuzzy models chosen

It is possible to solve the problem of the inconsistency by the choice of possibility distributions all having a certain degree of consistency, like the ones of figure 10. Consider two sources with respective appreciations of "poor" and "good". The hachured shape shows the combination of the appreciations through the min operator. The centre of mass of the resulting shape gives the notation that combined the two sources (here 5), while the high of the resulting possibility distribution can be considered as a measure of the consistence between the two appreciation (here 0.64). In that way, it is possible to have a notation on the result of the project, but also a measure of the reliability of this result.

The problem is a little bit more difficult in the case of more than two sources, because the combination of two identical appreciations would give the same possibility distribution. In order to show that the same appreciation from different sources reinforce the precision of the appreciation, we have used as a convention that:

- when one source gives an appreciation, it is modelled by the corresponding possibility distribution as described in figure 9. All the triangles have the [0,10] interval as a base. Figure 11 gives an example for the "good" appreciation, described by the possibility distribution noted (1) if only one source gives this appreciation.
- if two sources agree on an appreciation, the base of the triangle that describes the corresponding possibility distribution is divided by two, which corresponds to a decrease of the uncertainty on the "real" appreciation. The possibility distribution corresponding to "good" given by two sources is noted (2) in figure 11.

- if the three sources agree on an appreciation, the corresponding possibility distribution is transformed to a precise number, e.g. a possibility of 1 for the centre of mass of the shape, and 0 everywhere else (noted (3) in figure 11).

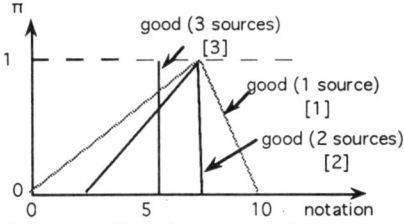

Figure 11. Reinforcement of an appreciation

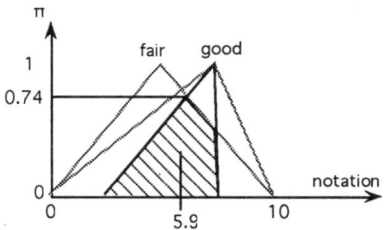

Figure 12. Example of combination

In figure 14, the hachured shape represents for example the combination of an appreciation of "good" by two sources, and "fair" by the other one. The resulting notation is 5.9, which is the abscissa of the gravity centre of the hachured shape, with a possibility of 0.74 which is the height of the shape.

4. FUTURE RESEARCH

We have now the tools which are necessary in order:
- to model the main characteristics of a team, through the team profile,
- to evaluate the project which have been realised by a team.
Teams need a consistent method of evaluation otherwise their selection will remain subjective and their success anecdotal. "What gets' measured, gets improved". Similarly teams need a measurable profile such as the one described here and which is a robust research unit. This work has so far given a stable platform for which to proceed. Further work will latest be needed to be done to refine and investigate both the Driver Questionnaires and the Team Profile Questionnaire.

The collection of data is now in progress in various companies. The next step of our study will be then to correlate the team profile to the performance evaluation in order to be able to select a team with the expected performance. Three methods will be tested in order to make this correlation:
- design of experiment. The goal of the design of experiment method is to obtain the best tuning of parameters without exhaustive tests. In order to do so, experimental models are designed; they allow a mathematical estimation of the effects of each parameter independently from the others through the definition of orthogonal arrays [12],
- design of an expert system, if it is possible to translate the expertise between the links between drivers and the performance of the project,
- use of a neural network, if a sufficient number of results are available. The main characteristics of neural networks is that they allow to constitute a knowledge base on a problem on the unique base of some experiments, while it is necessary to have a precise idea on the causal links between inputs and outputs in order to design an expert system. These three methods have for example been tested on the problem of the tuning of scheduling parameters in [13], with the conclusion that the neural

network approach is very interesting when strong expertise is not available on the problem, but when a lot of experiments can be made.

5. CONCLUSION

Effective teams need to be defined and evaluated. The teams here are defined in terms of the combination of our main Drivers or internal motivations, and the evaluation of each team is by a Team Rating by three managers. The uncertainty and imprecision of subjective measures is reduced by using fuzzy logic methodology, and is being applied to the data being collected.

Project Managers will have to pay some attention to Team Design rather than development in the new lean production quality paradigm, especially where there is a model to use. This will also encourage them to collect their own data to refine their own team compositions, and help the shift to a cooperative culture by being able to assess teams rather than individuals. The aim is to increase the efficiency and productivity of a team through better design.

REFERENCES

1. M. Belbin, Management Teams: Why They Succeed or Fail, Heineman, London, (1981).
2. R. Lewis, P. Lowe, Individual Excellence, Kogan Page, London, (1992).
3. T. Kahler, Process Communication Model, Little Rock, USA, T. Kahler Ass., (1979).
4. A. Stickley, Team Design for effective projects, 2nd International Conference on Concurrent Engineering and Electronic Design Automation, Bornesmouth, Dorset, UK, April 7-8, (1994).
5. H. Thamhain, D. Wilemon, Building High Performing Engineering Project Teams, IEEE Transactions on Engineering Management, EM-34 N°3, pp.130-137, (1987).
6. W. Falkowski, K. Munn, Interrater Agreement on Driver Questionnaire Items, Transactional Analysis Journal, N°19, vol. 1, (1989).
7. E. Berne, What do you say after you say hello?, Corgi Books, (1988).
8. J. Hazell, Drivers as mediators of Stress Response, Transactional Analysis Journal, vol.19, N°4, (1989).
9. L. Zadeh, Fuzzy Sets, Information and Control, N°8, (1965).
10. L. Zadeh, Fuzzy Sets as a basis for a theory of Possibility, Fuzzy Sets and Systems, N°1, (1978).
11. D. Dubois, H. Prade, Possibility Theory, an Approach to Computerized Processing of Uncertainty, Plenum Press, New York, (1988).
12. G. Taguchi, System of Experimental Design, Unipub/Kraus International Publication, (1987).
13. B. Grabot, L. Geneste, A. Dupeux, Multi-heuristic scheduling: three approaches to tune compromises, to be published in the Journal of Intelligent Manufacturing, (1994).

PART FIVE

Performance Indicators and Measurement

35

Indicators for the Evaluation of Organizational Performance

Gert Zülch, Thomas Grobel, Uwe Jonsson

ifab - Institute of Human and Industrial Engineering (Institut für Arbeitswissenschaft und Betriebsorganisation), University of Karlsruhe, D-76128 Karlsruhe, Kaiserstraße 12, Germany.

Abstract

Due to increasing international competition, productive enterprises are currently searching for measures to improve their order processing. As the major potentialities for further improvements are strongly connected to the organization of production systems, the organizational performance has to be investigated and evaluated.

The following paper presents four different approaches for evaluating the organizational performance and for revealing potentialities for improvements, namely: controlling, branch comparison, benchmarking and simulation. Based on a comparison of these approaches, a conceptive concept for evaluation will be developed.

Considering certain prerequisites, four measures are defined in the shape of degrees of goal achievement. Such degrees of goal achievement compare the real performance of a production system to a theoretically optimal performance. The four measures refer to lead time, due-date dependability, utilization and work in progress of a production system.

The concept for evaluation is connected to a simulation-aided approach in order to improve the organizational performance. This concept is demonstrated by using a case study of an enterprise of mechanical engineering as an example. The case study is performed with the simulation tool FEMOS and results in a proposal for the reorganization of the departments for design and operations planning.

1. NEED FOR INDICATORS

In order to maintain their competitiveness, productive enterprises have to estimate their position in the market compared to competitors on one hand. On the other hand they must search for approaches to improve their over-all performance in order to stay in the market.

Thereby, the organization of order processing becomes a dominating influence on performance, as technical means which might be used to support the performance of certain tasks are insufficient to meet all requirements [1, p. 133 ff.]. Regarding the order processing, it is of special interest to reveal possibilities for an improvement and to quantify potentialities by all possible means.

In order to evaluate the quality and discover such potentialities of an organizational structure, special indicators can be used. In the following, different approaches to evaluate the organizational performance will be presented. Afterwards, an adequate concept for the registration of relevant aspects of organizational performance will be developed.

2. DIFFERENT METHODS TO REVEAL POTENTIALITIES FOR IMPROVEMENTS

The revelation of improvement potentialities is a major topic for every company, and therefore a wide range of methods have been developed. A feasable approach for their classification is the distinction of various sets of solutions which are systematically taken into consideration by these methods. In figure 1 the methods are divided into four classes "controlling", "branch comparision", "benchmarking" and "simulation". Before selecting one method, the characteristics of these classes will be described.

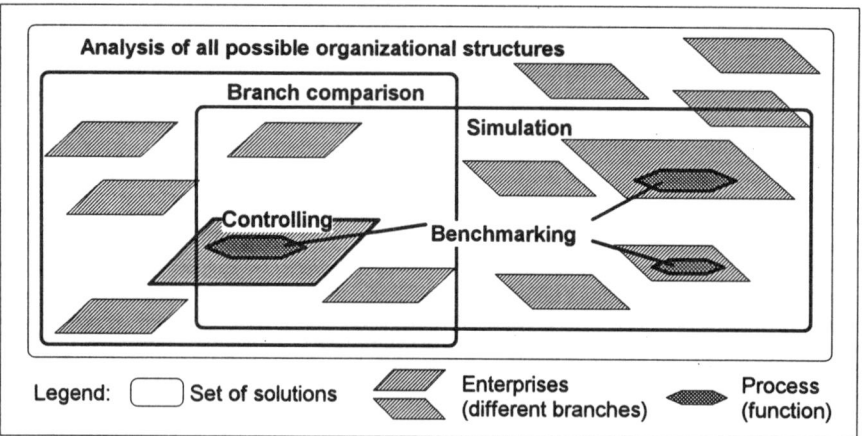

Fig. 1:   Solution sets of different methods to reveal potentialities of improvement

Controlling

The most common way to gain improvement potentialities is the performance measurement of the production system under investigation, usually performed by pre-defined indicators. This method is relatively easy to perform and the procedure of measurement is well-known. But because of the missing possibility of a comparison to other enterprises the systematic search for better solutions is rather limited. Often, the only available data for comparison is derived from preceeding measurements and improvements are made, more or less against an empirical back-ground. Therefore, controlling relates only to one special set of solutions, respectively one single company. A direct comparison of indicators to other enterprises is nearly impossible, unless the organizational structure standing behind is known as simlar and comparable.

An approach to overcome the described deficits is the comparison of the indicators to a theoretical optimum. Although this may lead to more distinctive key data the definition of the theoretical optimal values turns out to be a problem.

Branch comparison

For decades, there already has existed another possibility with which to compare the situation of one enterprise to that of other companies working in the same economic branch. This is the comparison of typical economic and logistic characteristics with mean values achieved by other productive enterprises within this branch. These data is based on inquiries which are usually initiated by industry associations (i.e. [2]).

But obviously, this approach bears a certain disadvantage. Even if the averages of the report-ed characteristics are enhanced by their standard deviations, the reason for a difference between the values of one company and those of the whole branch can hardly be discovered. Such proof demands a detailed description of the underlying organizational structures which certainly influ-ence the values on both sides,whereas this data is usually not available.

Benchmarking

When comparing existing production systems by using comparable key data, e.g. the measur-ed performance time of a standardized task, distinctively transferable results, which may lead directly to potentialities of improvement, can be achieved. Unfortunately, suitable companies for such a comparison are usually competitors on the market, and therefore their willingness to take part in an analysis and disclose their internal data is, in practice, a great obstacle.

Therefore, another approach has been established using the term "benchmarking" that avoids the mentioned problem by comparing only certain processes or sub-units of companies (cf. [3]). The aim is to recognize a process or sub-unit in another company that represents best practice in the fulfillment of the assigned task and then to improve the own process or sub-unit according to this "best of class".

By referring only to sub-units, also non-competing companies can take part in a benchmarking analysis and learn from each other. Nevertheless, gathering the required data and its adequate evaluation is in many cases a great burden within the benchmarking process.

In figure 1 this approach is represented by its set of possible solutions. This set includes all real existing systems and in addition all solutions that are reachable by combining existing processes or sub-units.

Simulation

The major problem of the preceeding methods is that they are based only on existing systems and thereby only seldom reveal organizational solutions which are really new. This is related to the need of proving the economic value of a new solution, before it is realized. For solving this problem, a model of the real system has to be be built which allows the prediction of the system's behaviour by simulating its dynamic flow of events. In a simulation based analysis the processes inside the production system can be tracked in detail and different approaches of improvement can be tried without any monetary risk.

Of course, the expressiveness of a simulation based analysis depends strongly on the quality of the used model, which has to be built very carefully by experienced personnel. But this effort is justified by the great potential usage of such an analysis. Achieveable solutions of existing systems dominate the benchmarking approach. On the other hand, a simulation based analysis can, potentially, consider every possible solution.

In practice, the set of possible solutions is limited by the features of the simulation tool. In figure 1 the simulation approach is represented by a subset of all possible solutions. But nevertheless, the solution set of simulation is much wider than the set of benchmarking.

3. CONCEPT FOR THE EVALUATION OF THE ORGANIZATIONAL PERFORMANCE

Regardless what specific method is applied for evaluating the organizational performance, an evaluation and a comparison of different systems is performed by quantifying a certain set of measures. The definition of measures has to be based on a suitable model and has to be embedded in a comprehensive system's concept (cf. [4, p. 260]). Such a concept will be presented in the following section (cf. [5, p. 82 ff.]).

3.1 Conceptual Frame

When evaluating a system's behaviour, this system has to be marked off by a definition of its system boundaries (cf. fig. 2). Herewith, the evaluation has to take into consideration all relevant influences on this system. As the organizational performance regards the winding-up of orders, the system boundaries can be fixed with the order entry in the production system and the finish of the last step of the processing of the order.

Due to these boundaries, measures can be divided into exogene and endogene measures (cf. e.g. [6, p. 36; 7, p. 330]). Exogene measures quantify the relation between system and environment and can be designated as market goals, as they are related to orders. The endogene measures describe the behaviour of the resources within the system and are also named as cost goals (cf. e.g. [4, p. 2]).

In order to get a deeper insight into the system's behaviour it is insufficient to quantify goal achievement only by a mean value. Besides measures with the character of a mean value, such measures should additionally be applied for a complete description of all effects, which quantify the variation from the mean values (cf. [5, p. 55 ff.]; for example cf. [8, p. 726; 9, p. 53]). This differentiation of key data between mean values and deviations implies their independency.

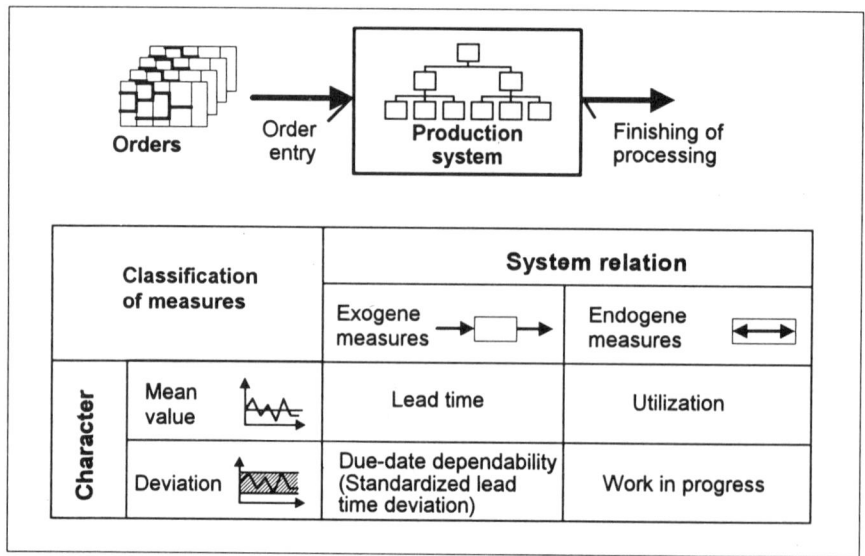

Fig. 2:   Conceptual frame for the evaluation (following [5, p.83])

3.2 Selection of Measures

Following this idea, the concept of evaluation has to hold four different measures. For the evaluation of the organizational performance, the logistic key data utilization, work in progress, lead time and due-date dependability are regularly applied (cf. e.g. [10, p. 8 ff.]). Supposing that there is a suitable definition, these measures can be integrated into the conceptual frame which has been introduced in the previous chapter (see fig. 2).

Lead time and due-date dependability quantify the quality of order processing and therefore are exogene measures. Utilization and work in progress are endogene measures and cover the use of the resources of the production system. Lead time and utilization can be characterized as mean values, while due-date dependability and work in progress signify the deviation from the respective average. The due-date dependability worsens with an increasing deviation of the lead times of orders; the work in progress increases, if the utilization varies predominantly over time.

A major problem connected to the use of these logistic indicators is the fact that no generally accepted definition for the quantification of the due-date dependability is available (cf. [11, p. 1093]). Furthermore, a dependancy often exists between lead time and due-date dependability, as long lead times imply poor due-date dependability. Therefore, this key indicator does not contain additional information for a logistic evaluation.

As a consequence, the due-date dependability has to be replaced by a suitable measure. This measure should be independant from a pre-defined due-date and independent from lead time (cf. also the approaches of [12, p. 322; 13, p. 78; 14, p. 592]). Such a measure will be defined in the following chapter using the standardized lead time deviation.

3.3 Definition of the Measures

When defining the measures used in a concept for evaluation, some requirements have to be met. In order to simplify the interpretation of results, the measures should be standardized and monotonous (cf. [15, p. 287 f.; 16, p. 20]). The definition should be dimensionless and imply a uniform range of value. An increasing value should quantify an increasing approximation to an ideal state.

In order to fulfill these prerequisites, so-called degrees of goal achievement have proven to be a suitable approach (cf. [4, p. 69 ff.; 17, p. 133 ff.]). Degrees of goal achievement relate the performance of a system to a theoretically optimal performance. If the optimal performance is achieved, the value of the degree of goal achievement will be 100 %.

Due to the conceptual frame developed above, the concept for evaluation consists of four different measures. These measures are the degrees of goal achievement of lead time, of standardized lead time deviation, of utilization of the organizational units and of work in progress (measured as demanded capacity). In contrary to most of the available concepts, these measures are independent because of an appropriate definition. Below, the different measures will be defined and explained (for a detailed description see [5, p. 82 ff.]; for other approaches cf. e.g. [4, p. 70 ff.; 18, p. 96 ff.])

As a measure for *lead time* the degree of goal achievement of lead time degree (GLT) will be used which represents the average lead time degree of all the processed orders of a system within a certain period. The lead time degree LTD of an order k is calculated by a division of the minimal theoretical lead time MLT with the real resp. simulated lead time RLT:

$$LTD_k = \frac{MLT_k}{RLT_k}.$$

The lead time of an order is defined as the time needed to process an order from order entry until delivery (the finishing of the last step of order processing). The minimal lead time of an order can be calculated based on the function network which represents an order (cf. [19, p. 181]). A function network consists of all steps and their relations needed to process an order. The minimal lead time of an order equals the length of the critical path in this function network when exclusively processing and transportation times are considered (cf. [20, p. 15 ff.]).

When using degrees of goal achievement, some authors define *due-date dependability* using a date which is calculated directly dependent on the minimal lead time of an order (cf. e.g. the approaches of [4, p. 70; 18, p. 100]). As a consequence, low values of a lead time degree imply a low value of the due-date dependability.

In order to avoid such a dependancy, a standardized lead time deviation is used here which quantifies the variation of the lead times based on the average lead time instead of on the minimal lead time. For that purpose, a corrected calculated lead time is used. This corrected calculated lead time can be achieved, if the minimal lead time of an order is elongated by an average prolongation factor of the orders. This factor can be calculated as the reciprocal value of the degree of goal achievement of lead time degree. Following this line of reasoning, the standardized lead time deviation SLD for an order k can be calculated by comparing the corrected calculated lead time with the deviation of real lead time from this corrected calculated lead time:

$$SLD_k = \frac{RLT_k - \dfrac{MLT_k}{GLT}}{\dfrac{MLT_k}{GLT}} = \frac{RLT_k}{MLK_k} \cdot GLT - 1.$$

As the optimal value of the standardized lead time deviation equals zero, a transformation for the calculation of the degree of standardized lead time deviation DSL has to be performed. As earliness should be equally considered to delay, the absolute value of SLD is used:

$$DSL_k = \frac{1}{1 + |SLD_k|}.$$

The degree of goal achievement of the standardized lead time deviation is calculated similar to the degree of goal achievement of lead time degree as an average value considering all finished orders.

The *utilization of a production system* quantifies the mean use of the capacity within a certain period. In order to calculate a degree of goal achievement of utilization the theoretical capacity stock is compared to the amount of capacity processed during the period of evaluation. The processed capacity covers set-up and processing time of all processed orders.

In order to evaluate the organizational performance, the *work in progress* has to be calculated in relation to the capacity stock of orders processed in the system. The work in progress gives an impression of the average capacity of the waiting orders. As minimal value a minimal work in progress can be defined which considers the fact that the orders have to be at least within the system when they are processed (cf. [4, p. 77 ff.]). In order to quantify the degree of goal achievement of work in progress GWP, this minimal work in progress MWP is compared to the average waiting amount of capacity WAC:

$$GWP = \frac{MWP}{MWP + WAC} \ .$$

The concept presented here follows the idea of comparing the system's behaviour with a theoretically optimal system. This approach was already used with connection to the simulation tool FEMOS in order to reveal and compare different solutions for the organizational design of several production systems [5, p. 166 ff.]. This procedure will be demonstrated here using an enterprise of mechanical engineering as an example.

4. EVALUATION OF AN ENTERPRISE OF MECHANICAL ENGINEERING

This case study treats a middle-sized enterprise of mechanical engineering which is a located in the South-West of Germany. The enterprise has about 270 employees and a turnover of about 60 Mio DM. 200 persons were employed in parts production and assembly, the remaining number in the pre-productive sectors such as design, operations planning and sales.

4.1   Production Situation

The enterprise produced components for material flow automation, e.g. conveyor belts and devices for machine interlinkages. The product spectrum consisted mainly of more or less standardized components, which were tailor-made to customer specifications. A customer order regularly covered a complete conveying system consisting of various components. Each system had to be planned and realized considering the special production environment of the customer, in particular with regard to the layout of the production system and the interfaces of the existing machines. Therefore, the production situation of this enterprise could be characterized as a specialized production with a strong customer-orientation.

The enterprise was traditionally organized according to functions. Basically it consisted of the following organizational units: the sales department, the purchase department, the design department, the department of operations planning and scheduling, parts production and assembly and the shipping section.

The winding-up of orders could be described as follows: Starting from a request the sales department run the negotiations with the customer. After order entry, the design department developed the drawings of the complete system and its components. The purchase parts were ordered by the purchase department. The operations planning and scheduling could be initiated, if the construction drawings and bill-of-materials were completed. Supplied by job papers and wiring design, parts production and assembly could start subsequently. After a final control the conveying system was delivered and assembled at the customer's site.

The major problem of the enterprise was its reliability to delivery dates promised to the customer, as overlapping of this date caused conventional penalties. Compliance with due dates proved rather difficult, as the capacity of the workshops was loaded at a relatively high rate. The existing organizational structure did not allow any immediate reaction to the customer's desires for a change in the product design, for example, if the customer wanted to change the layout of the conveying system after order delivery.

4.2 The Simulation Tool FEMOS

The evaluation concept was realized and applied in connection with the simulation tool FEMOS for a systematical analysis of the influences on the organization of production systems (cf. [5; 21]). FEMOS has been developed for the analysis such systems at the ifab-institute of the University of Karlsruhe since 1988. It has been successfully applied solving various types of problems in real production systems (see e.g. [1; 17; 19, p. 177 ff.; 22]). In these applications e.g. the shaping of the organizational structure of production systems, structuring of workshops and strategies of production control were analyzed supported by the FEMOS-program (cf. e.g. [23]).

The simulation tool can be characterized by an organization-oriented approach. The model which the program is based upon, allows the differentiation between personnel and work places on one hand and the condensation of different workplaces to departments on the other hand. Furthermore, sophisticated levels of the capabilities of a department's personnel can be considered. Orders are described by function networks. Thus, parallel processing of different functions can be modelled rather easily.

4.3 Evaluation of the Initial Situation

In order to analyze and develop approaches for organizational improvements, the enterprise was modelled with the simulation tool FEMOS. As the organization of parts production and pre-assembly were not to be changed, the study concentrated on the pre-productive functions. Parts production and pre-assembly were treated in the following discussion as a black box with fixed lead times.

Using the order entry to the design department and the delivery to the shipping department as system boundaries, nine steps of order processing could be identified: After checking the material need, the construction drawings and bill-of-materials for the various components were created and the switches and motors had to be selected. Based on this information, the electronic design could be performed for the entire conveying system. The last steps of the order processing in the pre-productive sector were the creation and the electronic documentation of the job papers. Parts production and assembly were treated as black box with fixed lead times. These steps had to be performed for all of the orders which combined various, more or less, standardized components (cf. the example in fig. 3).

In order to control the variety of components, 40 product groups were defined, such as different types of chain conveyers, auto guided vehicles, delivery stations and line portals. These groups could be characterized by equal function networks and similar work contents for the different functions.

Based on these product groups, for the simulation study 143 orders of 50 different order types were considered. The order types covered the complexity of the orders processed during the one year, on which the simulation study was based on, and varied from orders enclosing just single components, up to entire conveying systems.

In the initial situation, 26 employees of the pre-productive functions were considered in the simulation study. Six groups of mechanical designers with three to five members are specialized for different product groups. A seventh group of designers with three members performed the electronic design. The first group of two groups of operations planners was responsible for the documentation of components, the other created the job papers. All considered groups, including the department of electronic design, worked without direct linkage to the customer order. Hence, a customer order was processed in different departments of mechanical design and all orders passed the departments of operations planning.

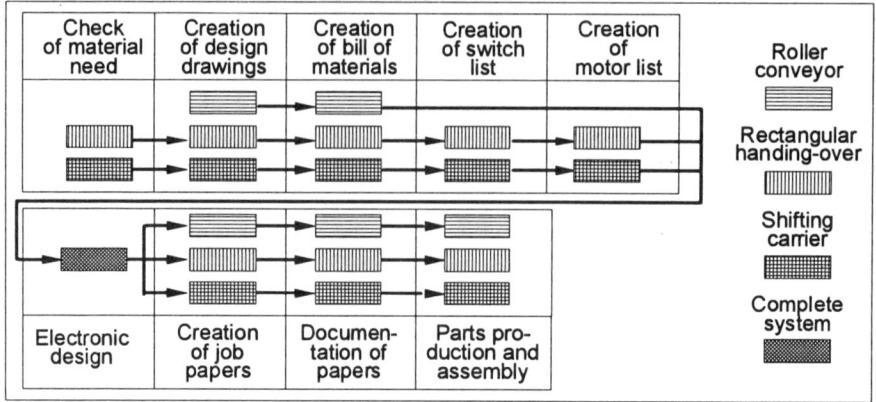

Fig. 3:   Example of a function network for a customer order

Based on this data, the initial situation was analyzed with the FEMOS program. In order to assure the validity of the model, the results of a first simulation run concerning the existing system were compared to the values derivated from the real system. As the results revealed the lead times and the utilization of the personnel as comparable to the real system, the validity of the model of the initial situation could be stated: The average lead time of the orders was 80 days, with a standard deviation of 64 days. A specific characteristic was the remarkable variation of the dynamic utilization of the different design groups, where - similar to the real situation - values between 67% and 92% could be measured.

When analyzing the degrees of goal achievement (cf. fig. 4), the major problems of the initial situation and potentialities for improvements could be identified. The utilization of the system was not too bad, because about 80% of the theoretical capacity of the system was used. This value equals the dynamic capacity which can be normally achieved using a functional structure of a production system (cf. [5, p. 137]). The value of the dynamic capacity is always below 100%, as in every real production system disturbances occur due to dependencies of order processing.

Criteria	Measure	Unit	Value
Lead time	Simulated lead time (average)	Time units	37.523
	Theoretical lead time (average)	Time units	14.071
	Degree of goal achievement of lead time	%	35,7
Lead time deviation	Average prologation factor	-	2,80
	Degree of goal achievement of lead time deviation	%	66,1
Utilization	Capacity processed	Time units	$3,99 \cdot 10^6$
	Theoretical capacity	Time units	$4,99 \cdot 10^6$
	Degree of goal achievement of utilization	%	80,0
Work in progress	Minimal work in progress	Time units	$8,3 \cdot 10^6$
	Simulated work in progress	Time units	$26,1 \cdot 10^6$
	Degree of goal achievement of work in progress	%	24,2

Fig. 4:   Results of the simulation of the initial situation

The value of the standardized lead time deviation indicated that the orders are late in a controlled and regular manner. All orders had more or less the same deviation from the ideal lead time. Based on this fact, it was possible to predict due dates with a certain reliability. The problem of the enterprise with its fixing of due dates was not caused by a high deviation of lead times but by long lead times themselves. As a consequence, attempts for major improvements should focus on a general reduction of lead times. The values of the other degrees of goal achievement gave hints for such improvements. The value of the lead time degree was as poor as the degree of work in progress. Thus, as a subject for further improvements the departmental organization could be identified, because its traditional functional orientation was responsible for high lead times and work in progress.

4.4 Development of Suitable Solutions

A possible reorganization was analyzed with four different variations. The first idea was to organize the functions following the mechanical design in a more order-oriented way. Therefore, operations planning and electronic design were condensed into one department that performed all the functions for an entire customer order. In order to realize flexible and small units, this new order centre was divided into three sub-units which treat different order types, but was able to process all of the components.

In the second variation, the idea of order-orientation was enlarged: The six design groups were replaced by three groups which were structured according to the order centres of the electronic designers and the operation planners.

Among the arriving orders there were occasionally some that comprise the production of a complete conveying system. Such orders tied up the capacity over a long time. In order to achieve a clearer division of the regular business with such projects, a special project group was installed in the third variation. This group consisted of five designers and two persons for operations planning.

The last variation was assumed to be the most flexible one. The division between designers and personnel for operations planning was abandoned. Each person had to perform all tasks allocated to the respective order centre. This measure implied a higher education and enlarged capabilities of the personnel, especially of those persons who performed the documentation of the papers in the initial situation. For economic reasons, the total number of persons could be reduced by one compared to the other solutions.

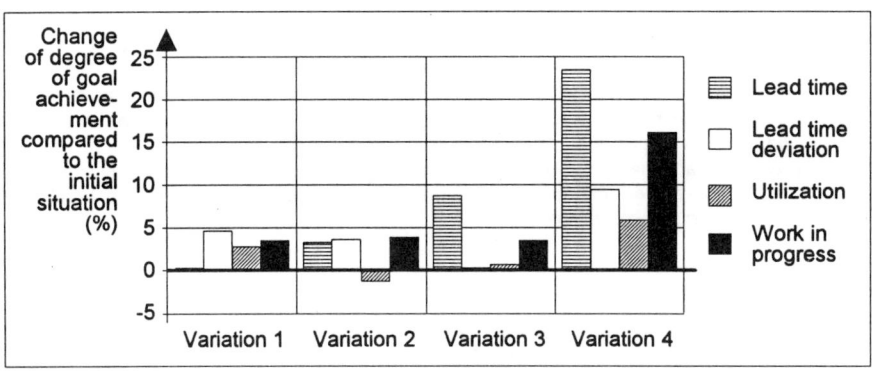

Fig. 5:   Results of the simulation study

Regarding the simulation results of the different variations, probable effects revealed. Figure 5 shows the change of the degree of goal achievement for the planning solutions mentioned above when compared to the original situation. According to the simulation results, only minor improvements concerning the degrees of goal achievement can be forecasted when restricted organizational measures as in variation 1 or 2 should be realized. All changes of the values are smaller than 5%. That means, the behaviour of the system cannot be approximated significantly towards an ideal state.

This situation changes with the introduction of a specialized organizational unit for the processing of projects. This measure allows to increase the transparency of the entire order processing, as such large orders cause bottlenecks. If a small unit of, say, just a few design engineers is occupied with a large project, it is not able to process other orders which could be wound-up quickly. The project itself is processed comparatively slowly, as too few persons are working on it. As a consequence, the bottleneck would exist over a long time.

Without any doubt the best results are achieved by the most flexible solution which implies an equal level of capabilities of all employees within the order centres. The degrees of goal achievement of lead time and work in progress rise significantly by more than 15%. Even lead time deviation and utilization increase remarkably. Regarding the utilization this improvement implies a higher dynamic capacity of the system. This fact is supported by the results of [5, p. 137], who stated a dynamic capacity of about 92 % for an object-oriented form of organizational structure. However, this value cannot be achieved by variation 4, because object-orientation is realized here not in a strict way implying an organizational unit for each order type: Due to the allocation of several order types to one organizational unit, the possibility of disturbances during the order processing increases.

5. CONCLUSIONS AND FURTHER DEVELOPMENTS

This study shows the suitability of simulation tools for the evaluation of the organizational performance in production systems. Apart from controlling, branch comparison and benchmarking, simulation is one possible approach to estimate the position of an enterprise in the market and to reveal possible further improvements. The advantages of this approach are as follows: No information from other companies is needed and data regarding only the own production system has to be collected. In search for improvements only theoretical restrictions limit possible solutions. But, realistic boundaries of improvements are hard to define.

Therefore, neither simulation nor benchmarking represents the only solution of the problem. Both approaches should be used in parallel. Information on good solutions for the processing of certain tasks which have already been realized can only be achieved by analyzing these processes in other companies, if possible. The idea of benchmarking is based upon the discovery of such successful processes.

In order to estimate the performance of an enterprise, different independent indicators should be used. Otherwise the danger of deteriorating the performance of an entire system can emerge. A single indicator or dependent indicators cannot consider the complexity of a production system in an adequate manner. Consequently, a simulation aided procedure seems to be a suitable concept for the evaluation of production systems because this approach fulfills the requirements of a comprehensive analysis.

REFERENCES

1. Zülch, Gert; Grobel, Thomas: Simulating Organizational Structures. In: Production Planning & Control, London, 4(1993)2, pp. 128-138.
2. VDMA - Verband Deutscher Maschinen- und Anlagenbau, Abteilung Betriebswirtschaft (Ed.): Kennzahlenkompaß. Frankfurt/M.: Maschinenbau, 1992.
3. Spendolini, Michael J.: The Benchmarking Book. New York: Amacom, 1992.
4. Wedemeyer, Hans-Georg von: Entscheidungsunterstützung in der Fertigungssteuerung mit Hilfe der Simulation. Düsseldorf: VDI, 1989. (Fortschritt-Berichte VDI, Reihe 2, Nr. 176)
5. Grobel, Thomas: Analyse der Einflüsse auf die Aufbauorganisation von Produktionssystemen. University of Karlsruhe, Diss. 1993. (ifab Schriftenreihe des Instituts für Arbeitswissenschaft und Betriebsorganisation der Universität Karlsruhe, Band 6, ISSN 0940-0559)
6. Weck, Manfred (Ed.): Simulation in CIM. Berlin u.a.: Springer; Köln: TÜV Rheinland, 1991. (Reihe CIM-Fachmann)

7. Schulte, Helmut; Stanek, Werner; Wirth, Siegfried: Ganzheitliche Unternehmensplanung. In: Zeitschrift für wirtschaftliche Fertigung und Automatisierung, München, 86(1991)7, pp. 328-331.
8. Fryer, John S.: Organizational Segmentation and Labor Transfer Policies in Labor and Machine Limited Production Systems. In: Decision Sciences, Atlanta GA, 7(1976)4, pp. 725-738.
9. Sculli, Dominic: Priority Dispatching Rules in an Assembly Shop. In: Omega, Oxford, 15(1987)1, pp. 49-57.
10. Augustin, Siegfried; Jahn, Siegfried: Entwicklungslogistik: Der Weg zum schnellen "time-to-market". In: Logistik-Synergie zwischen Handel und Industrie. Ed: Bäck, Herbert. Köln: TÜV Rheinland, 1990, pp. 7-17.
11. Baker, Kenneth R.: Sequencion Rules and Due-Date Assignment in a Job Shop. In: Management Science, Baltimore MD, 30(1984)9, pp. 1093-1103.
12. Adam, Nabil R.; Bertrand, J. Will M.; Surkis, Julius: Priority Assignment Procedures in Multi-Level Assembly Job Shops. In: IEE Transactions, Norcross GA, 19(1987)3, pp. 317-328.
13. Dar-El, Ezey M.; Wysk, Richard A.: Job Shop Scheduling - A Systematic Approach. In: Journal of Manufacturing Systems, Dearborn MI, 1(1982)1, pp. 77-88.
14. Goodwin, Jack S.; Goodwin, James C.: Operating Policies for Scheduling Assembled Products. In: Decision Sciences, Atlanta GA, 13(1982)5, pp. 585-603.
15. Grünwald, H.; Striekwold, P.E.T.; Weeda, P.J.: A Framework for Quantitative Comparison of Production Control Concepts. In: International Journal of Production Research, London, 27(1989)2, pp. 281-292.
16. Amon, Markus: Instandhaltungs-Controlling mit Kennzahlen. In: REFA-Nachrichten, Darmstadt, 44(1991)1, pp. 17-20.
17. Grobel, Thomas: Simulation der Organisation rechnerintegrierter Produktionssysteme. University of Karlsruhe: Institut für Arbeitswissenschaft und Betriebsorganisation, 1992. (ifab Schriftenreihe des Instituts für Arbeitswissenschaft und Betriebsorganisation der Universität Karlsruhe, Band 3, ISSN 0940-0559)
18. Graf, Karl-Robert: Systematische Untersuchung von Einflußgrößen einer Fertigungssteuerung nach dem Zieh- und Schiebeprinzip. University of Karlsruhe, Diss. 1991. (ifab Schriftenreihe des Instituts für Arbeitswissenschaft und Betriebsorganisation der Universität Karlsruhe, Band 2, ISSN 0940-0559)
19. Zülch, Gert; Grobel, Thomas: Simulating the Departmental Organization for Production to Order. In: 'One-of-a-kind' Production: New Approaches. Eds.: Hirsch, B.; Thoben, K.-D. Amsterdam u.a.: North Holland, 1992, pp. 177-193. (IFIP Transactions B: Applications in Technology, B-2)
20. REFA - Verband für Arbeitsstudien und Betriebsorganisation (Ed.): Planung und Steuerung. Part 2: Chapter 5-9. München: Hanser, 1991. (Methodenlehre der Betriebsorganisation).
21. Grobel, Thomas: Influences on the Organizational Structure of Production Systems. In: Human Factors in Organizational Design and Management. Ed.: Bradley, Gunilla; Hendrick, Hal. Amsterdam u.a.: North-Holland, 1994, pp. 29-36.
22. Zülch, Gert; Grobel, Thomas: Shaping the Organization of Order Processing With the Simulation Tool FEMOS. In: 8th Working Seminar on Production Economics. Pre-prints. Volume 1. Hrsg.: Grubbström, R.W.; Hinterhuber, H.H.; Lundquist, J. u.a. Igls, 1994, S. 319-332.
23. Zülch, Gert; Grobel, Thomas: Suitability of Selected Strategies of Production Control. In: Advances in Production Management Systems. Eds.: Pappas, I.A.; Tatsiopoulos, I.P. Amsterdam u.a.: North-Holland, 1993, pp. 313- 321.

36

TOPP - A METHOD FOR COMPANY SELF ASSESSMENT

Knut S Stokland
SINTEF Production Engineering, N-7034 Trondheim, Norway

1. TOPP SELF ASSESSMENT (TSA)

This handbook has been written by NTH/SINTEF on the authority of TOPP. The TOPP program offers several methods and tools for productivity measurement; TOPP Self Audit, Extended audit and Benchmarking; Handbook in TOPP Self Assessment (TSA) is a supplement to these tools.

The TSA system was developed to meet the needs in industry for measurement and evaluation of results of the business processes in the company. Identification of the business processes is a central part of designing the company's assessment plans. Understanding of these processes will be the basis in management of manufacturing enterprises in the future, and measuring and assessment are thus important tools for the company management.

Business processes represent the conversion of resources, capital, materials and manpower into internal or external products through a set of activities.

TSA focuses on productivity and on the companies performing the assessment themselves. Comparisons to other companies or any other defined reference are not of primary interest. TSA should provide trends that show how the results from selected business processes progress over time.

TSA should be a tool for continuous improvements of the business processes. The target groups for TSA are all manufacturing enterprises in the technological industry including the electronics/computer industry, the textile/clothing and woodware industry. The TSA primarily addresses the top management of the company, but every employee will benefit from the information procured by TSA.

Participation in the TOPP programme is not a requirement for utilizing the TSA.

The handbook makes suggestions as to *what* may be analyzed (analyzing areas and indicators) of the business processes, and how (measurement techniques) these areas may be analyzed. TSA is a management tool for evaluation of the most important business processes. All processes given in the handbook are *not* supposed to be part of every evaluation system in each company.

The TSA handbook is consistently avoiding the terms "revision" and "audit", thus taking the focus away from comparisons with a defined reference or standard.

The handbook is tentatively giving a step by step plan for designing and accomplishing self assessment in a company.

2. THE TSA CONCEPT

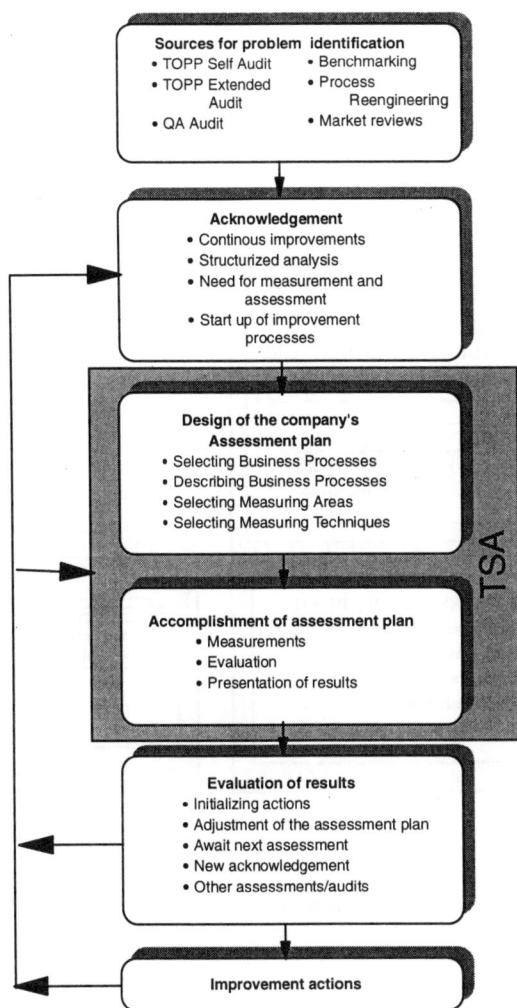

Figure 1 The TSA Concept

TOPP Self Assessment is a tool to support the management in increasing the productivity and the competitiveness of the company. The tool is not trade dependent, and it is meant to be supportive in designing and accomplishing the company's assessment plans. It is based upon measuring the results of own work on a higher level over time. Through repeated measurements a set of data showing trends is built up. The data is presented in an assessment report, adjusted to the different decision levels in the company. Hiring external expertise for design and accomplishment of the assessment plans is not necessary.

3. DESIGN OF THE COMPANY'S ASSESSMENT PLAN

Design of the company's assessment plan consists of:

- selecting business processes
- accomplishment of the assessment plan

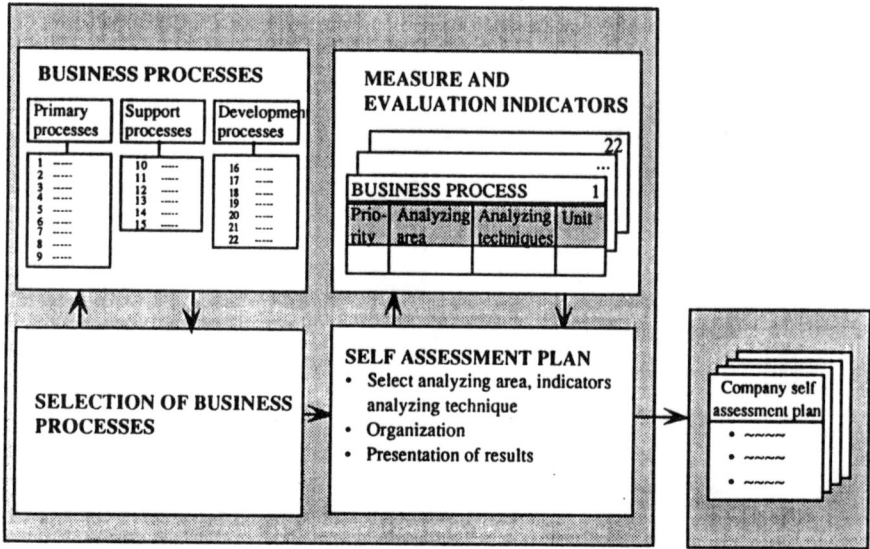

Figure 2 Procedure for designing a company self assessment plan

Business processes are selected acknowledging the need for performing a self assessment. The company's top management makes the selection. In TSA a set of "standard business processes" is available.

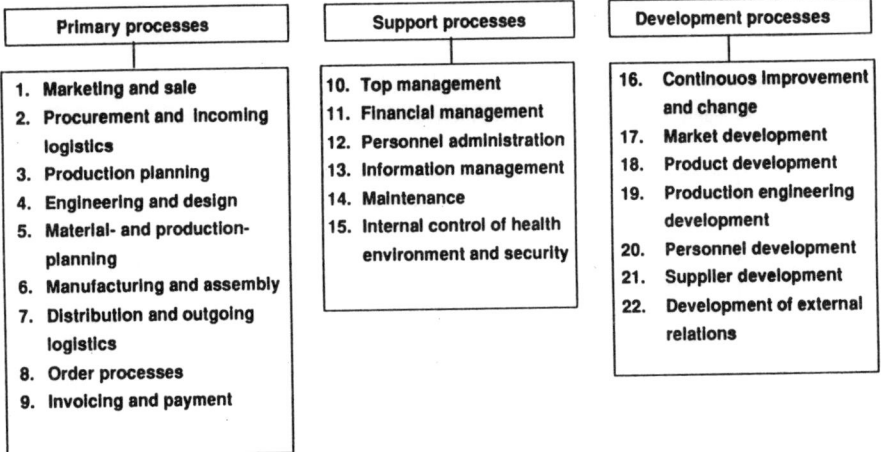

Figure 3 Standard business processes, Mechanical industry

The assessment plan consists of a set of assessment documents in which each measuring area and each measuring technique is described. The assessment team is chosen, and the handling and presentation of the results is decided upon.

The TSA handbook offers a lot of analyzing areas, each one with indicators and measuring techniques. The company has to select what is suitable, convenient for them to analyze and measure. Example of such analyzing areas, indicators and measuring techniques are given in the following figure.

	Business process 2		
PROCUREMENT AND INCOMING LOGISTICS			Page 1 of 2
By procurement and incoming logistics we mean all activities concerning the obtainment of raw material.			

PRIORITY	ANALYZING AREA	ANALYZING TECHNIQUES	UNIT
	Stock		
1	• Turnover	- Measure the turnover in stock of raw materials	
1	• Total value	- Measure the total value of raw materials	NOK
2	• Scrap	- Measure the share of scrap in stock of raw materials	NOK, %
2	• Response time	- Measure the time from confirmation of order to delivery	Time
	Security of delivery		
1	• Share of too late deliveries	- Make a system to registrate arrival	%
2	• Volume (NOK) of too late deliveries	- Measure the value and number of too late deliveries	NOK, time
2	• Accumulated delays	- Measure and accumulate too late deliveries	Time
	Deviation by delivery		
2	• Number of deficiencies pr delivery	- Measure and group	Number
2	• The consequence	- Measure the consequence of deviation in delivery - time/cost	Time, NOK
	Volume		
2	• Volume of purchase	- Volume of categorized purchases	NOK
	Resources		
1	• Man-hours	- Share of total man-hours spent on the process	%
1	• Value	- Value of purchased goods compared to total sales	%
2	• Persons involved	- Percent of the working staff involved in the process	%

Figure 4 Example of analyzing areas, indicator and analyzing techniques

4. ACCOMPLISHMENT OF SELF ASSESSMENT

The accomplishment of Self Assessment consists of the following main activities:

- collecting, arranging, sorting and editing of data
- report

When the assessment plan is worked out and the assessment team is chosen, the assessment may start with collecting and arranging/sorting the data.

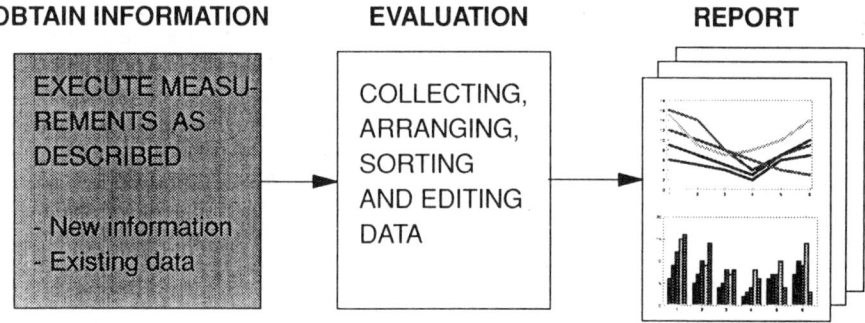

Figure 5 Accomplishment of Self Assessment

37

Performance Measurement - An Important Tool in Change Process

T. Laakso[a] and J. Karjalainen[b]

[a] Senior Researcher, Institute of Industrial Automation, Helsinki University of Technology, Otakaari 4 A, 02150 Espoo Finland.
[b] Senior Researcher, Department of Industrial Management, Helsinki University of Technology, Otakaari 4 A, 02150 Espoo Finland.

Key words: performance indicators, performance measurement, management of change, performance management

Abstract

The increased interest in performance measurement (PM) has caused the Federation of Finnish Metal and Engineering Industries to include PM as one of the subject areas of its CIM development program. The Helsinki University of Technology took part in this program by means of a questionnaire study concerning the present state, a number of case studies in various companies considered best in class in performance measurement, and building PM systems for Finnish companies. The questionnaire revealed that present PM systems seem to overemphasize financial measures, whereas measures that support modern manufacturing initiatives' of importance for change management are missed. The case studies helped to establish procedure for implementing PM. One of the most important lessons has been that the human aspects of PM must be emphasized in order to achieve successful result.

1. Motivation

At the same time when enterprises are applying new manufacturing initiatives such as JIT, TQM, Lean, etc. there is an urgent need for new management tools to implement these initiatives. The capability of a company to implement new strategies distinguishes a high-performing company from average companies. A properly defined and implemented performance measurement system is a tool that facilitates the implementation of the strategy of a company. Besides providing feedback on business and production process performance to operators and management, performance measurement facilitates learning and drives continuous improvement.

As the Finnish industry has become more and more aware of the importance of PM, the Helsinki University of Technology (HUT) has directed some research effort on developing PM systems. During the past two years HUT has been carrying out a number of studies concerning PM. One of the studies explored the use of present and planned PM among Finnish manufacturing enterprises. Studies were also carried out in companies that had been using PM for some time. In addition to this we assisted some companies to design and to implement partial PM systems. The PM research work has been carried out in co-operation with the Federation of Finnish Metal and Engineering Industries (FIMET) and a number of industrial enterprises as a part of the CIM development program for Finnish metal and engineering industry (SIMSON).

2. The Questionnaire Study

The aim of the study was to clarify the current situation of PM in the Finnish industry. The study seeks to establish the target areas of measurement and needs for further development of measurement systems. Furthermore examples of actual measures used in companies have been gathered and the importance of the measures have been explored.

We have used the Performance Measurement Questionnaire (PMQ) developed by Prof. Nanni et. al. at Boston University [1]. The questionnaire study was carried out among members of FIMET. FIMET consists of companies representing the Finnish metal and electrical industry. 1350 questionnaires were sent to 400 companies. 123 responses were received from 82 companies.

Originally the PMQ was designed to be applied internally in an organization. The objective was to provide a means by which an organization can articulate its improvement needs with regard to performance measures. It determines the extent to which the existing PM system of the organization supports necessary improvements and establishes guidelines for designing the measurement system. As mentioned we have used the PMQ method to analyze the status of PM systems use among several businesses and not within an individual company.

The PMQ is composed of four major parts. *The first part* consists of some general data of the individual respondents and the companies they are working in. The respondent has been asked about his or her position, organizational level, and about the functional area in which he or she works. Also the name of the company, turnover and total number of employees in the company has been asked for.

The second part of the PMQ focuses on competitive priorities and the current PM systems of the companies. In this section 27 improvement areas have been listed. Examples of listed improvement areas are quality, labor efficiency, volume flexibility, information systems, etc. The form of the questionnaire has been presented in figure 1.

Long-Run Importance of Improvement			Improvements Areas	Effect of Current PM system on Improvement			Not used
None	>>>>	Great		Inhibit	>> >>	Support	
1 2 3 4 5 6 7			QUALITY	1 2 3 4 5 6 7			0
1 2 3 4 5 6 7			LABOR EFFICIENCY	1 2 3 4 5 6 7			0
1 2 3 4 5 6 7			VOLUME FLEXIBILITY	1 2 3 4 5 6 7			0
1 2 3 4 5 6 7			INFORMATION SYSTEMS	1 2 3 4 5 6 7			0

Figure 1. The form of the PMQ.

On the left-hand scale, respondents have been asked to evaluate the importance of the improvement area for the long-term competitiveness of the company. The number 1 indicates that maintaining the present level of the area is adequate to meet future requirements, while the number 7 indicates that major improvements are required in that area in the future. On the right-hand scale, respondents evaluate the influence of the PM system of the company in supporting or inhibiting improvements in that area.

The third part of the PMQ is being constructed similarly than part two, but the focus is on performance factors that could also be called performance measures. In the PMQ 38 performance factors have been listed. Examples of the factors are inventory turnover, cost of quality, manufacturing lead time, vendor quality, etc.

On the left hand scale, respondents give their opinion as to the importance of achieving excellence in this factor for the long-run health of the company. The scale is from none (1) to great (7). The right-hand scale measures the extent to which they believe the PM system of the company emphasizes that factor.

The fourth part of the PMQ clarifies individual and company specific performance measures. Respondents are asked to record measures that are used in their companies. Measures have been evaluated as in part three of the PMQ.

3. Results of The Performance Measurement Questionnaire Study

Alignment analysis provides a general overview of the consistency between strategy, actions, and measures. Analysis has been made by averaging both left-side and right-side responses as to improvement areas and measurement factors. After the responses have been sorted in descending order, the top and bottom quartile are listed. In table 1 are the results of the alignment analysis of the improvement areas and in table 2 are the results of the analysis of performance factors.

Table 1. Alignment analysis of the improvement areas of all respondents.

	Long-run Importance		Effect of the Current PM system
1	Customer satisfaction	1	Direct cost reduction
2	Manufacturing throughput time	2	Overhead cost reduction
3	Integration with customers	3	Inventory management
4	Quality	4	Capital use
5	Capital use	5	Manufacturing throughput time

23	Machine efficiency	23	New product introduction
24	Information systems	24	Information systems
25	Product technology	25	Product mix flexibility
26	Product mix flexibility	26	Product technology
27	Environmental control	27	Environmental control

Table 2. Alignment analysis of the performance factors of all respondents.

	Long-run Importance		Emphasis of the PM system
1	On-time delivery	1	Sales margins
2	Sales margins	2	Deviation from budgets
3	Cost reduction	3	Return on investment
4	Return on investment	4	Inventory turnover
5	Customer survey	5	Cost reduction
6	Manufacturing lead times	6	On-time delivery

	Long-run Importance		Emphasis of the PM system
33	Process R&D costs	33	Indirect labor productivity
34	Number of engineering changes	34	Number of material part numbers
35	Number of suppliers	35	Productivity of middle management
36	Education and training budgets	36	Environmental monitoring
37	Number of material part numbers	37	Productivity of supervisors
38	Environmental monitoring	38	Productivity of specialist

When looking at the above results some interesting facts can be seen. "Integration with customers" have been considered one of the most important areas of improvement in the future, but when one looks at the top performance factors customer surveys are not mentioned at all. Respondents were satisfied with the present level of cooperation with suppliers. However, quality and throughput time of the manufacturing have been seen as an area of improvements on which supplier quality and on-time delivery have a major influence.

The greatest differences between left-side and right-side scores will be the main ingredients of what is called the gap analysis . The bigger the gap is the more the performance factor calls for increased support by the measurement system. If the sign of the difference is negative then the performance factor is called a false alarm. The tables 3 and 4 summarizes the biggest gaps and false alarms that have been found.

Table 3. The gaps and the false alarms of the improvement areas.

	Gaps		False alarms
1	Customer satisfaction	1	Capital use
2	Flexibility of capacity	2	Environmental control
3	Integration whit customers	3	Overhead cost reduction
4	Marketing	4	Inventory management
5	Multiskilled workers	5	Direct cost reduction

Table 4. The gaps and the false alarms of the performance factors.

	Gaps		False alarms
1	Productivity of specialists	1	Deviation from budgets
2	Productivity of middle management	2	Inventory turnover
3	Productivity of top management	3	Education and training budgets
4	Productivity of supervisors	4	Sales margins
5	Productivity of indirect labor	5	Return on investment
6	Supplier performance	6	Production goals

The individual measures that the respondents have listed, can be classified into financial and non-financial. The ratio between these is 50%/50%. Because the measures listed varied so much, we decided to classify them into nine categories. The categories of the measures are shown in figure 2.

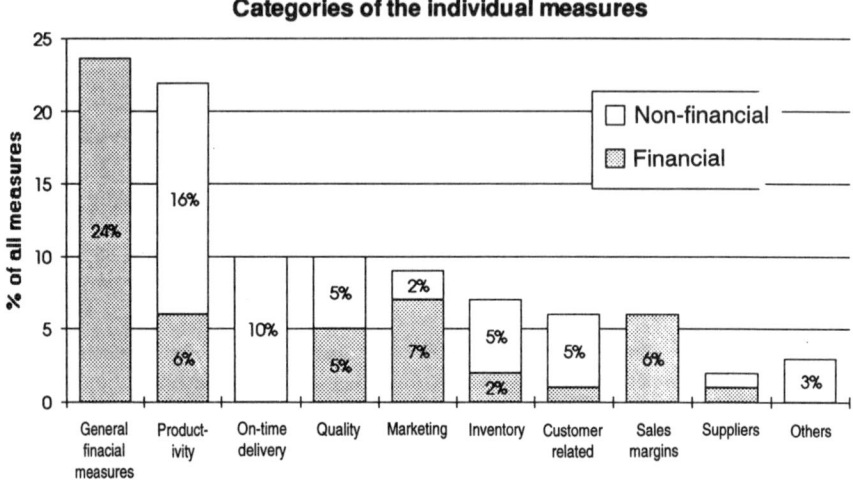

Figure 2. The distribution of the performance measures used in the companies.

In the gap analysis of the individual measurements, it has been pointed out that the most difficult categories of measures are related to suppliers and customer, and quality. It did not pass unnoticed that most of the measurements are focusing on results instead of causes. Emphasis is put on ROI, inventory turnover, variances, etc. instead of delivery precision, response times, machine failures, etc.

PMQ is primarily a diagnostic tool. There is also a need to extend the model to encompass such factors as management style, the structure of the organization and the procedures for strategy deployment.

4. Case Studies

In the best practice studies four Japanese, four American, one French, one German, one Swedish, and two Finnish companies have been included. In addition to carrying out the case studies more than a dozen small scale performance measurement systems has been installed at a number of companies.

From the cases we have learned that PM is a very versatile tool for management. Among the enterprises involved in our studies the following application types have been found:

1. Problem solving and problem spotting
2. Process control
3. Improvement of all important operational aspects
4. Measurement as a basis for bonus payment
5. Introduction of new management paradigms
6. Emphasis on human factors

In the Japanese companies studied by us, the abbreviation Q.C.D.S. was most often quoted as the essence of PM. Q.C.D.S stand for Quality control, Cost management, Delivery management, and Safety control in that order of importance. Q.C.D.S is the key of strategy deployment of the company and it is also the base of self management. All case companies had improvement (Kaizen) programs that contribute to all these four areas. In figure 3 the structure of the PDCA (Plan Do Check Act) management system is shown in one of the Japanese case companies.

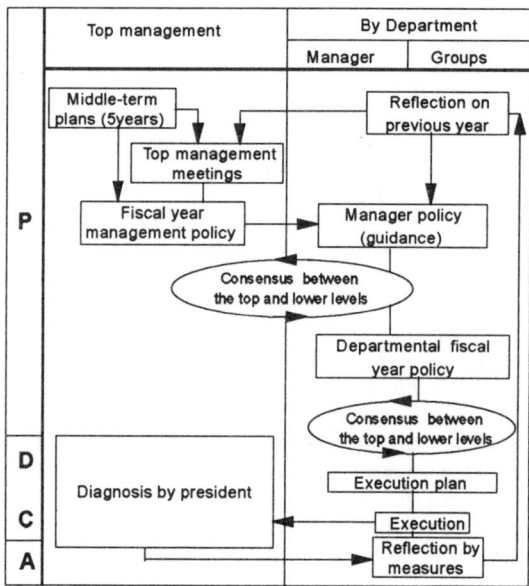

Figure 3. Management system of the NEC Shizuoka.

Japanese companies are forerunners in taking advantage of visual management. Graphs of the measures can be seen along the hallways and around the work cells. Photographs are also used widely to show working procedures and standard levels. The group or the team, whose performance is measured, is responsible for maintaining these measurements. It is one way of getting the operator level involved and empowered.

In the Swedish case a new human dimension has been added to the traditional cost and quality improvement dimensions. Human dimension measurements have been extended to encompass such features as skill, motivation, participation, and empowerement. The PM system of this case company was a byproduct of a large improvement project. The primary aim of the project was to reduce the number of staff personnel and management, and to distribute the responsibility to the heads of manufacturing cells and to self managed teams, and to simplify the lay-out. However, to work effectively towards a common goal the new organization needed a new PM system. This new PM system has three levels described by different colors: red, yellow, and green. The red measures are indicating performance at the factory level. The yellow measures depict efficiency at departmental level. Performances at cell level is shown by green measures. Depending on the organizational level the reporting interval is monthly, or weekly, or daily. Each level determines and maintains the measures indicating its performance. The performance measures must cover all three dimensions: costs, quality/delivery, and human factors. More and more challenging targets are constantly attached to the measures because one of the aims of the operation is continuous improvement.

In the French case the PM system has been developed as a part of a target setting and problem solving mechanism, aiming at pushing process improvement decisions down the hierarchy line, in order to utilize all available knowledge. By means of the PM system management should learn how to share responsibilities and to understand cause/effect relationship of the operations.
According to the management philosophy of the French case company continuous improvement and elimination of waste are supported through:

1. Action plans that are formally managed and which recognize the cross-
   functional interdependencies of the operations.
2. Measurements that are formally defined are shared at factory level.
3. Rolling forecast that is prepared monthly and which commits employees to the
   operational targets.
4. Regular management meetings that are focused to set and re-set new targets and in
   which action plans are initiated and reviewed.

A large number of indicators (measurements) is used in the company. Totally there are around 300 indicators of which 90 % are non-financial and 10 % are financial. Most of the indicators are reviewed in an action group that concentrates on problem solving activity. Problem spotting concerns only few key indicators. The new PM supported management procedure of the company is described in figure 4.

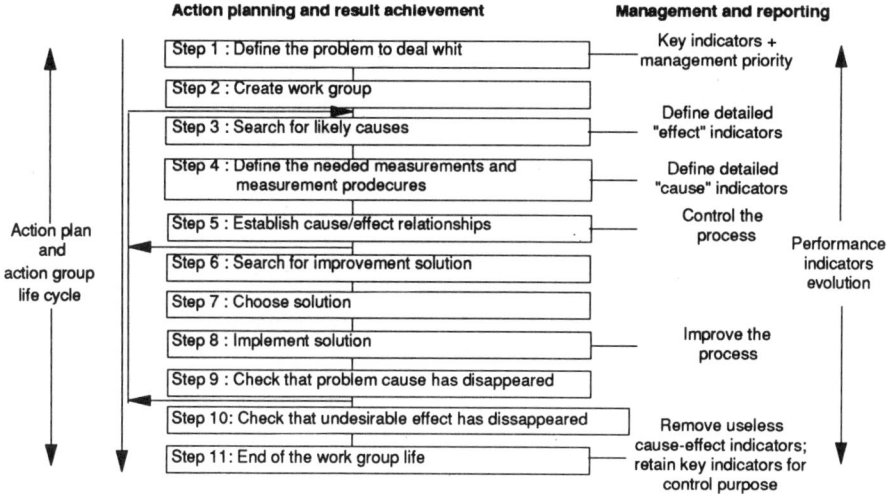

Figure 4. The management procedure of the French case company.

In the Finnish case the PM was a byproduct of establishing a new personnel appraisal system forming the basis of individual bonus payment. The new system covers both staff and manufacturing floor personnel. The part of the PM that is applied to staff personnel, is of special interest. It is an area of PM that often is referred to as the most difficult to put into practice. The four bonus connected areas of employee performance are profitability, quality, delivery, and development targets.

The projects which aimed at installing small scale PM systems in a number of industrial companies covered the following areas: internal and external delivery precision, supplier quality, logistics, productivity, throughput, quality, purchasing, and information systems. The experience gained from these projects has been mainly positive. We learned that the first step of the PM system project need not be a gigantic one. Small steps can lead to substantial improvements without using a lot of manpower, money and time.

What have we learned from our studies? First of all we have identified three main company categories. In category one companies performance measurement is abundantly and widely used as a universal tool. PM systems are created by a bottom-up-design and top-down-checking. Each organizational level has a number of individual measures designed by the level personnel itself. In category two companies PM is successfully used for specific purposes only. In category three companies the performance is measured mainly for routine reasons, such as accounting, reporting or controlling. In such companies the idea of PM is easy to sell to top management but the lower down in the organization you go, the more resistance you will encounter. In this category top management often views PM as an additional tool for control instead of self management.

5. Typical difficulties encountered

Although most people accept the general ideas of a good performance measurement system there are a number of dangers that can harm the well-started development project. The following list is collected during the various stages of our PM-project.

- The role of the measures in self management, empowerment and learning is not fully understood. Performance measurement is only considered to be a management control tool.
- Measures are dictated, although in a civilized manner, top-down. People are not allowed to include their own measures or measures that have an understandable meaning for their daily work.
- The visualization of the measures is ignored. Measures remain numeric, and they do not motivate people for continuous improvement.
- Too much effort is used to describe the present situation, and the project team is worn out before the creative work begins.
- Measures reflect too much the organization chart. Thus, they isolate different functions instead of uniting them into business processes.
- Measures are chosen subjectively, and they are not explicitly linked to the strategies and goals of the business.
- The participation of those who will be measured is ignored. An inpatient, but talented individual can dominate the process. The measures become his measures, and the true owners of the measures get frustrated.
- Measures do not become a part of the daily activities. After the first enthusiastic implementation period, measures remain unused because those who were not involved in the development process do not understand the meaning of measures.

To avoid these usual setbacks of performance measurement, a normative process of implementing performance measures is described below [4].

6. How to implement performance measures

Our general framework of performance measurement (fig. 5) includes strategies, business process to be measured, set of measures, situational characteristics and the continuous implementation process. There must be a link between the overall objectives and the objectives of any single unit. Performance measures are tools to establish this link and synchronize people's actions.

Examples taken from other companies and checklists from the literature are commonly used practical aids during the implementation phase. All the applicable measures should be tied to the conceptual framework of the company. So, the battery of metrics is a structured catalog of reasonable measures, not just an ad-hoc list. Every measure is linked to a performance factor or an improvement area, such as quality, customer satisfaction, human resources or waste. Thus, any chosen measure is a part of a larger entity.

Situational characteristics, such as present problems, former experience or work-force attitudes, are strongly dependent on the business unit that requires new or revised performance measures. Connection to overall objectives is important in order to achieve correctly directed actions. However, if the measures ignore the local realities, no true improvement is activated.

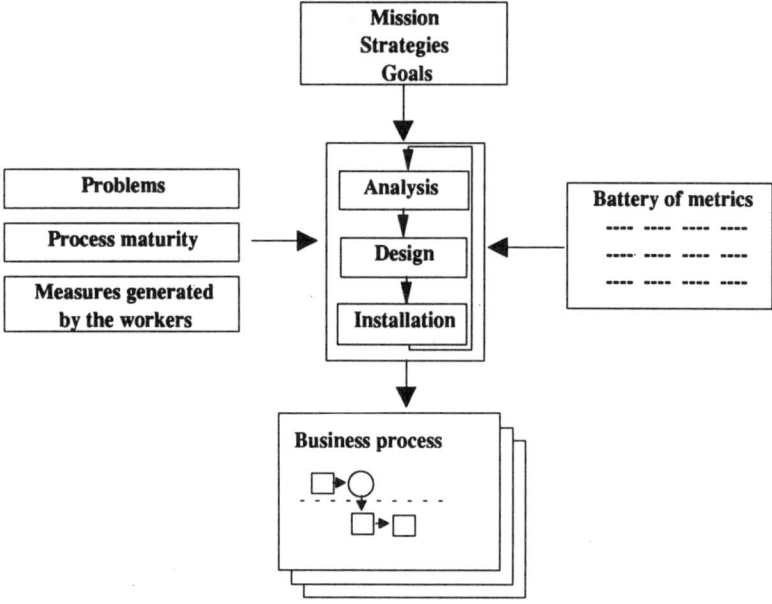

Figure 5. The general framework of establishing performance measurement

The implementation process includes the following main phases: analysis, design and installation. Especially in a change process, a major PM system implementation project can be identified as a part of the improvement project. If the old measurement system requires radical changes, it is easy to see the project approach, too. However, one should understand that the once implemented measures need to be continuously revised, because companies do not operate in a static environment.

Figure 6 summarizes the necessary steps included in the three main phases. The order of the steps is not rigid, and in some cases all steps may not be seen so clearly. Some steps are discussed more closely in the following paragraphs. It is important to notice, that the questions that need to be solved become more detailed when moving from analysis to installation. In a practical implementation process, there is always some iteration.

The analysis phase can be relatively long, starting right after the previous installation and continuing until the need for redesign is clarified. Naturally, some modifications to the installation may take place during the analysis period. The most important results that must be achieved before the design phase can begin are following: the reason for measuring and the activation of the right people.

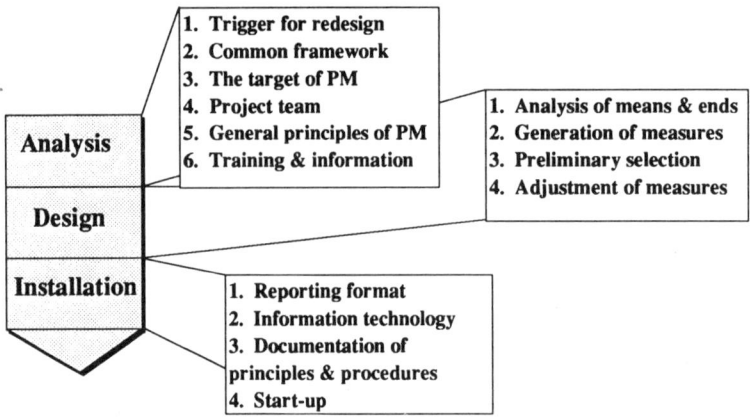

Figure 6. The steps of implementation

The reason for measuring identifies the target group of the measures (business unit, process, design team) and the link from measures to business objectives. In a change process, the need to reshape the way people think and act is essential.   So, the concepts and frameworks of communication should not be ignored. The modern initiatives may be clear for the management élite, but the focal point is to get everybody to understand what, for example, JIT or Lean means in their own work. Therefore, process mapping, team work techniques (e.g. storyboards) and questionnaires (e.g. the Performance Measurement Questionnaire) are suggested tools for this step.

The team that makes the design and installation must have a right mixture of people who set overall business objectives (top management), those who are responsible for achieving the objectives (middle management, supervisors and operators) and those who are familiar with the technical aspects of measurement (IT experts, consultants). The role of the supervisors and operators is the most important, and the participation of these people must be supported by training and information. The role of top management is strongest in the early stages of the implementation project; whereas IT experts are mostly needed in the installation phase.

The design phase includes the analysis of means and ends, selection of measures, and refinement of selected measures. There are some widely accepted objectives, like good quality or short throughput time. The basic measures for these objectives can be easily chosen from a list of measures. Even for benchmarking purposes, it may be relatively easy to find a measure that is meaningful in different organizations. However, each organization must examine the cause and effect relations behind any general metric. Fishbone analysis is a simple but effective tool for this. The analysis of means reveals new factors to measure, and these measures are usually closer to people's daily operations than the original end-measure. Another important result of the fishbone analysis may be that some unfavorable means are identified.

The adjustment of measures includes analysis of possible conflicts of interests. Performance measures must be close to the people, and so, each work group must have independence in choosing the measures. There is a danger, however, that measures drive actions that move problems from one group to another, not actions that terminate problems.

Another aspect of refinement is more psychological by nature. Same factors can be measured using slightly different measures, but one of the possible ways may be more inspiring and motivating. Positive measures are usually more effective than negative ones; so, one should measure successes instead of failures.

In addition to various aspects related to information technology, the installation phase includes design of reporting format and documentation of measurement principles and procedures. The best practices companies seem to have paid great attention to the visualization of measures. Using different kinds of graphs, critical aspects can be high-lighted, unimportant details can be faded, and various factors can be compared or combined.

The documentation of measurement principles and procedures is necessary to make sure that the ideas collected earlier are not forgotten. It must also be assured that those who were not directly involved in the process will understand the measures correctly. Important points in the document, besides the issues considered earlier, are the following: the procedure of setting and updating target values and control limits, suggested actions if control limits are broken, responsibilities and procedures of gathering and transforming data, and frequency and delivery of measures.

7. Future research activities

We have learned that PM is a universal technique that can be applied to practically speaking all regular as well as change management situations of a company. This is shown in figure 7.

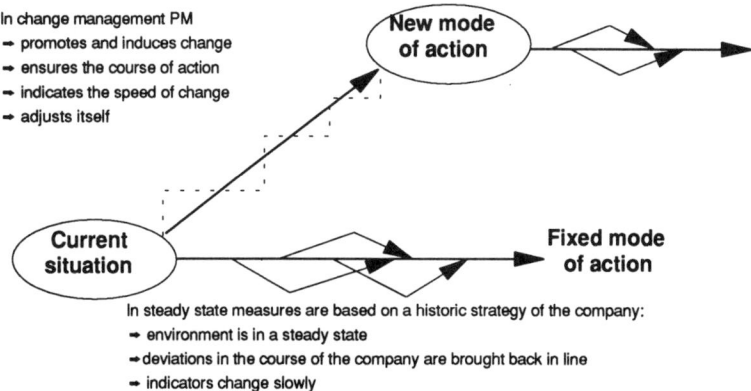

Figure 7. The different areas where performance measurement can be applied

The path we are following in our further research is to clarify the problems of managing change. Managerial attention must shift from maintaining current operations to managing strategic evolution. The capability to implement rapidly and successfully new manufacturing practices is the key issue when we try to maintain the competitiveness of the enterprise. In order to manage the evolution of the enterprise many new tools and methods must be developed.

The result of the project will be a tool kit that will guide companies through the phases of change identifying which areas/processes to measure, developing suitable measurement criteria for the chosen processes, analyzing results, and making comparisons.

References

1.   Nanni,A. & Vollman, T. & Dixon, J., The New Performance Challenge; Measuring Operations For World-Class Competition. Homewood, Illinois 1990, Richard D. Irwin Inc. 199 pages.

2.   Laakso, T., Performance Measurement: a Tool for Management, Licentiate Thesis, Computers in Industrial Production, Helsinki Univesity of Technology, Otaniemi, 1993 127 pages. (in Finnish)

3.   Andersin, H. & Laakso, T., Performance Measurement as a tool for management and selfmanagement - cases and conclusions. 12th ICPR seminar, August 16-20, 1993, Lappeenranta.

4.   Andersin, H., Karjalainen, J. & Laakso, T., Performance Measurement as a Driver of Actions. Publication of FIMET 1994. 121 pages (in Finnish).

38

BENCHMARKING AS PERFORMANCE REFERENCE IN A PERFORMANCE MANAGEMENT MODEL

H. Bredrup and R. Bredrup

Department of Production and Quality Engineering, University of Trondheim, N-7034 Trondheim, Norway

1. INTRODUCTION

Nobody has the fortune to know exactly what future will bring. Countless attempts are made to foresee the future, but the number of different opinions is enormous. Some common characteristics could be drawn by looking at the most popular predictions. The 1992 U.S. Manufacturing Futures Survey (Kim and Miller, 1993) revealed the following anticipation's among the managers regarding changes in business environment:

- Increasingly globalizing market competition, and cooperation (37%)[1]
- More focus on customer's expectations for quality and time (24%)
- Changing nature of work force: their tasks, attitude, expectations, and capabilities (19%)
- Increasing concerns and regulation for environmental issues (13%)
- Declining or non-growing domestic market (12%)
- Rapid change in technology and shortened product life cycle (10%)
- Increased level of competition (9%)

A clear trend from this survey is that competition will increase due to more globalization and more demanding customers. As customers are exposed to better products and services their expectations for better quality, service and value will increase. Higher expectations have to be met by improved performance to obtain customer satisfaction.

Performance management is a response to this development to manage improvement of matters of importance for competitiveness as defined from an external point of view. Competitive benchmarking is an applicable technique to identify and decide performance gaps.

---

[1] Percentage of respondents that mentioned the issue

2. PERFORMANCE PRIORITIES FOR BENCHMARKING

We will provide a methodology to define performance from an external point of view as shown in Figure 1. By defining performance based on the stakeholder model and the vision and strategy, we achieve a direct link to competitiveness and business achievements. However, it is important to know the existing priorities and indicators to develop a common understanding of priorities. Priorities are a part of the organizational culture, and implementation based on a top-down approach without involvement, are likely to fail. A combination between the top-down and bottom-up approach is then preferable.

Unless a performance planning process is forced by a serious crisis, it is hardly possible to develop completely new priorities with a supporting measurement system immediately. Our methodology is general and could be applied in total reengineering of management, but we strongly emphasize use of pilot cases to make a participative process possible. Achieving a common understanding of priorities is the most critical task for future efficiency.

Our model integrates existing and future requirements. Efficiency and effectiveness are covering existing requirements whereas future requirements are covered by adaptability. Looking at performance requirements from an external point of view is then the only feasible approach to balance all three dimensions of performance.

2.1 Stakeholders performance requirements

Long term survival of a company depends on satisfying stakeholders needs. Competitiveness depends on attractiveness towards the different stakeholders where customers have an exceptional position. Stakeholders may have interest in special aspects of performance and conflicting interests especially regarding distribution of improvements exist.

Stakeholder satisfaction is determined by the difference between expectations and perceived performance. Perceived performance is related to the interfaces between the company and its stakeholders. Definition of performance requirements from a stakeholder's perspective could follow the succeeding sequence:

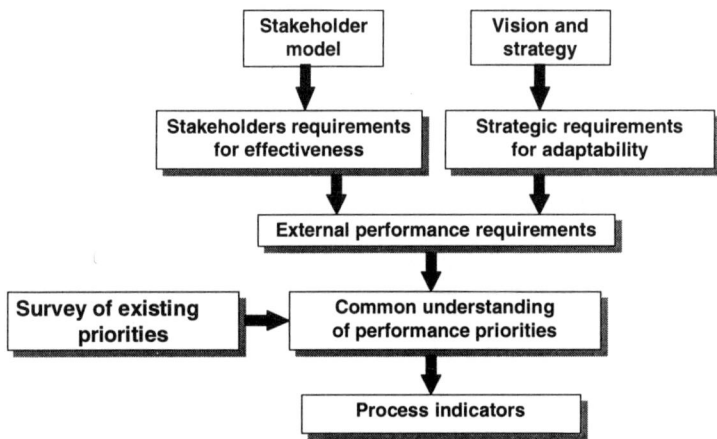

Figure 1 Model for definition of performance

1. Define stakeholders model with relative importances
2. Identify interfaces betwen company and stakeholder with relative importance
3. Identify performance requirements for interfaces
4. Define the relative importance of each performance requirement

Define the stakeholder model with relative importance

The list of stakeholders includes groups like customers, suppliers, authorities, owners, financial institutions, employees, management, environment and alliance partners. Within each group there may exist several subsets with different characteristics. Application of the stakeholder model means accepting the fact that a company and its stakeholders have mutual interest and dependency concerning spesific business processes.

Most stakeholders are easily identified due to direct business relations whereas more indirect stakeholders could easily be forgotten. Brainstorming or Nominal Group Technique are useful tools to provide a representative set of the most important stakeholders. In some cases it could be wise to involve groups in different departments to ensure that all important stakeholders are identified.

The different stakeholders have different impact on competitiveness. Assessment of importance is necessary to achieve a balanced understanding of the priorities for the business and maximize cost/benefit of investments in performance improvement. Figure 2 shows that it is waste to provide high performance of characteristics of low importance. The most efficient strategy is naturally to improve performance of characteristics of high importance to competitiveness. Rating the different stakeholders is a step in the process of identifying the most important characteristics.

Assessment of importance is a difficult task in complex companies. However, to desist from giving importance, is to give equal priority to everything. Objectiveness is impossible, but a well-prepared process of subjective evaluation is sufficient for our purpose. It should be emphasised that this evaluation of importance is individual for each company and a necessary step to understand performance. Transformation of performance improvements into business achievements depends on the ability to identify the best prospects.

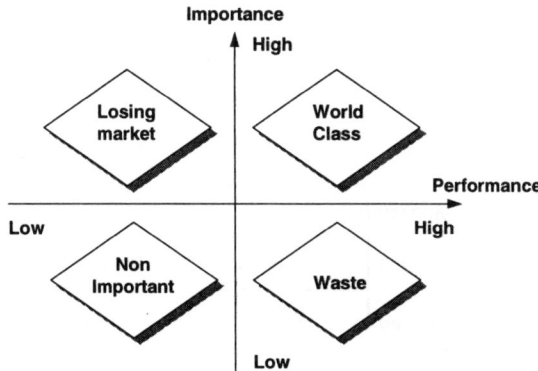

Figure 2 Relation between importance and performance

Identify interfaces between stakeholders and company

A stakeholder may have several interfaces to a company. These interfaces could have different characteristics and be of different importance. Identification of interfaces from the most important stakeholders' points of view will serve as a checklist for development of performance requirements. Three different kinds of interfaces may exist:

- Stakeholder and company are partners
- Stakeholder have direct interest in the output or the process
- Stakeholder have indirect interest in the output or the process

Creative techniques should be applied to develop an extensive list of interfaces. Afterwards, the set is reduced by rating the interfaces by relative importance.

Identify performance requirements

Performance has to be defined from the stakeholders point of view since their perception of performance determines the competitiveness of the company. Some stakeholders may have well-known requirements for performance as for instance finance institutions demanding a given interest rate. Others may have dynamic expectations of performance level.

Negotiations between company and stakeholders often regulate the financial aspects of the relationship, whereas more operational performance requirements often are left out. Access to continued education and training are examples of factors for employees satisfaction often neglected in negotiations with unions. A stakeholder will typically have some outspoken and some implied requirements to performance. Figure 3 shows a model of the different categories of match between the stakeholders and the company's perception of performance.

Traditional surveys are often not intended to collect this kind of information, but could easily be adjusted to suit our purpose. However, most knowledge is already available within the company. A common problem is that this information is distributed and related to individuals within the organization. A systematic approach is required to benefit from the knowledge of each individual.

The list of most important interfaces works as a checklist for development of stakeholders' performance requirements. Those involved in each interface have to systematize their knowledge. Additional information could be collected with surveys or less formal inquiries. Other sources for information both internally and externally may be tried.

		Company	
		Included in performance	**Not included in performance**
	Outspoken	Obvious	Ignored
Stake-holder	**Implied**	Identified	Not identified
	Not intended	Waste or surprise	Unknown

Figure 3 Match between stakeholders' intention and performance

Define importance of each requirement

By assessing importance of each requirement, a rating of performance priorities from the stakeholders' point of view is obtained. This importance rating is the key input to the performance planning model. Information about these priorities is already available within the organisation, but additional information have to be collected. Surveys and different informal inquires could provide the necessary supplement.

The Kano-model (Akao, 1990) in Figure 4 relates satisfaction to the degree to which performance requirements are achieved. This could be helpful in the assessment process. The straight line represents the expressed performance requirements from the stakeholder. In general, only the expressed requirements would be explicitly mentioned if the stakeholder is asked for input. However, there exist a set of basic requirements that are so obvious that they are not explicit mentioned. The degree of perceived achievement decides the satisfaction level. Identification and fulfilment of expecations is necessary to achieve satisfaction. Offering unspoken features could enhance satisfaction. However, it depends of achievement of basic expectations. Extra service at the dealer will not equalize a scratch in the paint of a car.

Satisfaction of expectations are required before thinking of providing extra benefit. This is a guiding star for the importance evaluation process. Even if all of the basic requirements are met perfectly, we would not achieve real satisfaction. We would only eliminate dissatisfaction.

Performance planning is a continuous process and not a sequence of events. A new step in the process could provide new knowledge about previous steps, that should be included. New knowledge from other sources should also be included. The systematic approach makes it easy to update previous steps in the model. Importances set for requirements is not fixed and is a matter of tactic. Tactics have to be frequently updated. The performance planning model will help making better choices and to see the consequences of changes.

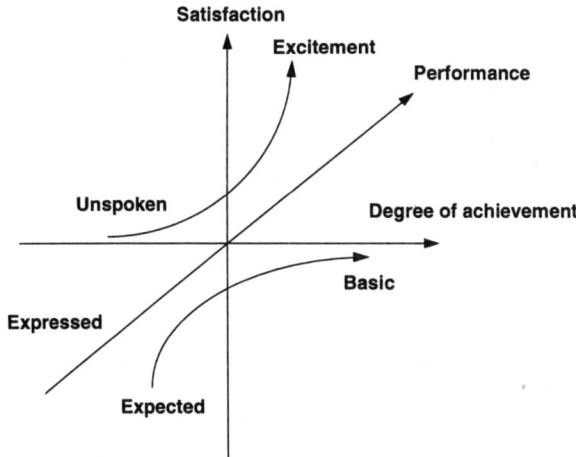

Figure 4 The Kano-model

2.2 Performance requirements based on vision and strategy

Performance planning has to be integrated with the traditional planning hierarchy. Development of a long term vision, indicating what the company should look like in the future, will provide useful input to performance planning. Strategic plans concerning new markets, new products, new marketing strategies, new technologies, etc., means new performance requirements. These requirements have to be integrated with those from the stakeholder model so that both short term and long term business achievements are ensured.

Performance requirements have to be outlined both in vision and strategic plans. Time horizon decides how detailed specifications are. It is not controversial to emphasize a need for integration of all kinds of strategies. Problems arise concerning the variety of characteristics of the different strategies and lack of integrative tools. Performance planning is an integrative methodology, but depends on the ability to develop compatible performance requirements.

Development of performance requirements should be a major concern in development of different strategies. By applying the stakeholder model, we have developed a methodology to identify performance requirements of the existing stakeholders, including customers. However, strategic decisions often imply dramatic changes in opportunities and new requirements for performance. Identification of these is essential to achieve improvement.

New opportunities could include investments in new technology, suppliers with new characteristics, new systems. Utilization of new opportunities depends on a mutual adjustment process with the existing system. Research (Bredrup *et al.*, 1994) has shown that investments in modern manufacturing often fail to provide expected benefit because existing production management systems remain unchanged. Expected improvements in flexibility and reduction of lead times are not achieved due to lack of support in existing systems.

Entering new markets and introducing new products often implies changes in performance requirements for the different business processes. Change to more tailor made products requires a better flexibility, better skilled operators, new performance indicators, more responsive production management systems, etc.

2.3 Integration of performance requirements

Performance requirements is a matter of integrating the set of stakeholder requirements and strategic requirements. They should cover all important aspects of competitiveness. Performance requirements could then be defined as:

> *Qualities that a company is expected to possess to fulfil the needs of present and future stakeholders in order to sustain or enhance competitiveness.*

Lacking qualities could be compensated by other qualities, but possessing the right qualities is assumed to be the most cost effective strategy. However, there will probably exist several conflicts of interest. A summation of requirements is then impossible. Revealing trade-offs is important because hidden trade-offs are potential sources for dissatisfaction. This process will provide valuable information about the nature of competitiveness. It is often said that the planning process is more important than plans itselfs.

What we are really looking for, is to develop a common understanding of priorities within a company. Indicators for measurement and benchmarking are derived from these priorities. However, priorities may vary within a company. In order to develop a common understanding of priorties within the company, the priorities must match the performance requirements. However, developing of common understanding is a process that includes organizational development, and involving people in the process is necessary to succeed. Gaps between current and desired priorities serve as a basis for managing the cultural process.

Identify possible performance priorities for the business

Performance planning has to be based upon the real driving forces of the business. Identification could be based on experience or by creative approaches like brainstorming, Surveys are also valueable sources for input. It is important to be openminded and not allow evaluation in the creative process. Awareness of the influence from existing values and beliefs is important to avoid ending up at old tracks. Inertia against change is considerable.

Evaluation of the different forces is achieved by auditing them against the stakeholder and strategic requirements previously identified. Based on the given importances we are able to suggest a rating of different performance requirements. However, the final rating is a matter of performance planning. Important aspects to consider in performance planning, is performance gaps, strategic and tactical decisions, the internal relationship between different requirements, investments, organizational difficulties, etc.

Analysing the match between 'is' and 'should be'

An essential activity in all processes involving organizational development, is to define current status. We have to know the existing priorities to choose appropriate approaches to change. An internal survey including the different departments and organizational levels could reveal the existing set of priorities and identify gaps between current and desired priorities. Lynch and Cross (1991) suggest an approach where each department manager indicates existing priorities. This is done by distributing a given number of points on different predefined priorities. A systematization of all forms from the complete survey and comparison with previously defined priorities provides answers on two important questions:

- Degree of convergence or divergence among the different functions
- Size of the gap between 'is' and 'should be'

Develop a common understanding of priorities

A common understanding of priorities is necessary to develop an efficient performance plan and to achieve integration with business strategy. This is a very difficult exercise, because existing paradigms and values are challenged. Managers from different departments seldom agree upon how to respond on changes. Our model provides some important characteristics that will ease this process:

- Participative and involves employees based on their qualifications
- Visual and easy to apply
- Boosts existing functional silos
- External view that pushes the discussion above internal prestige
- Try and fail approach that makes 'simulation' possible
- Forces the important questions on the agenda

Applying the model is no guarantee for acceptance of common priorities, but these elements provides a leap compared to traditional methods. Traditional methods are often too specialised to achieve real involvement of those responsible of implementing the plans. Involvement is perhaps the most important factor to achieve acceptance by all affected parties. Understanding and acceptance of the performance priorities is absolutely necessary for development of a consistent set of action plans throughout the organisation that will result in real business achievements. Otherwise, contradicting efforts may cause a loss in efficiency and hence competitive position.

Tools like Nominal Group Technique (NGT) could be applied in the process to achieve a common understanding. It is important to look at this as a part of organizational development. Performance planning is only one of the activities within performance management, and they have to be seen as a whole. Training and information is essential to benefit from the tools and the methods described.

3. PERFORMANCE PLANNING WITH QFD

The QFD methodology offers a favorable tool for transforming stakeholder and strategic requirements into performance requirements for a given business unit. The methodology can be applied for any kind of business units as business process, organization units and limited problem areas. Figure 5 shows the 'House of Quality' for the relationship between some performance parameters for production management and a sample of stakeholder and strategic requirements. The methodology guides us through the process of defining requirements.

The QFD-chart includes benchmarking with two competitors. A desired target level is decided to define the performance gap. The chart indicates that conformance quality is both important and has a large performance gap. Therefore, we should focus on improving parameters that influence on conformance quality. Performance targets must be as measurable or descriptive as possible.

Comparison of the performance parameters level with other companies provides an effective tool for recognition of the performance gap and establishment of possible targets. The QFD-chart shows that the defect level has high influence on conformance quality and a high performance gap. Therefore, reducing the defect level seems to be a reasonable strategy.

It is obvious that some performance parameters are easier to improve than other due to different implementation problems, knowledge, organizational obstacles and so on. QFD takes into account organizational difficulties. Improvement of lead time, defect rate and more efficient production, is the main means for improving the stakeholder and strategic requirements. From the roof, we also remember that improvement of these factors can gain advantages of concurrent improvement at the sacrifice of resource utilization..

References

Akao, Y. (editor) (1990) *Quality Function Deployment: Integrating Customer Requirements into Product Design*, Productivity Press, Cambridge, Mass.

Bredrup, H. (1994) Measurement systems based on standard costing inhibit JIT implementation, *Proceedings for Eighth International Working Seminar on Productions Economics*, Igls/Innsbruck, Austria.

Bredrup, H., Bredrup, R. and Estensen, L. (1994) Factors Influencing Effectivenness of investments in Modern Manufacturing Technology, *Proceedings for Automation '94*, Taipei, Taiwan, R.O.C, July 6-9, 1994.

Harrington, H. J. (1994) The Collapse of Prevailing Wisdom, Presented at *EOQ'94*, Lisbon June 13-17, 1994.

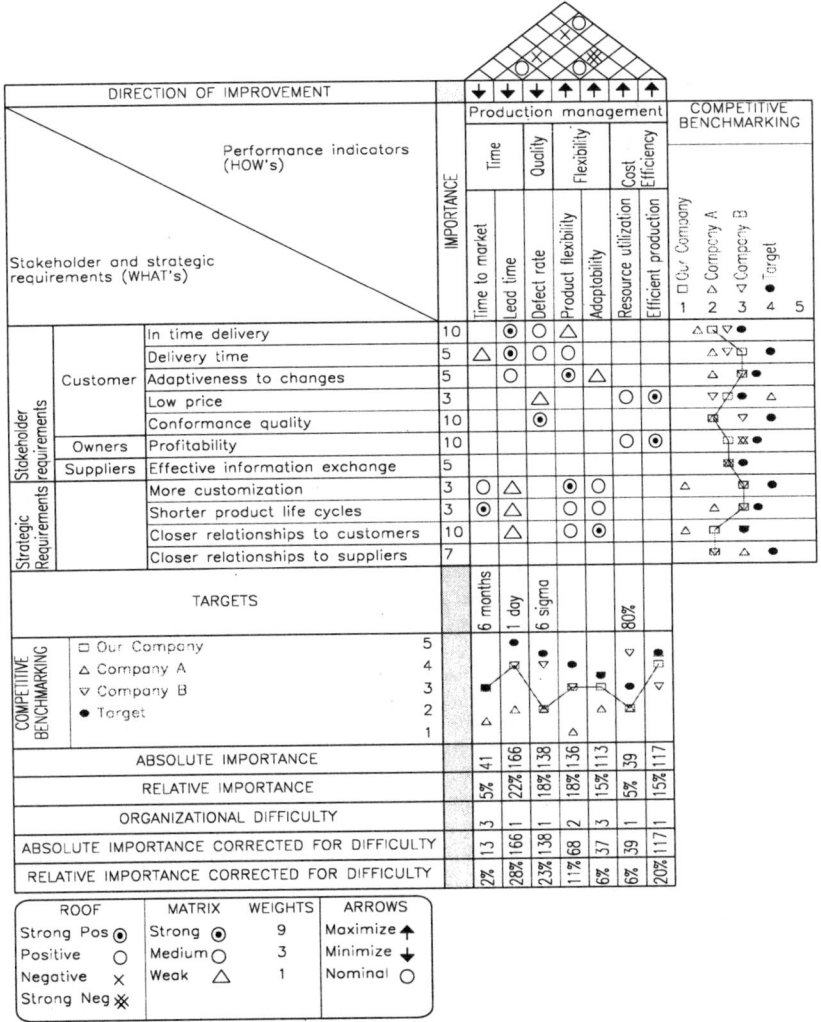

Figure 5 An example of a 'House of Quality' for production management

Kim, J. S. and Miller, J. G. (1992) Challenges for building the value factory: Key findings from the 1992 U.S. Manufacturing Futures Survey, *Operations Management Review*, **9** (3), p. 1-21.

Lynch, R. L. and Cross, K. C. (1991) *Measure up!: yardsticks for continuous improvement*, Cambridge, MA : Blackwell Business.

39

ECOGRAI - A method to design and to implement Performance Measurement Systems for industrial organizations - Concepts and application to the Maintenance function

G. Doumeingts [a], F. Clave [a] and Y. Ducq [a]

[a] GRAI/LAP - Université Bordeaux 1 - 351 Cours de la Libération 33405 Talence cedex - FRANCE -Tel 56 84 65 30

Abstract:

ECOGRAI is a method to design and to implement Performance Indicator Systems (PIS) for industrial organizations. Based on the GRAI model, it can be applied on all the production functions (Engineering, Manufacturing, Quality, Maintenance, delivery...) with a global approach or more specifically on only one function. The results of the PIS design is a coherent set of specification sheets describing each Performance Indicator. The implementation and the operating of this system is supported by an EIS (Executive Information System) tool.
This paper is presenting first the main characteristics of the ECOGRAI method, the six phases of its structured approach, the various actors involved and the ECOGRAI implementation. The second part shows an example of application to one production function : the Maintenance function.

Key words :

Performance Indicators, method, modelling, structured approach, maintenance, EIS

1. INTRODUCTION

The industrial performance measurement is today one of the basic tools for the production unit control. This performance measurement becomes more and more important because it is used in the "Benchmarking" approaches and in the "Self Auditing Procedure".
Based on the GRAI method users request, the GRAI Group has been working on this domain since 1988. Indeed, the GRAI method allows to elaborate specifications in order to choose or to develop Production Management Systems. But when the system is chosen and implemented, the user wants to know its performance. It is why ECOGRAI method was developed.
ECOGRAI is a method to design and to implement Performance Indicator Systems (PIS) for industrial organizations and used by the decision makers of

the Production Management Systems to measure the achievement of their objectives. The results of the PIS design is a coherent set of specification sheets describing each Performance Indicator (indicators, concerned actors, required information and processing...). The implementation and the operating of the PIS is supported by an EIS (Executive Information System) tool.

First, the various phases and the various actors involved in the ECOGRAI structured approach are presented. In a second part, an example of application on one function of the production system, the Maintenance function, will be described.

2. THE ECOGRAI METHOD

The main characteristics of the ECOGRAI method are :

- a logical process of analysis / design using a top-down approach, and allowing to decompose the objectives of the strategic levels into objectives for operational levels,

- a concrete process of participative implementation, creating a dialogue between the various levels of the hierarchy, and favouring the identification of indicators by the future users involved in the study : it is a bottom up implementation,

- the use of a number of tools and graphical supports : GRAI grids, GRAI nets, splitting up diagrams, coherence panels, specification sheets,

- the search of a limited number of Performance Indicators by an original approach (figure 1) : identification of the objectives assigned to the decision makers (target situations which have to be reached inside the functions, depending on the management level which are considered), identification of the variables on which the decision makers can act to reach their objectives (we can call these variables "Decision Variables" or "Drivers"), and identification of the Performance Indicators, quantified data which measure the efficiency of an activity or a set of activities of a function in the process to reach the objectives. The originality of the ECOGRAI method is not defining the Performance Indicators, but the search of Decision Variables or Drivers on which decision makers can act to reach their objectives. The Performance Indicator is a consequence of the preliminary choice,

- a coherent distribution of Performance Indicators covering the various functions and the various decision levels (strategic / tactical / operational).

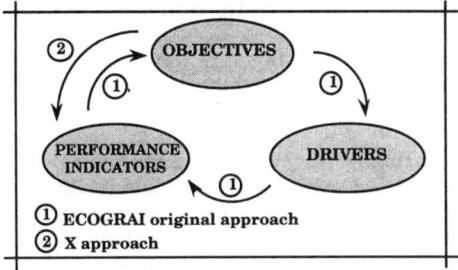

Figure 1. The ECOGRAI original approach

2.1 The six phases of ECOGRAI method

The logical structured approach of the method is decomposed into six phases (figure 2). The first phase (phase 0) consists in analysing the Production Management System and in determining in which decision centers the Performance Indicators will be defined. The two following phases (phase 1 and 2) aim at identifying the basic elements which are required : the objectives and the drivers. The fourth phase (phase 3) consists in identifying the Performance Indicators, the fifth (phase 4) in designing the information system of the Performance Indicators, and the sixth (phase 5) in implementing it on the Production Management Information System.

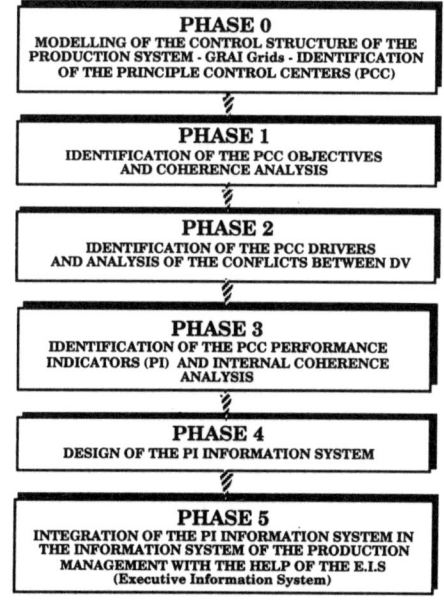

PHASE 0
MODELLING OF THE CONTROL STRUCTURE OF THE PRODUCTION SYSTEM - GRAI Grids - IDENTIFICATION OF THE PRINCIPLE CONTROL CENTERS (PCC)

PHASE 1
IDENTIFICATION OF THE PCC OBJECTIVES AND COHERENCE ANALYSIS

PHASE 2
IDENTIFICATION OF THE PCC DRIVERS AND ANALYSIS OF THE CONFLICTS BETWEEN DV

PHASE 3
IDENTIFICATION OF THE PCC PERFORMANCE INDICATORS (PI) AND INTERNAL COHERENCE ANALYSIS

PHASE 4
DESIGN OF THE PI INFORMATION SYSTEM

PHASE 5
INTEGRATION OF THE PI INFORMATION SYSTEM IN THE INFORMATION SYSTEM OF THE PRODUCTION MANAGEMENT WITH THE HELP OF THE E.I.S (Executive Information System)

Figure 2. The six phases of the structured approach

2.1.1 Phase 0 : Modelling of the Production System Control Structure and Identification of the PCC

The objective of this phase is to determine the Principal Control Centers of the Production Management System in which the Performance Indicators will be designed.

Phase 0.1- Modelling of the Production System Control Structure.
As written above, ECOGRAI uses the GRAI tools (grids and nets) to model the Production System control structure in order to identify the set of the decision centers, their activities, their links (decisional and informational) and the basic elements which are taking into account to design a Performance Indicator : the objectives and the drivers.

This paper is not aiming at describing the GRAI method. Therefore, for a better understanding of ECOGRAI, a short description of the GRAI tools used (the grid and the nets) is presented below.

The GRAI grid takes up the hierarchical and functional approach. It allows to identify the set of decision centers of the studied system, as well as their links.

The GRAI grid is presented in the form of a matrix :

- the managerial axis or control axis which represents the various levels of decision which can be found in a Production System. Traditionally, this axis is decomposed hierarchically in several levels, according to the nature of the decisions : strategic, tactical, and operational levels.

- the production axis which describes the various activities required to the product life cycle. It is decomposed into several functions which group a set of activities having a same identified finality (Engineering, Manufacturing, Quality, Maintenance, Delivery, Recycling...).

Each function of this axis is decomposed in : to manage the products (internal or external, it means supplying and purchasing), to manage the resources (human or technical) and to plan (to synchronize at each level product and resource management).

A decision center is defined by a function and decision level cross.

The GRAI nets are aiming at describing in details all the activities identified inside each decision center of the GRAI grid.

Then, in this phase, the production system control structure studied is modelled with a global GRAI grid (all the functions). It is split up into specific GRAI grid by function (figure 3).

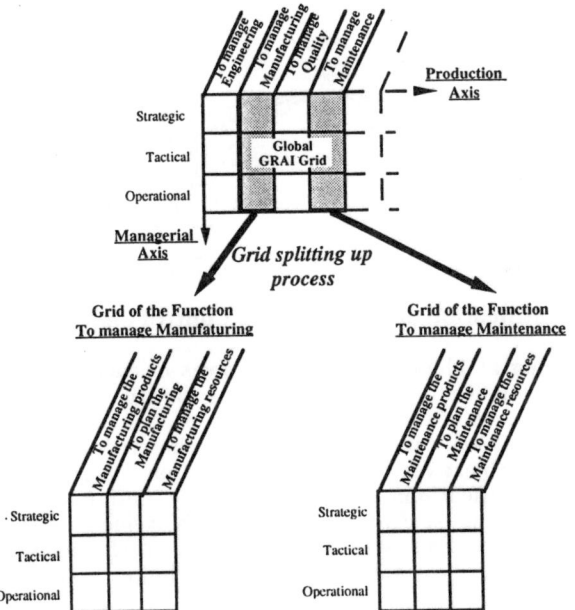

Figure 3. Global Grid split up into various grids for each function of the Production Axis

Phase 0.2- Identification of the Principal Control Centers (PCC)
The objective of this phase is to "scrutinize" the grids in order to determine the
Principal **C**ontrol **C**enters. By definition, they are the decision centers which
have, as a consequence of their activity, a principal influence on the system
control. We search a set of PCC allowing to cover the various functions of all
the grids and the various decisionnal levels (figure 4).

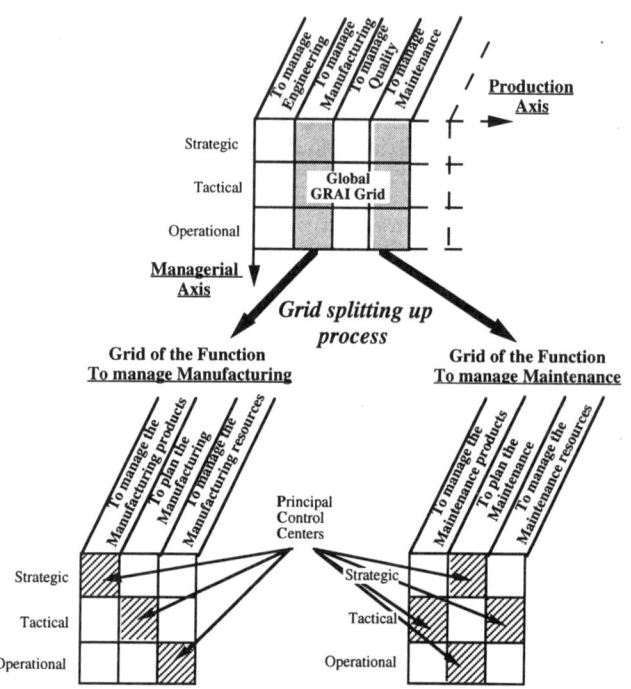

Figure 4. Phase 0 results - Identification of the Principal Control Centers

2.1.2 Phase 1 / Identification of the Principal Control Center objectives and coherence analysis

This phase is aiming at identifying the objectives of each Principal Control
Center. We follow a top-down approach : it means that the first step consists in
identifying the objectives of the Production System, the second in identifying
the global objectives of each function belonging to the production axis, and the
third in defining the objectives of each Principal Control Center inside each
function. These identifications are based on the notion of contribution.
Actually, each objective must contribute to the achievement of the objectives
identified at a upper level. Each step is supported by graphic tools (splitting up
diagrams) allowing to verify if a sub-objective contributes to an objective at a
upper level.

Phase 1.1- Identification of the Production System objectives
These objectives come from the Business Planning. They are often expressed in term of optimisation of the triplet Quality / Lead Time / Cost.

Phase 1.2 - Identification of the global objectives for each production axis function and coherence analysis
The same performance criteria (Quality / Lead Time / Cost) appear but more in detail and more specific to the studied function. The global objectives of the production axis functions must contribute to the achievement of the production system objectives (figure 5).

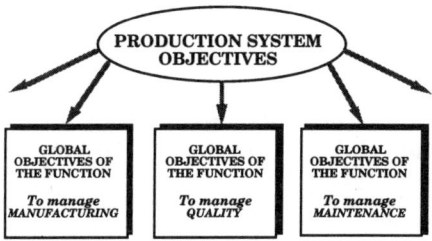

Figure 5. Splitting up of the Production System objectives

To verify these contributions, a splitting up diagrams is used (figure 6).

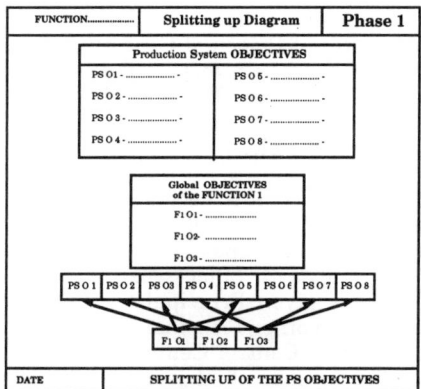

Figure 6. Example of Splitting up diagram

Then, the inter-function coherence between the global objectives is analysed. The links which exist between the global objectives of the various functions are identified in order to check there are no perverse effects (the objectives assigned to a function do not prevent another function to achieve its objectives) (figure 7).

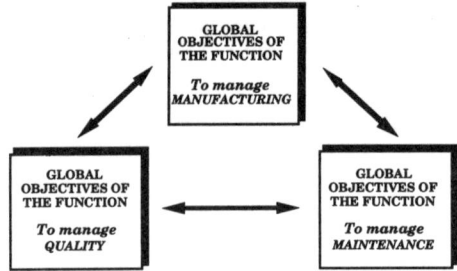

Figure 7. Inter-function coherence analysis between three functions

Phase 1.3 - Identification of the Principal Control Center objectives for each function and intra-function coherence analysis
The last step of the phase 1 consists in identifying the objectives for each Principal Control Centers (figure 8).
This identification is validated by the intra-function coherence study. The principle used is the same : to verify if the objectives of the Principal Control Centers contribute to the achievement of the function global objectives. The same tools as on phase 1.2 are used to show these contributions.

2.1.3 Phase 2 : Identification of the PCC Drivers and analysis of the conflicts
As already said, if it is necessary to know the objectives in order to build relevant performance indicators, it is not sufficient. Actually, the drivers, corresponding to each objective of Principal Control Centers must be identified (figure 8). This identification must be interpreted as one of the steps leading to the building of the triplets {Objectives / Drivers / Performance Indicators}. This notion of triplet is another valuable characteristic of the method, and it expresses the controllability principle.

During the identification of the DV, it is necessary to put in evidence the intra-function and inter-function influences of the Drivers. The aim here is to evaluate the relationships which appear into a function and between the functions. Indeed, the proposed objectives (and by consequence the Performance Indicators) for a given Principal Control Center are sometimes related to drivers that belong to other Principal Control Center. In this case, we must evaluate, for the Principal Control Center concerned, the degree and the origin of the influence. The notions of "driver with direct effect and indirect effect" refer to this phenomena. A direct effect is assigned to the driver which has a dominating influence on the considered objective .

2.1.4 Phase 3 : Identification of the PCC Performance Indicators and internal coherence analysis

Phase 3.1: Identification of the Performance Indicators for the Principal Control Centers.
The previous phases allow, for each Principal Control Center, to identify one or several objectives (coherent with the global objectives of the function themselves

coherent with the production system objectives) and the associated drivers. The determination of the Performance indicators is performed during this phase 3 (figure 8). The approach uses the knowledge of all the people involved in the study and this identification is validated by an internal coherence analysis inside each Principal Control Center.

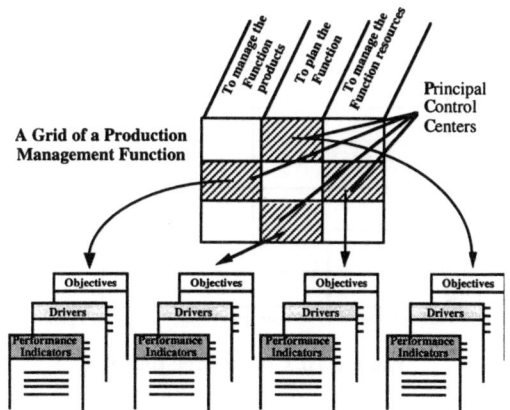

Figure 8. Identification of the PCC Objectives, Drivers, and Performance Indicators

Phase 3.2 : Internal Coherence Analysis of the PCC
This study consists in verifying the internal coherence inside the Principal Control Center in terms of triplet {Objectives / Drivers / Performance Indicators}. A triplet is coherent if :
   - it is composed of one objective, one or several drivers and one or several performance indicators,
   - the performance indicators allow to verify the achievement of the objective, and are influenced by actions on the drivers (figure 9).

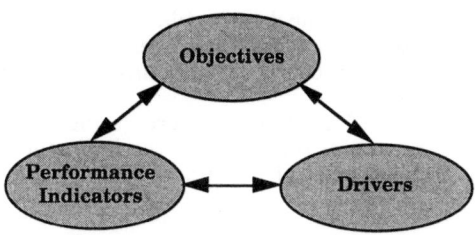

Figure 9. Internal coherence

In order to verify this coherence, coherence panels are built. They allow to identify the various links between the elements of the PCC as well as their weight.

Then, "coherence panels" are filled in (figure 10). The links between the PCC elements are classified according to the connection (strong link / weak link / no link).

Function	Decision Centι Level	INTERNAL COHERENCE ANALYSI			
OBJECTIVES O 1		**	**	**	*
O 2			*		**
PERFORMANCE INDICATORS	IP 1	IP 2	IP :	I P4	
DRIVERS DV 1			*		**
DV 2		**		**	
DV 3			**		*

Strong link (**) / Weak link (*) / No linl

Figure 10. Coherence panel

2.1.5 Phase 4 : Design of Performance Indicator information system

An indicator is basically a measure which will become more and more sophisticated : "measure -> raw information -> process -> review -> statiscal process" can be an example of a possible chain. ECOGRAI is completely oriented towards the phase of specification, which is preliminary to any possible automation of the performance evaluation system. Two aspects are considered : the data aspect (which information is necessary ?) and the processing aspect (the processings which are necessary to build the indicators, starting from basic information). Whatever the case of study, it is always necessary to define clearly each indicator with fundamental parameters. The tool which guides these definitions is the specification sheet for each indicator which contains (figure 11) :

     - the identification of the indicator (name, decision center, horizon, period),

     - the objectives and the drivers related to the indicator,

     - the perverse effects which have been identified,

     - the identification of the data required for the implementation of the indicator,

     - the definition of the associated processings,

     - finally, the way of representing the indicator, determined by the future users (using graphics most of the time).

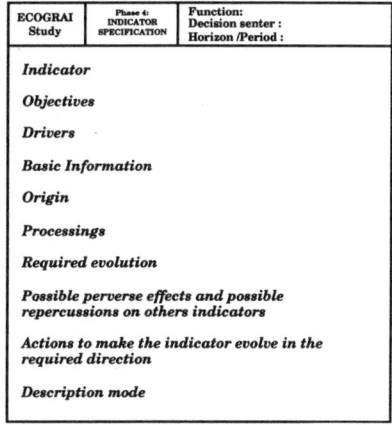

ECOGRAI Study	Phase 4: INDICATOR SPECIFICATION	Function: Decision senter : Horizon /Period :

Indicator

Objectives

Drivers

Basic Information

Origin

Processings

Required evolution

Possible perverse effects and possible repercussions on others indicators

Actions to make the indicator evolve in the required direction

Description mode

Figure 11. An example of specification sheet

2.1.6 Phase 5 : Integration of the Performance Indicator information system in the Production information system

The phase five consists in integrating the performance measurement system in the production information system. This implementation phase is supported by the use of an EIS tool (EIS : Executive Information System). It is used to converse with the existing data bases. The data on Production information system are located and the extraction frequencies, the processings and the visualisation choice to exploit the Performance Indicator are then specified into the EIS tool. This work is performed from the specification sheet.

2.2 The implementation of the ECOGRAI method

One cannot define a method to design and to implement PIS without defining also the way to implement it. So the ECOGRAI structured approach is based on the GRAI structured approach. In particular, the notions of involvement (creation of various working groups), and top-down / bottom-up approach have been kept.

Three kinds of group are defined (figure 12) :

- the project board, is composed of the people in charge of the studied production unit. They define objectives and orientations, and they check the results presented by the analysis group during the project board meetings. They also structure the information and locate them in their context,

- the synthesis group, is composed of the person in charge of each studied function of the production system, of the supervisors, and if necessary, of the operators. During the synthesis meetings with the analysis group, this group is involved in the organisation analysis of the function they are in charge of, and in the definition of the indicators,

- the analysis group, composed of an analyst and of a person in charge of the considered production unit. This group is in charge of ensuring the operational part of the study. They organize meetings and collect information.

They also formalize the results, and propose solutions to the problems which may appear during the study.

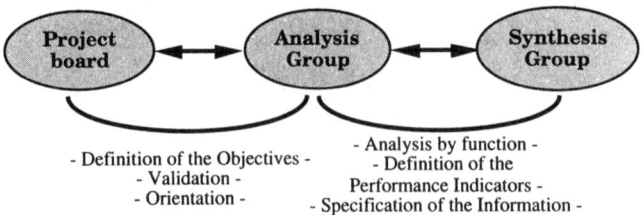

Figure 12. The various groups involved in the structured approach

The information technology department involved in the design, with the help of an EIS specialist, ensures the physical implementation of the performance indicators.

Thus, after having constituted the various groups, the structured approach follows the six phases of the method. The main remark which can be made is the continuous iteration between the synthesis group and the project board all along the study.

3. EXAMPLE OF ECOGRAI APPLICATION TO THE MAINTENANCE FUNCTION

ECOGRAI can be applied on all the production functions (Engineering, Manufacturing, Quality, Maintenance, delivery...) with a global approach or more specifically on only one function. This paragraph is aiming at presenting the results of the ECOGRAI method application on a specific function of a production system : the Maintenance function.

3.1 Results of the phase 0 : " Modelling of the Maintenance control structure and identification of the PCC"
The GRAI grid presented below (figure 13) models the maintenance control structure. The grey tint decision center correspond to the identified PCC.
Indeed, nine Principal Control Centers have been identified inside this grid. They correspond to the decision centers who have a principal influence on the maintenance control system.

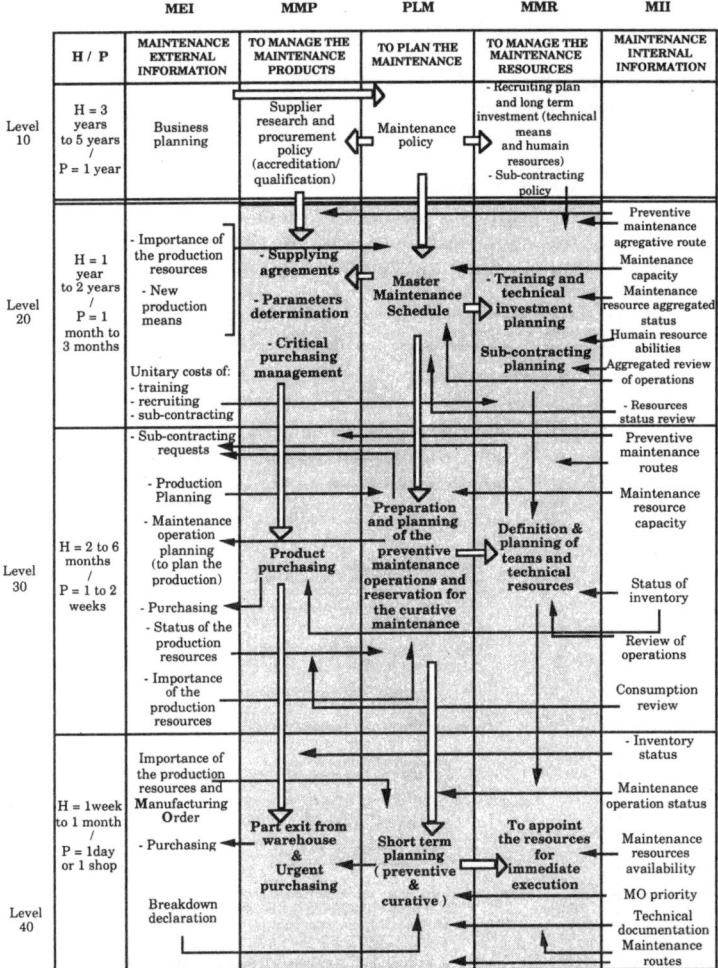

Figure 13. GRAI grid of the Maintenance function

3.2 Results of the phase 1 : " Identification of the Principal Control Center Objectives and coherence analysis"

3.2.1 Results of the phase 1.1 "Identification of the Production System Objectives"

According to the triplet {Quality, Lead time, Cost} (Q, Lt, C), three Production System Objectives (PS O_n) have been identified :

> PS O_1 : To respect the required Quality level

PS O_2 : To respect the production Lead time
PS O_3 : To minimize the production Costs

3.2.2 Results of the phase 1.2 "Identification of the Maintenance Global Objectives and coherence analysis"

Three global Maintenance function Objectives (M_F O_n) have been identified :

M_F O_1 : To maximize the Maintenance operation Quality
M_F O_2 : Maximal availability of production resources
M_F O_3 : To minimize the maintenance Costs

In order to analyse the coherence between the Production System objectives and those of Maintenance function, the following splitting up diagram is used (figure 14) :

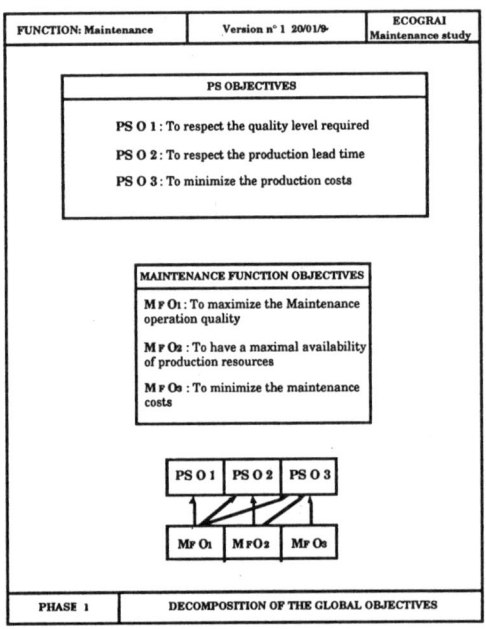

Figure 14. Splitting up diagram of the maintenance function (M_F)

3.2.3 Results of the phase 1.3 "Identification of the PCC Objectives and coherence analysis"

In order to have an explicit presentation and to avoid redundancies in the explanations, we present only one Principal Control Center. We have chosen PLM $_{20}$: function "To plan the Maintenance" (PLM) Level 20, because it determines the Master Maintenance Schedule.

Master Maintenance Schedule : Function "To plan the Maintenance" (PLM) Level 20

Activity : To plan the Maintenance activities by resource sub-set (team, specialist...)

Objectives : - PLM $_{20}$ O_1 : To optimize the appropriateness maintenance
load / maintenance internal capacity.

- PLM $_{20}$ O_2 : To minimize the maintenance cost for an equal availability.

This decision center receives its decision frame from the PLM $_{10}$ decision center. The PLM $_{10}$ objectives are in fact the same as the global objectives of the Maintenance function.

Then, in order to analyze the coherence between the PLM $_{20}$ objectives and the PLM $_{10}$ objectives, the following splitting up diagram is used (figure 15) :

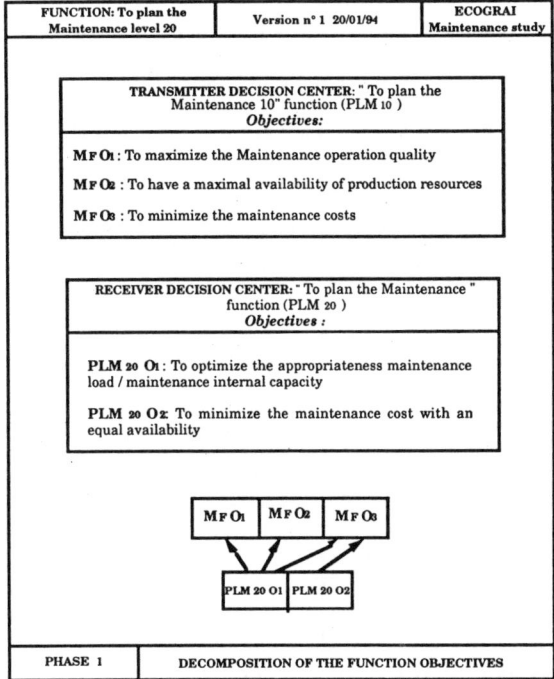

Figure 15. Splitting up diagram of PLM $_{20}$

3.3 Results of the phase 2 : " Identification of the Principal Control Center Drivers and analysis of the conflicts"

We can remind here that the Decision variables are the variables on which the decision makers can act in order to reach the objectives. So, according to the PLM 20 objectives, the following Drivers have been identified :

- PLM $_{20}$ DV_1 : Choice of the maintenance type (preventive, predictive or curative) ratio by resource sub-net.
- PLM $_{20}$ DV_2 : Maintenance activity priority.

The analysis of the conflicts has been performed between the various PCC of the maintenance function. In this example, we have not found conflicts between the various Drivers.

3.4 Results of the phase 3 : "Identification of the Principal Control Center Performance Indicators and internal coherence analysis"

3.4.1 Results of the phase 3.1 "Identification of the PCC Performance Indicators"

We can remind here that the Performance Indicators allow to the decision makers to measure the achievement of the objectives and that they are influenced by actions on the previous drivers. So, according to the PLM $_{20}$ objectives and drivers, the following Performance Indicators have been defined:

- PLM $_{20}$ PI_1 : Maintenance internal capacity use rate,
- PLM $_{20}$ PI_2 : Maintenance cost (salary + training + investments + parts + inventories + sub-contracting + infrastructure).

3.4.2 Results of the phase 3 (2) "Internal coherence analysis"

The internal coherence analysis is performed in filling in the coherence panel of the PCC. This panel is presented below (figure 16) :

Function Maintenance	Decision Center: PLM Level: 20	INTERNAL COHERENCE ANALYSIS	
OBJECTIVES O1 : To optimize the appropriateness maintenance load / maintenance internal capacity	**		
O2 : To minimize the maintenanc cost with an equal availability	*	**	
PERFORMANCE INDICATORS	PI 1 Maintenance internal capacity use rate	PI 2 Maintenance cost	
DRIVERS DV 1 : Choice of the maintenance type (preventive, predictive or curative) ratio by resource sub-set	**	**	
DV 2 : Maintenance activity priority	**	*	

Strong link (**) / Weak link (*) / No link ( )

Figure 16. Coherence panel of PLM $_{20}$

Regarding to this panel, we can see :
- that each Objective is connected at least to one PI and one DV,
- that each DV is connected at least to one Objective and one PI,
- that each PI is connected at least to one objective and one DV.

So, this decision center is coherent inside.

The GRAI net built from all the results is presented below (figure 17). It describes the PLM $_{20}$ decisional activity, the involvement of each support, defined in the GRAI grid and in the previous results (information, constraints, objectives, drivers, performance indicators), in the decision making and the final result of this PCC.

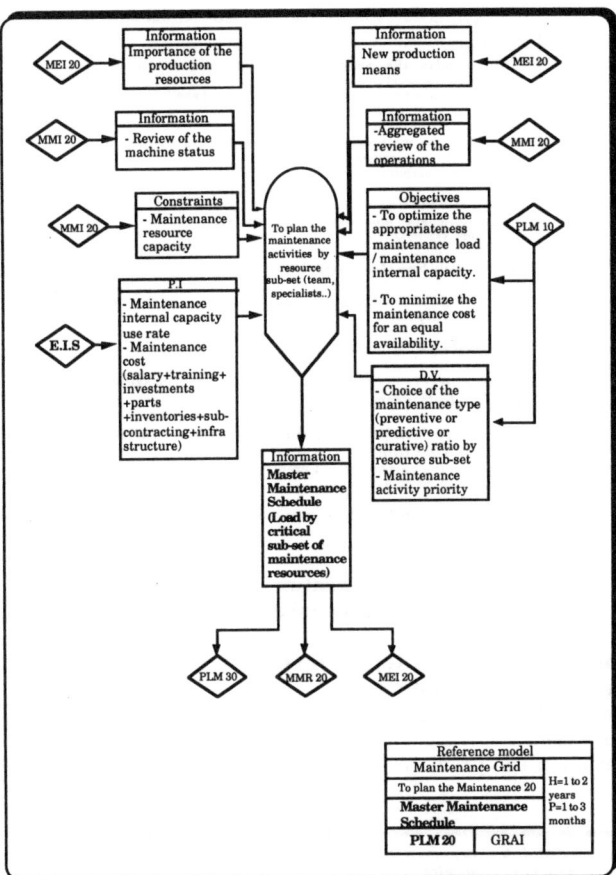

Figure 17. PLM ₂₀ GRAI net

3.5 Results of the phase 4 : " Design of the PI information system"

This phase consists in building the specification sheet of each PI. The specification sheet of the PI "Maintenance internal capacity use rate" is presented below (figure 18) :

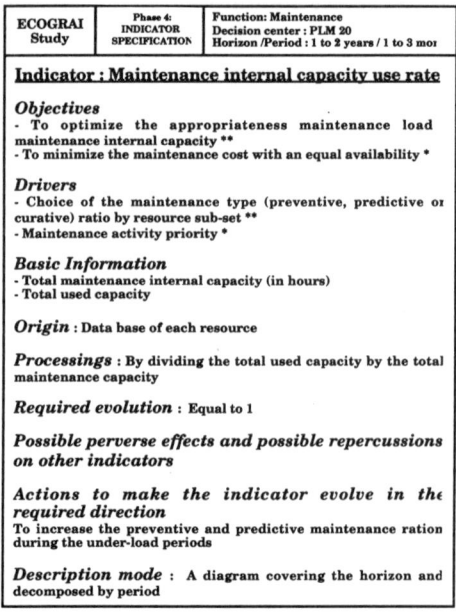

ECOGRAI Study	Phase 4: INDICATOR SPECIFICATION	Function: Maintenance Decision center : PLM 20 Horizon /Period : 1 to 2 years / 1 to 3 mo1

Indicator : Maintenance internal capacity use rate

Objectives
- To optimize the appropriateness maintenance load
maintenance internal capacity **
- To minimize the maintenance cost with an equal availability *

Drivers
- Choice of the maintenance type (preventive, predictive o1
curative) ratio by resource sub-set **
- Maintenance activity priority *

Basic Information
- Total maintenance internal capacity (in hours)
- Total used capacity

Origin : Data base of each resource

Processings : By dividing the total used capacity by the total
maintenance capacity

Required evolution : Equal to 1

*Possible perverse effects and possible repercussions
on other indicators*

*Actions to make the indicator evolve in th(
required direction*
To increase the preventive and predictive maintenance ration
during the under-load periods

Description mode : A diagram covering the horizon and
decomposed by period

**Figure 18. Specification sheet of the PI
"Maintenance internal capacity use rate"**

3.6 Results of the phase 5 : "Integration of the PI information system in the maintenance information system"

This phase consists of implementing the concerned PI in the Maintenance
information system.
The visualization of the previous PI from the EIS tool is presented below (figure
19) :

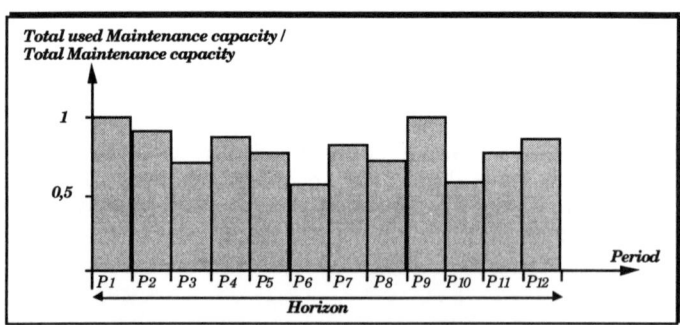

Figure 19. PI visualization

4. CONCLUSION

This application allows to illustrate the various possibilities of ECOGRAI method to design and to implement Performance Measurement Systems. The original approach, including a participative demarch, assures a real involvement of the future users. Actually, many methods proceed with the following way : Objectives --> Performance Indicators, whereas ECOGRAI uses with a different approach : Objectives --> Drivers --> Performance Indicators. This original approach allows also to identify relevant PI. Thus, it reduces the number of PI with improving efficiency and performance measurement system adoption.

This model was used in the IMS Project Globeman 21 as a basic concept to elaborate Matrix in order to benchmark industrial practices.

At this time, the works on ECOGRAI are oriented towards software tool developments, to support the method application on the one hand (linked with the CAGIM software : Computer Aided Grai Integrated Methodology) and on the other hand to support the Performance Indicator System implementation (EIS).

BIBLIOGRAPHY

ANCELIN B., (1989), "Quels critères de performance pour les nouveaux ateliers", *Revue Française de Gestion Industrielle*, N°1, 1989, pp. 66-84.

BERADA M., (1992), "Mise en place des tableaux de bord à la SAPSO", ADEPA, Rapport interne au GRAI.

BITTON M., (1990), "Méthode de conception et d'implantation de systèmes de mesures de performances pour organisations industrielles", *Thèse d'automatique*, Université de Bordeaux I, 220 pages.

BITTON M., DOUMEINGTS G., (1990), "Conception de systèmes de mesures de performances : la méthode ECOGRAI", ECOSIP, "Gestion industrielle et mesure économique, approches et applications nouvelles", *Economica* , pp. 251-274.

DOUMEINGTS G.,(1984), "Méthode GRAI, méthode de conception des systèmes en productique", *Thèse d'état es-Sciences,* Université de Bordeaux I.

DOUMEINGTS G. , CLAVE F. (1993), "Méthode pour concevoir et implanter un système d'indicateurs de performances en production - Concepts et exemples d'application -", CETIM, "Mieux gérer la production : les outils économiques d'aujourd'hui".

EVRAERT S., MEVELLEC P., (1990), "Calcul des coûts : il faut dépasser les méthodes traditionnelles", *Revue Française de gestion*, Mars-Avril-Mai 1990, N°78, pp. 12-24.

FRAY C., GIARD V., SYBORD T., (1988), "Méthodologie d'analyse et d'évaluation économique des décisions en production", *Communication aux journées "Productique" de l'AFCET*, 20 Juin 1988, 21 pages.

GIARD V., (1988), "Evaluation économique et prise de décision en gestion de production", *Revue Française de Gestion,* N° 67, Janv-Fév 1988.

GRADY M., (1991), "Performance measurement : implementing strategy", *Management Accounting,* June 1991, pp. 49-53.

GREENE A., FLENTOV P., (1990), "Managing performance : maximizing the benefit of Activity-Based Costing", *Journal of Cost Management,* Summer 1990, pp. 51-59.

KAPLAN R., JONHSON T., (1990), "Relevance lost : The rise and fall of management accounting", *Harvard Business School Press*, Boston, Massachussets.

LORINO P., (1991), "Le contrôle de gestion stratégique", la gestion par les activités", *Dunod Entreprise,* 230 pages.

40

Performance parameters to define "lean production level" in spanish home appliance cluster: from component suppliers to assembling factories.

I. Zugasti[a] & J.C. Beitialarringoitia[b]
[a]Design and Management of Production Systems Department, Ikerlan S.Coop.
[b]Robotics and Advanced Automation Department, Ikerlan S.Coop.
2 J.M. Arizmendiarrieta, 20500 Mondragon, Spain

This position paper presents the set of parameters that has been defined to obtain a quick diagnosis of the level of fitness of a group of companies that are part of the total chain from raw materials and component manufacturers to assembly factories. These companies are part of the cluster of Home Appliance Finished Products* , some of them are suppliers of the other ones and a high level of consensus has been achieved to get a global competitive Finished Product. The objective of the project in the first phase was to accomplish a rapid evaluation of potential improvements to achieve a Lean Production of Finished Goods, define and prioritize pilot projects to eliminate waste.

1. MAIN STEPS OF THE PROJECT IN THIS FIRST PHASE.

CHANGE PROCESS TO LEAN PRODUCTION (LEANPRO)
1st. Phase - Year 1993

TASK	Month 7	Month 9	Month 10	Month 11	Month 12
Management and coordinate (Steering Committee)					
Presentation of the project to Directors					
Technical Audit of actual situation					
Joint discussion of the results					
Definition of the future value of parameters.					
Define method and tools in each business					
Definition of pilot projects and next planning					
MILESTONE			A	B C, D	E

2. THE BENCHMARKING: WHAT TO BENCHMARK.

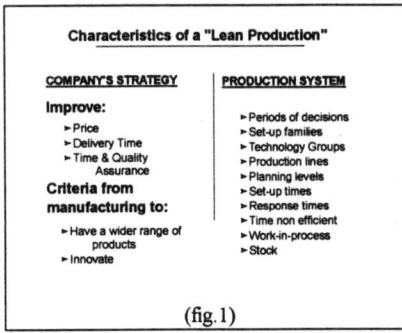

Characteristics of a "Lean Production"

COMPANY'S STRATEGY	PRODUCTION SYSTEM
Improve:	► Periods of decisions
► Price	► Set-up families
► Delivery Time	► Technology Groups
► Time & Quality Assurance	► Production lines
	► Planning levels
Criteria from	► Set-up times
manufacturing to:	► Response times
	► Time non efficient
► Have a wider range of products	► Work-in-process
► Innovate	► Stock

(fig.1)

The parameters to be benchmarked are defined in relation with the strategy of the company. In this case, we defined both the main characteristics of a Lean Production (see fig 1) and the activties mainly involved and, if case, redesigned (see fig2).. The figure 1 shows the external parameters (that mesure the performance of the company in this environment) and the internal parameters (that mesure the performance of the internal activities of the company to serve strategy).

The figure 2 shows the main activities to performe a lean production in this project.

* We would like to thank Mondragon Cooperative Corp. for its active participation as test bed and the Provincial Council of Guipuzcoa for its fianacial support of the project.

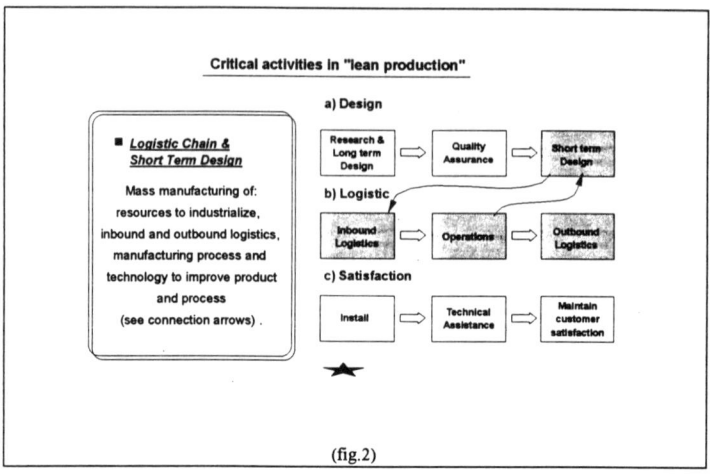

(fig.2)

3. THE BENCHMARKING: SELECTION OF THE KEY CONCEPTS.

The benchmarking thus, has been focused on measuring such parameters that would be able to show the present level and to drive the future evolution of the company towards the lean production concept as well. Wide experience of Ikerlan in the field of the use of the power of technology to design market oriented production systems lead to the following key characteristics:

- Synchronism. This concept is related with the control of the production rhythm, that implies a fine tuning between the decision making model (planning) and the physical system (product families and layout).

- Simplicity. Today, manufacturing with a production system simpler than the competitor's means better service to the customer at lower cost. To take advantage of this challenge we need a keen sense in the use of the power of the technology.

- Flexibility to the market. This means the widest range of satisfaction in a particular technology with the shortest delivery time.

- Flow tension. The potential of this production system to improve lead time.

- Waste In this project the waste refers to the time dedicated to activities that do not add value to the product.

4. THE BENCHMARKING: DEFINITION OF PARAMETERS.

These parameters are to be defined for every technology business. The defined parameters are listed below:. All the values are proyected to a range that defines five levels (from 1 to 5). The value 1 is the worst and 5 is the champion.

- **p1- Order Combination Ratio: Measures** the level of the synchronism of the production system. The higher it is, the higher the effect of programming changes is and the more complicate is the management of the order dispatching.

Definition:

$$\frac{\text{"Order Horizon"}}{\text{"Scheduling Period"}}$$

Level	1	2	3	4	5
Value	10	7.5	5	2.5	1

- **p2- Production Planning Levels: Measures** the complexity of the decision system.

Definition:

$$\text{"Planning Levels"}$$

Value	1	2	3	4	5
Range	10	5	3	2	1

- **p3- Order Set-up Ratio: Measures** the adequacy of the set-up time to the flexibility requirements of the market and the trade-off between the widest range of satisfaction in a particular technology with the maximum of flexibility.

Definition:
(at worst)

$$\frac{\text{"Number of Set-up families in every resource"}}{\text{"Number of Set-ups acceptable in the Scheduling Period"}}$$

Value	1	2	3	4	5
Range	10	7.5	5	2.5	1

- **p4- Response Ratio: Measures** the adequacy of the response time of the production system to the market needs. To be fast is the simplest way to be flexible..

Definition:

$$\frac{\text{"Lead Time"}}{\text{"The most exigent delivery time of the representative market"}}$$

Value	1	2	3	4	5
Range	10	5	1	0.5	0

- **p5- Flow Tension Ratio: Measures** the potential of this production system to improve lead time.

Definition:

$$\frac{\text{"Lead Time"}}{\text{"Minimum time for urgent orders"}}$$

Value	1	2	3	4	5
Range	20	10	5	2	1

- **p6- Supplier's Response Ratio: Measures** the adequacy of the response time of the suppliers production system to our needs..

Definition:

$$\frac{\text{Representative Delivery Time of Suppliers}}{\text{"Lead Time"}}$$

Value	1	2	3	4	5
Range	2	1.5	1	0.5	0

- **p7- Flow Control Ratio: Measures** the control level of the production flow. Every stop, every queue in the shop floor increases the WIP needed to manunfacture the final product The W.I.P.(Work-In-Process) is the best indicator of flow discontinuities.

Definition:

$$\frac{\text{"W.I.P. in time units"}}{\text{"Lead Time"}}$$

Value	1	2	3	4	5
Range	3	2	1	0.5	0

- **p8 to p10 - Stocks Ratios: Measures** the level of stocks needed to be used to cover the problems of synchronism.

Definition:

$$\frac{\text{"Stock in time units"}}{\text{"Lead Time"}}$$

Value	1	2	3	4	5
Range	2	1.5	1	0.5	0

5. THE BENCHMARKING: DATA COLLECTION AND CONCLUSIONS.

parameters	B1	B2	B3	B4	B5	B6	B7	B8	B9	B10	B11	B12	B13	B14	B15	B16	B17	B18
								Product businesses										
p1	3	4	3	3	5	4	5	3	5	5	4	1	4	5	4	5	4	5
p2	3	4	4	5	4	3	5	3	5	5	4	5	4	5	5	5	5	5
p3	2	5	5	5	4	3	5	4	2	4	4	3	3	5	2	4	3	5
p4	2	3	3	3	3	2	1	4	3	3	3	3	1	3	2	4	3	3
p5	4	4	4	5	4	3	3	3	5	5	3	3	3	3	3	3	4	4
p6	2	1	1	1	1	1	1	1	1	1	1	1	1	1	1	3	4	3
p7	2	3	1	2	1	3	2	3	4	3	3	2	3	1	3	4	3	3
p8	4	1	1	1	1	1	1	5	1	1	1	3	1	2	1	4	4	4
p9	1	1	1	1	1	4	1	1	1	1	1	1	2	1	1	1	1	3
p10	1	1	1	1	1	1	1	1.4	1	1	1	4.5	1	1	1	1	4	4
Total	26	27	24	27	26	24	24	27	27	28	24	27	23	27	23	31	35	39
Lead Time (days)	6	1	3	2	1	6	10	14	0.75	0.5	5	5	1-3	1	1	2	5	5

(fig.3)

This data (fig. 3) were collected in selected 18 product businesses in 17 companies of the Home Appliance Cluster. The procedure for data collection was as follows:

- A presentation of the key parameters was scheduled to the General Managers
- Half a day was spent in each product business. This time included:
  - A visit of the shop floor to get a real view of the production process.
  - A meeting with the product business manager and with the General Manager of the Company

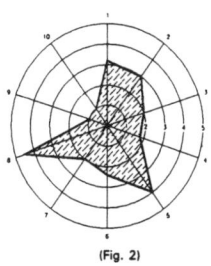

(Fig. 2)

The average time spent in each product business was 1 day/person. and the information collected was very useful in the discussions to define the global strategies to prioritize the set of actions and the expected benefits. Figures 2, and 5 show one example of the graphic used for communication purposes, one example to see the kind of actions defined (after more detailed data collection) and expected results. Finally the figure 6 makes a summary of the main conclusions.

	Improvement actions	Resp.	Lead	Waste	m2.	WIP	Stock	Labor
		EXPECTED IMPROVEMENTS						
1	Cleanness							
2	New management of the buffer n.2							
3	Redesign of the machinery area:							
4	Redesign the final assembly							
5	Set-up improvement in the die							
6	Improve flexibility (in the							
7	T.P.M. plan in critical machines							
8	Improvement of critical suppliers							
	Improvement Potential		**30 %**		**30 %**	**40 %**	**40 %**	**30 %**

(fig. 5)

Conclusions:

MAIN CONCLUSIONS

→ The defined parameters proved to be useful to know the distance to the "lean production ratios".

→ Rapid diagnosis and assumption of the real priorities to invest. The analysis allows the use of parameters easier and understandable to manufacturers.

→ Rapid launching of improvement programs.

→ Very useful as self diagnosis-tool (in current use)

PART SIX

Group Work Reports

41

How to implement Benchmarking

Report on group work

Jens O. Riis

Dept. of Production, University of Aalborg, Fibigerstraede 16, DK-9220 Aalborg, Denmark

1. INTRODUCTION

As a fairly recent development tool, only a few cases of industrial implementations of benchmarking exist, primarily located in the US. Nevertheless, issues related to implementation are essential for the continued success of benchmarking as a management tool.

With the weak empirical basis in mind and a humble attitude to the complexity of implementation the group[1] focused on (i) developing a model for placing a benchmarking activity in a larger context, (ii) defining participants in a benchmarking activity, and (iii) identifying critical issues related to implementation of organizational changes in general.

2. THE CONTEXT OF BENCHMARKING

A brief discussion of what benchmarking actually is led to the following definition

> Benchmarking is an instrument for improvement by providing a reference point.

This rather specific definition, it was felt, would help to make explicit *the role of benchmarking* as an important means for creating attention and momentum for an organizational change process, i.e. shock therapy. Furthermore, the definition would allow for a constructive dialogue with other methods, such as Total Quality Management, Kaizen, and Lean Production. Other definitions stress the development of a systematic and continuous process, and that the object of applying benchmarking should be selected processes of the enterprise. In a broad sense,

---

[1] The participants of the group were: Bill Baker, John L. Burbidge, J. Bräuner, Maya Daneva, J. Favrel, Kjetil Jacobsen, Georg Naeger, I. Pappas, Ingvar Persson, Gunnar Siebert, Riitta Smeds, Adrian Stickley, Fred Swift and Markus Weber. The group leader wants to express his appreciation of their active participation and contribution.

benchmarking is part of the continuous improvement process and insures that internal goals are calibrated to achieving market success.

The area of applying benchmarking should not be limited to engineering and manufacturing, but was seen as applicable also to areas such as sales, marketing, purchasing, and service.

The discussion led to a distinction between benchmarking activities at two levels, respectively the company and the specific improvement project, cf. figure 1.

At the company level top management has the responsibility to formulate the corporate policy with respect to improvement activities. To assist management in realizing its objectives, a champion may be appointed to pave the way for the application of benchmarking. It is important that at least one person knows the methods and techniques and can assist individual improvement projects. A general orientation program is located at the company level. As the company gains experience from improvement projects, the experience should be collected in a corporate memory, to be available for new projects.

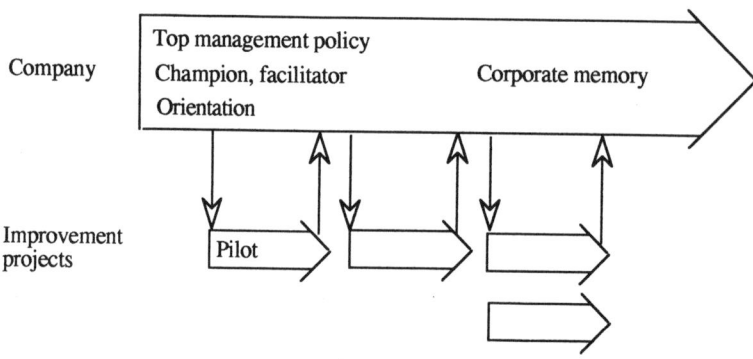

Figure 1.   The context of an improvement project

The actual benchmarking activities are carried out in improvement projects of various kind. Often, a company will start with a pilot project selected in a smaller area with a relative high probability of success. The experience gained form a basis for subsequent improvement projects.

Benchmarking is seen as an element of an improvement project which may use methods and techniques from other programs, such as Total Quality Management, Total Productive Maintenance, Lean Production, Quality Function Deployment, etc.

3. PARTICIPANTS IN BENCHMARKING

Many persons in the company will somehow be involved in a benchmarking activity. They should be considered as stakeholders, because each person or group of persons will contribute to the improvement process in a unique way different from other persons, and has specific goals and requirements. To be able to successfully use benchmarking in a company contributions of various kind are necessary, and it is a challenge for project management to find a way of reconciling opposing interests on part of the stakeholders.

The following stakeholders were identified:

o Management sponsor. Management sponsors and leaders need to demonstrate support for benchmarking team successes. Recognition and reward should be an integral part of benchmarking studies.

o Functional managers and process owners. They will contribute to the realization of improvements identified, functional managers by providing necessary resources. Process owners may be foremen responsible for a part of the process, relevant staff members, and importantly operators directly involved in the execution of the process.

o Team leader (coordinator, speaker). This person will be the main driver of the improvement process. However, the role may be played with different management styles, depending on the team; for example, a leader, a manager, a coordinator, or a speaker merely representing the opinions of the team.

o Facilitator. It is important to have a person who knows the methods and techniques of benchmarking and who can play an active role in supporting the process in the various improvement projects. The role will be that of a facilitator, a champion or a coach.

o Team members. They will contribute with their specific knowledge and experience, and the team will be cross functionally composed.

o External partner. This company will play an important role by providing a reference for the internal measurements.

The way in which stakeholders will be involved, either formally in the project organization or informally, will vary from company to company and from project to project. The identification of stakeholders, thus, may serve as a check list for analyzing each stakeholder. Together with an analysis of the overall situation in the company a good basis may be provided for designing an appropriate project organization.

4. CRITICAL ISSUES

A brain storm method was used to identify critical issues which afterwards were grouped according to stakeholders and phases of the improvement process.

TOP MANAGEMENT
- Integration of benchmarking in the policy deployment process
- How to document successful profits for top management (benefits of benchmarking)?
- Benchmarking process must be formal (use of methodology)
- Benchmarking should be seen as a management tool and not as a quality tool
- Top management may display secretiveness, because of fear of results being publicized

FUNCTION AND MIDDLE MANAGEMENT
- Middle managers may feel insecure due to uncovered incompetence and fear of reorganization; this may lead to obstruction
- Opposition from partners along supply chain, e.g. wholesalers
- Consider the contribution of suppliers to important processes, and involve them in benchmarking activities if necessary
- People who are not used to have transparency will hesitate to publish their performance indicators, because they will be afraid of criticism
- Involve the process owners

TEAM
- Training of participants and facilitators, as well as management
- Poor team building may lead to lack of motivation and to poor results
- Full team representation "up front"
- Involve the people who will be affected by process changes (if any) in the benchmarking activity
- Education versus indoctrination in benchmarking issues
- An external facilitator will not make benchmarking a continuous process

PARTNER
- Create a win-win situation with partners with benefits for both parties
- Openness and respect on the part of both parties with regard to information interchange

COMMUNICATION
- Be sure that benchmarking results are spread and communicated to the whole organization
- Structured documentation of results for convincing presentation to management
- Link benchmarking activities to the company's vision, goals and strategy - assuming that they are realistic/appropriate (or benchmark first)
- Corporate knowledge repository as a basis for communication
- Success stories are good for demonstrating the value of benchmarking

PREPLANNING
- Choice of external process to benchmark against
- Look for a good occasion as a vehicle for creating attention and momentum for a benchmarking process
- How to apply benchmarking in engineering design and in the cooperation between engineering and manufacturing?
- Focus on customer and stakeholder-oriented needs for improvement
- Awareness of the company's environment, e.g. competition and new technological development
- Compliance with standards

PLAN AND DESIGN
- Begin with a critical, but relatively uncomplicated activity that may trigger new benchmarking activities
- Limit the number of benchmarking studies to "vital span"
- Adopt a use-friendly model for benchmarking
- Benchmarking should be managed as a project with stepping-stones
- How to combine a top-down and a bottom-up implementation process?
- Know your process; clarify what kind of tools and methods are appropriate
- Failure causes: Poor understanding of resources required, and poor linkage to other improvements activities, e.g. quality
- Let benchmarking activities be (i) customer oriented; (ii) internally sponsored; (iii) processes documented; (iv) pre-testing of quality; (v) process emphasis
- Gather information (quantitative and qualitative) prior to benchmarking site visit, and make site visits only when necessary
- Use EDI/on-line facilities to ease further benchmarking activities

CHECK AND CONTROL
- Advantages do not come automatic. Clarify what must be done to achieve them
- Study performance and results
- Monitor progress
- Procedures are necessary for continuous check of relevance, coherence and cost

ACT
- Integrate work already done (e.g. improvement results and analysis findings) and avoid unnecessary duplication of information gathering
- Beware that institutionalizing benchmarking may eventually kill it
- Obtain resources to sustain benchmarking efforts

42

Modelling for Benchmarking

Summary of group work edited by
A.S. Carrie[*], P. Higgins[**] and P. Falster[***]

ABSTRACT

The main findings of the group discussion on modelling for benchmarking can be summarized as follows:

- benchmarking should be about comparing processes, not about enterprises, therefore identifying common processes is the key point,
- a business process serves a customer - who is willing to pay,
- IDEF0 was proposed as an analysis tool suitable for process modelling, however, many people present believed that IDEF0 does not provide enough capability. Other modified IDEF approaches may be suitable.

SCOPE FOR DISCUSSION

The briefing note for the groups work was provided in the following problem statement:

One definition of benchmarking (among numerous others) is:

"Benchmarking is the process of continuously comparing and measuring business processes against business leaders anywhere in the world to gain information that will help the organization take action to improve its performance"

As can be clearly seen from this definition, benchmarking focuses on business processes. The unit of analysis in benchmarking is the business process. To ease the task of choosing and documenting the processes to benchmark, a model of the company based on business processes should be available. Such a model should be a comprehensive framework, covering most of the generic processes contained in a general company.

[*] A.S. Carrie, Department of Design, Manufacture and Engineering Management, University of Strathclyde, James Weir Building, Montrose Street, Glasgow G1 1XJ, Great Britain.

[**] P. Higgins, CIM Research Unit, University College Galway, Nun's Island, Galway, Rep. of Ireland

[***] P. Falster, Electric Power Engineering Department, Technical University of Denmark, DK 2800 Lyngby, Denmark

We will ask the group to discuss the feasibility of a such a framework, how to classify business processes, what processes should be represented in the framework, etc. Remember that the main task is to discuss modelling of the company for the purpose of benchmarking. The framework should be general enough to facilitate comparisons across companies and industries.

Based on the given problem statement above a very interesting open discussion took place and the remainder of this text attempts to synthesis this discussion.

INTRODUCTION

Modelling provides a basis for understanding, learning and development. The main task is to discuss modelling of a company for the purpose of benchmarking. It is said in the problem statement that such a model should be a comprehensive framework covering most of the generic processes. However, we would like to distinguish between a specific model for a specific company and a general model or framework covering as said most of the generic processes in a general company.

We prefer to use the term framework instead of model, because a framework does not require a theory in order to be used for design purpose. A framework may simply be a taxonomy or classification system.

We propose to use a Descriptive language, not necessarily a formal and executable language. For example IDEF0 is a descriptive language.

The following items have to be discussed:

* clarifying our definitions
* framework
* research needs
* organization impact

How to compare two completely different companies, for example a pizza company and a ship yard? We should be able to dentify a common set of processes that could be compared for benchmarking and then perhaps supply rules that could be used to provide a scaling mechanism to allow different enterprises to be composed. The point is therefore to find the business processes to compare, instead of trying to compare "pizza's and ships". Benchmarking should thereby be about comparing business processes i.e. P3A vs P3B from enterprises EA and EB, respectively (figure 1). In this figure two enterprises and a set of generic processes are illustrated.

Those processes, say P3A and P3B, from the two enterprises which share the same generic processes should be compared. Thereby, this technique would facilitate comparisons across companies. Accordingly, we need criteria to identify comparable processes. Another question is also how should we develop a framework to describe individual processes rather than classify business processes.

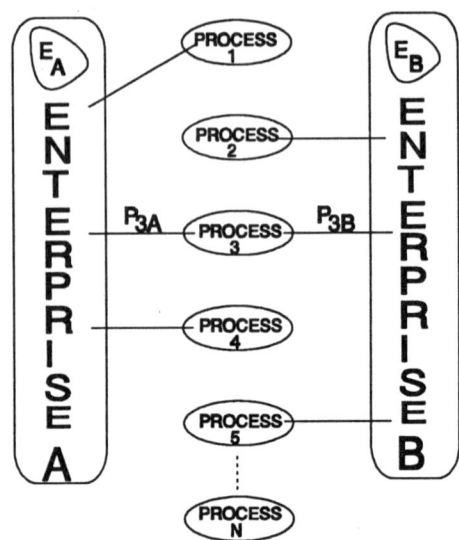

Figure 1. Relationships between Enterprises A and B and Business processes
1, 2, 3, ... , n.

The organizational structure of a company should not be important in the definition of
generic processes, however, since one the objective of benchmarking and modelling is
to change the organization to improve performance, a means of relating organization
structure to process definition must be developed. This leads to questions such as -
How can the organization be modelled? For each business process, where are the
organizational boundaries? How does the organization influence business processes?

WHAT IS A BUSINESS PROCESS?

What is a business process and how can we identify processes in the company? How do
we define them? Is there any difficulty in defining them? Do we need to classify them?
Does a process have certain definable characteristics? These are some of questions raised
during the discussion.

Various definitions of business processes exist. Some of the definitions are found in the
conference proceedings.

CIM-OSA [1]:
A Business Process is the business user's view of what tasks are required to achieve a
particular enterprise objective. A task is a general term which includes Business
Processes and enterprise activities.

CIM-OSA's [1] three main categories of Business processes is Manage, Operate and Support (see also Childe & Smart [7]):

Management processes (strategy planning, operations planning, performance monitoring and condition reporting) "Manage processes" as those which are concerned with strategy and direction setting as well as with business planning and control.

Operational processes (product development, manufacturing/product production) product marketing) "Operate processes" are viewed as those which are directly related to satisfying the requirements of the external customer, for example the logistics supply chain from order to delivery.

Support processes (installation, maintenance, set-up, repairing) "Support processes" typically act in support of the Manage and Operate processes. This include the financial, personnel, facilities management and Information Systems provision (IS) activities.

Burbidge's classification of industries using the material conversion classification system based on the ratio of material to product varieties (see this volume) is also a possibility to consider [5].

REQUIREMENTS TO MODELLING TOOLS AND FEASIBILITY OF A FRAME-WORK

Modelling the system should be in terms of 1) what the system is, 2) what is its environment and 3) the interaction between system and environment. Quality is the link between the system and its environment.

Several modelling methods were discussed. One view was that IDEF0 was inadequate as a technique because it modelled functions or activities, not processes. Some participants had developed alternative techniques. Several approaches to developing software tools for business process re-engineering may be found in [13]. On the other hand it was agreed that IDEF0 could be used successfully provided a business process viewpoint was kept in mind. IDEF0 could then be used to:

- identify processes (activities)
- identify the environment in a hierarchical structure

The decisioned system approach and the Grai Grid as possible means of benchmarking information and decisioned flows should also be mentioned in this connection [14].

The discussion reflected inconsistency in the use of the terms, functions, activities and processes. An analogy to the discrete simulation field may be useful to clarify some of the problems.

It is well known that a modeler could either take a machine-oriented or activity-oriented view (as in CSL) or a material-oriented or process-oriented view (as in GPSS). For the

Browne & Bradley [3]:
A Business Process can be viewed as a set of logically related tasks performed to achieve a definite business outcome. These business processes have two important characteristics:
- They have customers, and
- they cross organizational boundaries
- are independent of formal organizational structure

Davenport & Short [8]:
A Business Process is the logical organization of people, materials, energy, equipment and procedures into work activities designed to produce a specified end result.

Hickman [10]:
A Business process is a logical series of dependent activities which use the resources of the organization to create, or result in, an observable or measurable outcome, such as a product or service.

The definitions may seems different but they complement each other, where they do not overlap.

We can summarize the preceding definitions as follows:

A process is part of the value chain (according to Porter [12]). A Process is a sequence of operations or a series of steps. Could something that does not add value be a process? The answer is no.

There should be a customer because a process serves a customer, who is willing to pay. If it does not add value is it then a function? One person's function could be another's process (for example the personnel function). Let be added to the above definitions that there also should exist an External Event Trigger.

A process takes time and uses resources. It converts input to output. There are defined constraints, goals and objectives. Sample business processes include customer order entry, purchasing, product development, etc.

HOW SHOULD BUSINESS PROCESSES BE CLASSIFIED, AND WHAT PROCESSES SHOULD BE REPRESENTED?

There seems to be a variety of systems for classifying generic business processes. Burbidge defines an arterial flow system with nine generic functions: Design, Production planning, Production Control, Purchase, Market, Finance, Personnel, Secretarial and General management [4].

It was mentioned that for example Andersen Consulting defines 200 generic processes, IBM defines 17 generic processes, Childe & Smart [7] defines 7 generic processes, Mertins [11] defines 183 generic processes and Champy [6] defines 5 generic processes.

first viewpoint you place yourself on the machines or functions and watch the material input and output. For the second viewpoint you place yourself on the material, orders or information and watch the machines or functions you passes. The latter viewpoint is the dominant viewpoint as pointed out above for business process modelling.

However, it is also appreciated within simulation modelling that both viewpoint are equally useful depending of what you want to measure. This could explain the different viewpoints among the participants.

Analysis should be strictly goal-driven, modelling one level above the area of interest.

Required capabilities of modelling methods:

- hierarchical decomposition, defining the right granularity, level of detail,
- base level of primitive business processes in order to be able to compare similar procession different enterprises,
- reusability,
- transportability (to f.ex. EXPRESS),
- business processes are independent of how we implement them
- recursive or iterative,
- formalized structured methods,
- standardization,
- object-oriented,

The following dynamic business process model in figure 2 is proposed based on a descriptive model (not a formal or executable model):

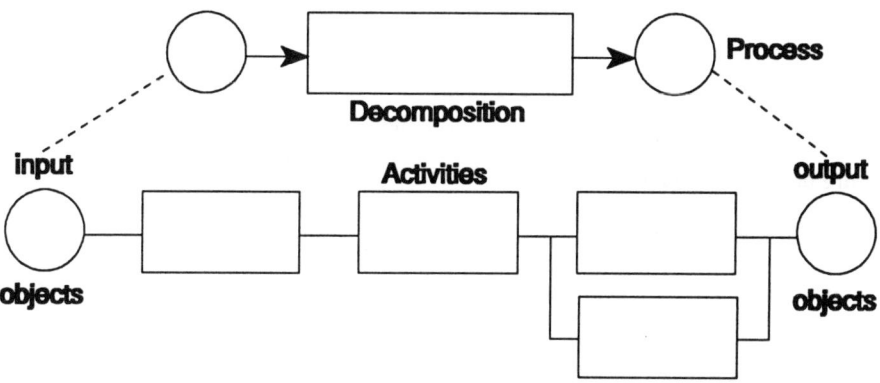

Figure 2. Dynamic Business Process Model.

The characteristics of a business model is summarized in figure 3.

A Business Process	
Use	Resources
Has	Customer
Takes	Time
Add	Value
Triggered by	External event

Figure 3. Characteristics of a Business Process

SUMMARY OF DISCUSSION AND OUTLOOK

The existing organizational structure is not important in defining processes, but it is necessary to have a way for representing the organization, such that changes to improve the organization, can be described.

Some of the questions and answers discussed during the group work are summarized below:

Model of the company based on business processes, identify the processes.

Essential capabilities of and requirements for modelling methods include:

- hierarchical decomposition, defining the right granularity,
- base level of primitive business processes in order to be able to compare,
- reusability,
- transportability (to EXPRESS for example),
- business processes are independent of how they are implemented,

Nature of the environment.

Criterias to identify comparable processes:

- how to redesign the system in order to do comparison with the purpose of benchmarking,

Which tools are needed to do:

- performance comparison,
- feasibility study

Develop a framework to describe individual processes rather than classify business processes.

Value-adding chain of business processes:

- have a customer to the business process,

Be strictly goal-driven, model one level above your interest.

Organization hierarchies is not of primary importance:

- for each business process, where are organizational boundaries?
- where does organization influence business processes.

Andersen consulting has developed 200 global standard business processes covering a variety of enterprises.

How can two completely different companies, f.ex. pizza company and a ship yard be comparable. Find the business processes to compare (not apples and pears).

Acknowledgement

The participants in the group discussion are thanked for their valuable contribution to this report.

REFERENCES

1. AMICE Vol1: Open System Architecture for CIM. ESPRIT Consortium AMICE (Eds.). Project 688, Springer-Verlag, 1989.

2. R.R. Bravoco & S. B. Yadeav: Requirement Definition Architecture - An Overview. Computers in Industry 6 (1985), pp 237-251.

3. J.Browne & P. Bradley: Business Process Reengineering (BRP) - A Study of the software tools currently available. Working paper at CIMMOD/CIMDEV meeting, Bordeaux, France May 9th & 10th 1994. CIMRU, University College Galway.

4. J.L. Burbidge: Production Control: A Universal Conceptual Framework. Production Planning and Control., Vol. 1, No. 1, pp 1-13.

5. J.L. Burbidge: The "Material Conversion Classification". IFIP 5.7 Working Conference on Benchmarking - Theory and Practice, Trondheim, Norway, june 16-18, 1994, Chapman & Hall (in this volume).

6. Champy, R.C.: Benchmarking - The Search for Industry best Practices that Lead to Superior Performance. ASQC Quality Press, Milwaukee, Wisconsin, 1989.

7. S.J. Childe & P.A. Smart: Benchmarking - An Approach Using Process Modelling for Redesign. IFIP 5.7 Working Conference on Benchmarking - Theory and Practice, Trondheim, Norway, june 16-18, 1994, Chapman & Hall (in this volume).

8. T.H. Davenport & J.E. Short: The new Industrial Engineering: Information Technology and Business Process Redesign, Sloan Management Review, summer, 1990.

9. M. Hammer & J. Champy: Reengineering the Corporation: A Manifesto for Business Revolution. Harper Business, 1993.

10. L.J. Hickman: Technology and Business Process Re-engineering: Identifying Opportunities for Competitive Advantage, British Computer Society CASE Seminar on Business Process Engineering, London, 29 june 1993.

11. K. Mertins: Benchmarking Techniques. IFIP 5.7 Working Conference on Benchmarking - Theory and Practice, Trondheim, Norway, june 16-18, 1994, Chapman & Hall (in this volume).

12. M. E. Porter: The Competitive Advantage of Nations. The Free Press. 1990.

13. K. Spurr, P. Layzell, L. Jennison, N. Richards (eds): Software Assistance for Business Re-engineering, J. Wiley, 1993.

14. G. Doumeingts: System Analysis Techniques. In Computer-Aided Production Management, A. Rolstadås (ed.): IFIP State-of-the-Art Report, Springer Verlag 1988.

43

PERFORMANCE INDICATORS

Group Work of IFIP W.G. 5.7

Eero Eloranta, Chairman
Steve Crom, Co-Chairman

INTRODUCTION

A central element in benchmarking is measurement and comparison. Companies must measure the performance of their business processes and practices to be able to compare themselves to others and to identify benchmark partners who are better than themselves. Furthermore, in order to recognize improvement based on benchmarking, the benchmarking company must be able to track performance over time. Therefore, good performance indicators are an essential ingredient of successful benchmarking. What follows is a summary of a group discussion on the topic of performance indicators which we hope provides a useful framework for performance measurement in benchmarking. This summary represents the collective thoughts of the 13 members of the group.[1]

The discussion is organized around questions frequently asked about performance measurement in benchmarking:

- What should I measure?
- What units of measurement have others successfully used?
- With whom should I compare myself?
- Who should do the measuring?

The dangers of over-focusing on the quantitative aspects of benchmarking are highlighted at the close of this discussion. We conclude with a check list to help you in choosing the right performance indicators to benchmark for continuous improvement.

WHAT SHOULD I MEASURE?

One way to think about what is useful to measure is to consider three groups of indicators: performance; practice/process; and, enablers. As depicted in Figure 1, performance is the result of work practices and processes which in turn are influenced by enablers such as leadership style, information technology infrastructure, human resources policies, etc. While more difficult to quantify, what separates average from world class companies is often the enabling elements. Benchmarking as a tool for taking action to improve performance, must therefore consider all three types of data. For any given company, what to measure in each category

---

[1] A special thanks to Bill Baker of Texas Instruments who contributed to the group's discussion from his experience in helping select and manage the use of global benchmark measures at TI.

depends on the company's business strategy and the areas most in need of
improvement.

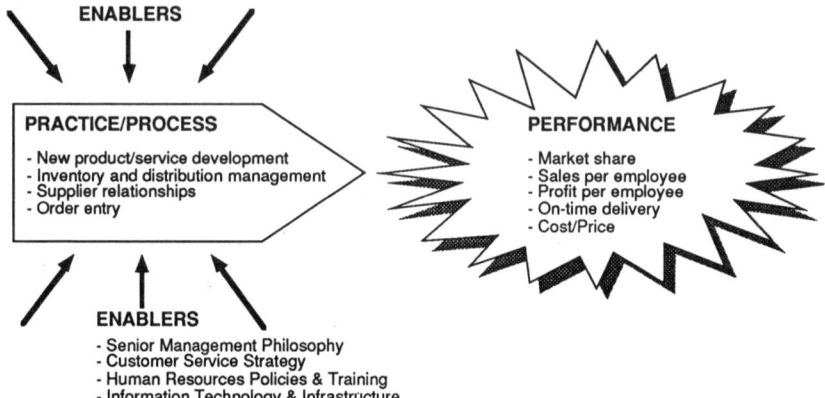

Fig. 1 *Performance, Practice and Enablers - Examples What To Measure*

WHAT UNITS OF MEASUREMENT HAVE OTHERS SUCCESSFULLY USED?

Since what to measure depends on a company's business strategy, the right
units of measure will also be company specific. The case of Texas
Instruments illustrates by example how to decide on the right measures
and units of measure.

At the highest level, Texas Instrument's (TI) business strategy is
"Customer Focus Through Total Quality." In deciding on the right
performance indicators to benchmark the questions they asked themselves
were:

1. What is our business strategy?
2. How can we measure if we are succeeding?
3. Since it is important that everyone in the company help implement
   the strategy, what measures apply to everyone?
4. What few global measures can everyone "buy into," help put into
   practice and use to drive improvement?

Senior managers at TI agreed that given the business strategy three areas
were critical to measure: people involvement; customer satisfaction; and,
continuous improvement. Agreeing on what to measure was more
difficult. Data on over 25 performance indicators was initially collected.

Seeing how time consuming it was to collect and use the information, senior managers worked until they agreed on four performance indicators:

People Involvement

- Hours of training (average and minimum)

Customer Satisfaction

- On-Time and Complete Deliveries (delivery of all components, including paperwork, to the right location at the right time)

Continuous Improvement

- Quality (6 sigma, defects in parts per million)
- Cycle Time

The above measures apply to support/service functions as well as to manufacturing. TI collects the information internally and uses it for internal and external benchmarking. In the area of training, for example, TI compares the average and minimum hours of training its employees receive with the level of training employees at other companies receive. TI expects every employee to receive a minimum of 32 hours of training per year and holds managers accountable for selecting the right training along with each employee and seeing that their employees devote the time to complete the training selected. One of the key criteria for selecting a performance indicator for a large organization is that the measure be easy to aggregate by level and department on a regular basis.

WITH WHOM SHOULD I COMPARE MYSELF?

Who to use as points of reference or benchmarks depends on the your reason or objective for benchmarking. If establishing an early warning system is the objective, focus your attention on the competition and on how well you are meeting your customers' expectations. If you want to create a sense of urgency for improvement, benchmark the impressions non-customers (e.g., recently lost accounts) have of you versus other suppliers. Have senior managers conduct the interviews. If you have selected a business process to improve and what to fuel the creativity of a team of employees involved in the redesign effort, look to companies with similar processes who are the best in the world at that particular process. The following is a framework for helping decide who to benchmark depending on your reason for benchmarking.

Whom To Benchmark

<u>Objective</u>	<u>Reference Points</u>
• Early Warning System	• Customers • Competitors
• Create Sense of Urgency	• Non-customers • Suppliers
• Process Re-Design	• Best in Class/World
• Find Benchmark Partner	• Benchmarking Conferences, Roundtables, Centers
• Continuous Improvement	• Own processes/practices, performance results and enablers

WHO SHOULD DO THE MEASURING?

The most successful benchmarking efforts are strategically driven with support from the top of the organization and involvement in benchmarking and performance improvement throughout the company. By level in the organization there is a hierarchy of measures that support one another:

<u>What</u>	<u>Who</u>
1. Strategic goal	• Senior Management
2. Core business processes	• Senior and Middle Managers
3. Process data	• Those closest to the process in question

As depicted in Figure 2 each of the above levels of data collection should be logically nested in each other. The right process data to collect depends on which processes are most important to the business, which depends on the business's strategy and strategic goals. Involving the right people at the right level of the organization in deciding what to benchmark goes a long way in creating support to collect benchmark data and convert the findings into action.

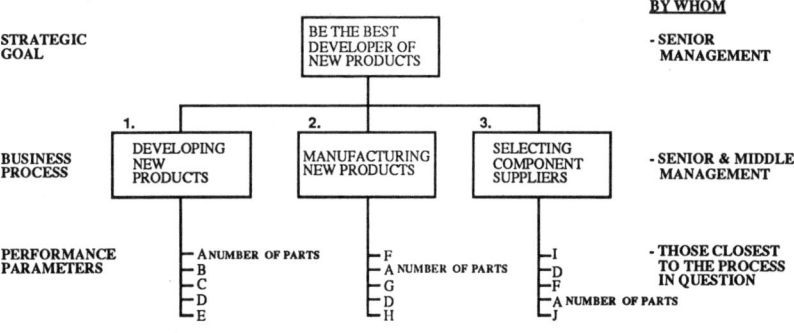

STRATEGIC GOAL — BE THE BEST DEVELOPER OF NEW PRODUCTS — - SENIOR MANAGEMENT

BUSINESS PROCESS

1. DEVELOPING NEW PRODUCTS
2. MANUFACTURING NEW PRODUCTS
3. SELECTING COMPONENT SUPPLIERS

- SENIOR & MIDDLE MANAGEMENT

PERFORMANCE PARAMETERS

A NUMBER OF PARTS
B
C
D
E

F
A NUMBER OF PARTS
G
D
H

I
D
F
A NUMBER OF PARTS
J

- THOSE CLOSEST TO THE PROCESS IN QUESTION

Fig. 2 *Example of Hierarchy of Benchmark Measures*

DANGERS IN IMPLEMENTING PERFORMANCE INDICATORS FOR BENCHMARKING

In deciding what to measure and how, do not lose sight of the spirit of benchmarking: *to gain information that will help the organization take action and improve its performance.* Avoid the temptation to use benchmarking to control, reward or punish performance. Information is power and the real power of benchmarking is in sparking people's enthusiasm to learn and improve.

CONCLUSION

Benchmarking as a tool for taking action to improve performance must consider three groups of indicators: performance; practice/process; and, enablers. What to measure depends on a company's business strategy and the right units of measurement are determined accordingly. In driving continuous improvement a number of criteria should be considered in selecting measures, ease of understanding and simplicity of data collection should be high on the list (see Figure 3). It may be appropriate to gather benchmark data on competitors, from customers, non-customers or those best in the world in performing a process that is strategically important to you. To determine with whom and what to benchmark get senior managers, middle managers and those closest to key processes involved in determining strategic goals, selecting core business processes and collecting/monitoring acting on performance data. Keep in mind that information is power and the real power of benchmarking is in sparking people's enthusiasm to learn and improve.

- **ARE THE PERFORMANCE INDICATORS (PI) CONNECTED TO THE STRATEGY?**

- **WILL THE PI's ENCOURAGE CONTINUOS IMPROVEMENT?**

- **ARE THE PI's EASY TO UNDERSTAND?**

- **IS IT EASY TO COLLECT THE DATA YOU NEED?**

- **CAN THE PI's BE AGGREGATED EASILY "ROLLED-UP" ON A PERIODIC BASIS?**

- **WILL THE PI MEASURES LEAD TO ACTION?**

Fig. 3   Performance Indicators - A Checklist

INDEX OF CONTRIBUTORS